Understanding Granites: Integrating New and Classical Techniques

Geological Society Special Publications
Series Editors
A. J. HARTLEY
R. E. HOLDSWORTH
A. C. MORTON
M. S. STOKER

GEOLOGICAL SOCIETY SPECIAL PUBLICATION NO. 168

Understanding Granites: Integrating New and Classical Techniques

EDITED BY

ANTONIO CASTRO
Universidad de Huelva, Spain

CARLOS FERNÁNDEZ
Universidad de Huelva, Spain

and

JEAN LOUIS VIGNERESSE
Université Nancy, France

1999
Published by
The Geological Society
London

THE GEOLOGICAL SOCIETY

The Geological Society of London was founded in 1807 and is the oldest geological society in the world. It received its Royal Charter in 1825 for the purpose of 'investigating the mineral structure of the Earth' and is now Britain's national society for geology.

Both a learned society and a professional body, the Geological Society is recognized by the Department of Trade and Industry (DTI) as the chartering authority for geoscience, able to award Chartered Geologist status upon appropriately qualified Fellows. The Society has a membership of 8600, of whom about 1500 live outside the UK.

Fellowship of the Society is open to persons holding a recognized honours degree in geology or a cognate subject and who have at least two years' relevant postgraduate experience, or not less than six years' relevant experience in geology or a cognate subject. A Fellow with a minimum of five years' relevant postgraduate experience in the practice of geology may apply for chartered status. Successful applicants are entitled to use the designatory postnominal CGeol (Chartered Geologist). Fellows of the Society may use the letters FGS. Other grades of membership are available to members not yet qualifying for Fellowship.

The Society has its own Publishing House based in Bath, UK. It produces the Society's international journals, books and maps, and is the European distributor for publications of the American Association of Petroleum Geologists (AAPG), the Society for Sedimentary Geology (SEPM) and the Geological Society of America (GSA). Members of the Society can buy books at considerable discounts. The Publishing House has an online bookshop (http://bookshop.geolsoc.org.uk).

Further information on Society membership may be obtained from the Membership Services Manager, The Geological Society, Burlington House, Piccadilly, London W1V 0JU (Email: enquiries@geolsoc.org.uk; tel: +44 (0)171 434 9944).

The Society's Web Site can be found at http://www.geolsoc.org.uk/. The Society is a Registered Charity, number 210161.

Published by The Geological Society from:
The Geological Society Publishing House
Unit 7, Brassmill Enterprise Centre
Brassmill Lane
Bath BA1 3JN, UK
(*Orders*: Tel. +44 (0)1225 445046
Fax +44 (0)1225 442836)
Online bookshop: http://bookshop.geolsoc.org.uk

First published 1999

The publishers make no representation, express or implied, with regard to the accuracy of the information contained in this book and cannot accept any legal responsibility for any errors or omissions that may be made.

British Library Cataloguing in Publication Data

A catalogue record for this book is available from the British Library.

ISBN 1-86239-058-4
ISSN 0305-8719

Typeset by WKS, Westonzoyland, UK

Printed by Whitstable Litho, Whitstable, UK

Distributors

USA
AAPG Bookstore
PO Box 979
Tulsa
OK 74101-0979
USA
Orders: Tel. +1 918 584-2555
Fax +1 918 560-2652
Email bookstore@aapg.org

Australia
Australian Mineral Foundation Bookshop
63 Conyngham Street
Glenside
South Australia 5065
Australia
Orders: Tel. +61 88 379-0444
Fax +61 88 379-4634
Email bookshop@amf.com.au

India
Affiliated East-West Press PVT Ltd
G-1/16 Ansari Road, Daryaganj,
New Delhi 110 002
India
Orders: Tel. +91 11 327-9113
Fax +91 11 326-0538

Japan
Kanda Book Trading Co.
Cityhouse Tama 204
Tsurumaki 1-3-10
Tama-shi
Tokyo 206-0034
Japan
Orders: Tel. +81 (0)423 57-7650
Fax +81 (0)423 57-7651

Contents

Preface

The modern view of the granite problem requires the application of many different theoretical, experimental and empirical resources provided by geophysics, geochemistry, experimental petrology, structural geology, scale modelling and field geology. Following this philosophy, we have edited this volume with the intention of providing an integrated approach in the study of topical granite-related problems, rather than to concentrate on individual facets. The edition of this volume was inspired by an international workshop entitled 'Modern and Classical Techniques in Granite Studies' that was held in the University of Huelva (Spain) in November 1997. The editors wish to acknowledge reviews by the following colleagues: A. Aranguren; M. P. Atherton; P. Barbey; C. G. Barnes; K. Benn; G. Bergantz; A. Berger; M. Brown; M. A. Bussell; L. G. Corretgé; A. R. Cruden; J. Dehls; J. de la Rosa; T. Dempster; D. Dingwell; E. Ferré; A. E. Helmers; R. Knoche; J. P. Lefort; K. McCaffrey; C. Miller; A. Nèdèlec; E. Rutter; M. Sandford; W. E. Stephens; C. J. Talbot; G. Watt; R. Weinberg.

Antonio Castro
Carlos Fernández
Jean Louis Vigneresse

Understanding granites: integrating new and classical techniques

ANTONIO CASTRO[1], CARLOS FERNÁNDEZ[1] & JEAN LOUIS VIGNERESSE[2]

[1]*Departamento de Geología, Universidad de Huelva, La Rábida 21819, Huelva, Spain*
[2]*CREGU, UMr CNRS 7566 G2R, BP 23, 54501 Vandoeuvre cedex, France*

Granite magmatism represents a major contribution to crustal growth and recycling and, consequently, is one of the most important mechanisms to have contributed to the geochemical differentiation of the Earth's crust since the Archaean times. The important role of granites has been acknowledged from the times of James Hutton, being an important part of Hutton's (1785) work *The Theory of the Earth*. Since that time, advances in Earth Sciences and the development of analytical instruments have produced a vast amount of data from granite terrains around the world (see, for instance, the *Proceedings of the Hutton symposia*, 1988, 1992, 1996). However, many problems still remain unsolved.

There have been several important controversies concerning granites during the last 50 years. One of the best remembered was that led by H. H. Read and N. L. Bowen around the 1950s (Read 1958). The end of this obscure controversy, full of misunderstandings and with some rhetoric, ended with the publication of the classic memoir by O. F. Tuttle & N. L. Bowen (1958) entitled *Origin of granites in the light of experimental studies in the system $NaAlSi_3O_8$–$KAlSi_3O_8$–SiO_2–H_2O*. The main conclusion of this study was that granitic liquids may be generated from melting of quartzo-feldspathic rocks in excess water, at depths of '... *12–21 km in geosynclinal areas where the initial gradient is on the order of $30\,°C/km$*' (Tuttle & Bowen 1958, p. 2). Detailed phase relationships in the granite system were determined in this seminal work. However, the conditions for the generation of granite liquids postulated by Tuttle & Bowen seemed unlikely to occur naturally: the composition of the source must be quartzo-feldspathic and water must be in excess.

It was not until two decades later that the first dry melting experiments on pelitic systems were carried out by Brown & Fyfe (1970) and Thompson (1982). They demonstrated that granite melts may be obtained by incongruent melting of hydrous phases, namely micas and amphiboles. Dry melting experiments explain satisfactorily the observed field relationships between pelitic migmatites and granites. Once more, the 'transformation' of metasedimentary rocks into granites, argued by Read, was set at the centre of the controversy. From the 1970s, many experimental works on the pelite system have been carried out, and today the crustal origin of granites is beyond doubt, at least for those of peraluminous composition. However, although these results greatly advanced our understanding of the granite problem, they are far from affording a definitive solution.

The recognition of several granite types introduced a new source of complexity to the granite problem. Bruce Chappell & Alan White proposed the distinction between S (sedimentary source) and I (igneous source) type granites (Chappell & White 1974). Very soon, the S–I classification scheme was applied to granites around the world. Wally Pitcher extended the classification to M (mantle fractionates) and A (anorogenic) type granites and established a correlation between granite types and tectonic settings (Pitcher 1982). However, these classifications have been shown to be inappropriate in many intrusions worldwide, and their tectonic implications have been questioned by many granitologists. Nevertheless, they promoted interesting discussion and argument, collection of new geological data, new experiments and new geochemical modelling. This in turn led to greater understanding of the granite problem and independent assessments of the validity of the S–I classification. The genetic model for so-called I-type granites is at the heart of this new and seminal controversy that still remains.

Isotope data seem to be in favour of a hybrid origin for these I-type granites. A hybrid origin implies that some mantle components were added to the granitic melt at the time of crustal melting. However, the way in which these hybrid granites acquire their mixed composition remains uncertain. Bowen (1922) postulated that the assimilation of pelitic rocks by basalts would give rise to enormous amounts of granitic fractionates. He proposed that the basalt–pelite

From: CASTRO, A., FERNÁNDEZ, C. & VIGNERESSE, J. L. (eds) *Understanding Granites: Integrating New and Classical Techniques*. Geological Society, London, Special Publications, **168**, 1–5. 1-86239-058-4/99/$15.00 © The Geological Society of London 1999.

reaction would produce a granitic (*sensu lato*) liquid and a noritic residue, based on the observed relationships in the rocks of Corlandt in New York. Seventy years later, experiments on assimilation by Patiño Douce (1995) finally demonstrated the validity of this hypothesis. These experiments explain many geochemical features observed in granodioritic rocks.

Another remarkable controversy is that of the so-called room problem: the question of how to generate the space required for a granite to find its place in the upper crust. This is associated with the question of the liberation of room caused by melt ascent in the intermediate to lower crust. This is a long-lived debate, as indicated by Pitcher (1997) who describes the lively field discussions between the Norwegian petrologist Baltazar Keilhau and Charles Lyell. The solution to the room problem must rely on the careful structural, geophysical and rheological description and modelling of granitic plutons and batholiths. Undoubtedly, the methods and results of Hans Cloos and co-workers inaugurated a new style in the structural description of granitic rocks, as evidenced by the widely cited memoir by Robert Balk (1937). In parallel with these meso- and megastructural descriptions, Bruno Sander (1930) gave a new impulse to the microfabric studies. Unfortunately, the two schools did not reach consensus, despite the complementary character of their two approaches. After a transition period, a renewed interest in the room problem occurred in the late 1970s and 1980s, based on new developments in structural geology. These successfully applied theories of flow and deformation to geological materials, allowing us to understand microfabric development. They have also led to the proposition of rheological models for magma behaviour and emplacement, through strain orientation and measurement in both host rocks and plutons. Analogue modelling of pluton emplacement, exemplified by the work carried out by Hans Ramberg and co-workers at Uppsala (e.g. Ramberg 1981), has been critical in reinvigorating the room controversy. Actual models relate to diapirism, ballooning, stoping, return flow, magma blistering, dyking, emplacement into shear zones, or even multiple coeval mechanisms in variable proportions in time and space. These are some of the elements that constitute the modern debate. Increasingly sophisticated theoretical models, both analogue and computational, coupled with field techniques, are being mobilized to answer the so-called room problem. They now occupy an increasing number of structural geologists all around the world. Whether or not an answer will be obtained in the future, there is no doubt that this topic will gain a place in the structural geology or tectonics textbooks of the next century.

Obviously, things are clearer than a century ago, and new paradigms have emerged, but important aspects of granite generation are still unsolved and several questions are pending. Are granitic melts pure melts? Are granites emplaced by dykes, diapirs or non-diapirs, in a single pulse or by multiple pulses? Are granites the products of crustal recycling or are they associated with continental growth from the mantle? Are the many different granitic types derived from the same number of varied sources, or are they related to different processes or different stages of the same process? One thing is clear—the main problems related to the origin of granites are independent of their setting and age, at least from the Proterozoic onwards. This means that their solutions cannot be local. Partial answers to some of these questions are proposed in this Special Publication. Although the answers are not definitive, we believe they contribute to a better understanding of the granite problem.

However, one major question remains: why are granites a problem? Different researchers from different schools would give very different answers to this question. Indeed, some colleagues might say that granites are not a real problem. In our opinion, granites remain problematic because our view has normally been restricted to a specific facet of the problem. During granite generation, distinct physical and physico-chemical processes control the complex systems. Some of these processes (for example, thermodynamics and kinetics at the site of partial melting, non-linear rheology during segregation and ascent of magma, interacting tectonic processes during emplacement of magmas) have been identified, but other processes, as yet unidentified, may also be involved. A geologist who has specialized in one particular facet may find it difficult to clearly envisage the remaining aspects of the problem. This is similar to the well-known joke about three or four blind people who investigate an elephant by touching its trunk, or leg, or tail. Each of them provides a partial description of the animal he touched, but none describe it correctly as an elephant. A related problem is the diversity of ways in which geological objects are studied. Hutton started the granite debate with field evidence: later on, we incorporated crucibles; and nowadays, we have a diverse set of objects, including batholiths, crucibles and geochemical/geophysical models. Over the years, we have not only incorporated new tools and new rationales, we have also developed the tools and made them increasingly sophisticated. The incorporation of these new methods and new rationales, as well as their increasingly technical nature, have raised the granite problem to extreme complexity. To understand a small part of granite generation, all these techniques must be integrated together.

Because of the complexity of the problem, we need to conduct our current research in a multi-disciplinary and collaborative way, integrating several techniques and corroborating our findings with field observations. This is the philosophy that guided us in editing this book: the integration of new and classical techniques in order to understand problems on the origin of granites. In this sense, many chapters of the present book are review papers dealing with the development and achievements of a particular technique. Other chapters deal with the application of a number of techniques to a specific problem.

Whether granites represent pure melts or, by contrast, are a mixture of restitic crystals with a low melt fraction, is still a matter of debate. It has important consequences on the rheological behaviour of the granite magmas and, consequently, on their ability to ascend through dykes or to infiltrate surrounding rocks. As discussed by **Dingwell**, considerable experimental progress has been made in the past years to quantify the physical properties that control the rheological behaviour of granite melts. A review of the different rates of orogenic behaviour and implications for crustal melt generation and magmatic ascent is given by **Thompson**. The influences of time and space scales are considered in this review, as well as the rheological layering and evolution of the lithosphere.

New aspects of the 'room problem', originally raised by Read (1958), are covered in some detail. The contribution of detailed geological maps of granite massifs, together with the application of geophysical techniques and scale-model laboratory experiments, have been of special relevance to understand how granites are emplaced and accommodated within the continental crust. **Améglio & Vigneresse** review geophysical surveys to provide data on the shape at depth of granitic intrusions. **Román Berdiel** compares the results of scale-model experiments with the geometry of natural plutons emphasizing the importance of extensional tectonics in favouring the intrusion of granite magmas into the continental crust. The question of the deep shape and relation with emplacement is also addressed by **Trzebski, Lennox & Palmer**, who present gravity data from the eastern Lachlan foldbelt, Australia.

A similar approach is also given by **Cruden, Sjöstrom & Aaro**, with application to the Gasborn granite in Sweden.

Another matter of debate is the proportion of crustal and mantle components involved in granite generation. **Patiño Douce** argues with experimental data that most granites, including the so-called S-type granites, have basaltic components in their compositions. The experimental contribution has been crucial in revealing new insights into the granite problem. Assimilation of crustal rocks by basalt magmas is a plausible hypothesis that must be taken into account.

Field examples of granite batholiths are the prime evidence for the interpretation of the emplacement mechanisms. One of the most important examples of passive emplacement of plutons in the upper continental crust is the Peruvian batholith in the Andes. **Cobbing** emphasizes the existence of a complex fracture control on the geometry of plutonic bodies from the Coastal batholith of Peru, a control that may be deduced from the observed field relationships at all scales. Another observation made by this author is the coalescence of basic and granitic magmas with symplutonic relationships. This is particularly important for the gabbro–granodiorite association commonly observed in many granite complexes. During ascent, basic magmas and granite may use the same conduit, allowing the possibility of mingling and mixing and favouring hybridization. This is illustrated by **Menéndez & Ortega** in their study of the Guitiriz granite complex in northern Spain. These relationships have important implications in understanding the generation of granite magmas in the continental crust because they imply that magmas from the mantle were intruded contemporaneously with partial melting of crustal materials. The existence of a relationship between mantle reactivation and crustal melting is a prerequisite to account for the energy budget in crustal recycling processes. The interaction between successive magma batches on the bulk disposition of facies is examined by **Hecht & Vigneresse** through two examples from granitic massifs of Cabeza de Araya, Spain and Fichtelgebirge, Germany. In both cases, the petrographical zoning appears to be dynamic, and related to a competition between magma ascent and crystallization.

Combination of geological inferences and theoretical analysis is also decisive in understanding the granite problem. **Fernández & Castro** investigate the rheological behaviour of granite melts. The result of this combined study is that granite melts may flow and fracture depending on their strength, the imposed stress and deformation rate, and this, in turn, may be dependent on local conditions. These results explain the observed field relationships in the Central System batholith (Spain) and may account the formation of large granite batholiths by a mechanism of repeated dyke-in-dyke intrusion. The textural evolution of megacrysts might be a complex process involving either growth and dissolution of crystals, according to a mechanism proposed by **Higgins** in his study of the Cathedral Peak granodiorite in California. His model combines igneous and metasomatic theories, and may explain the presence of early orientation fabrics in megacrysts, although very particular cooling histories are envisaged by the author to account for the crystal size distribution in granitoids.

Emplacement of plutons into crustal-scale shear zones is one of the main mechanisms invoked nowadays for the origin of large granitic masses. **Alonso, Carracedo & Aranguren** provide a multidisciplinary approach to this problem, combining field, petrographical, geochemical and geophysical (anisotropy of magnetic susceptibility) data to propose an emplacement model for the Campanario–La Haba pluton in central Spain. The outcome is a model for emplacement in extensional fractures, related to 'a NW–SE dextral shear zone parallel to the regional foliation. These results show how the tectonic evolution controls the location and shape of the granitic plutons that became emplaced into the anisotropic upper crust of the central Iberian massif. A different story is envisaged by **Dietl** for the Joshua Flat–Beer Creek pluton in Eastern California. He studied in detail the structures in the contact aureole, including field and petrographic descriptions, strain measurements, quartz c-axes fabric determinations and measurements of the anisotropy of magnetic susceptibility in the pluton. According to Dietl, stoping, dyking, downward return flow and assimilation are the mechanisms that acted during the emplacement of this pluton.

Processes whereby granite magmas are supplied from a migmatitic source to generate granite plutons are discussed by **Sawyer, Dombrowski & Collins**. Their study is based on a field example from the Wuluma Hills, Australia, and highlights the influence of coeval migmatization and deformation in granite magma movement. The relationship between migmatites and granites is reinforced by **Mouri & Korsman**, who analyse an example from the Svecofennian of the Tampere area (southern Finland).

In summary, multidisciplinary studies, applying a variety of techniques (geophysical and geochemical, mapping and experimental, geochemical and experimental, etc.), have resulted in significant advances in the understanding of granites. This is the philosophy of this Special Publication: showing by various examples how our knowledge of granites has advanced. However, many important questions still remain unanswered. These include the rate at which granites are produced and emplaced within the continental crust, and the role played by the mantle during granite production, both as a heat source and as a chemical reservoir for crustal contamination. Such questions are only partly addressed in the present volume and will be the focus of future investigations on the origin of granites. We hope that the philosophy behind this volume—the integration of new and classical techniques—will lead to further developments in the understanding of granites in the future.

References

BALK, R. 1937. *Structural behavior of igneous rocks.* Geological Society of America Memoirs, **5**.

BOWEN, N. L. 1922. The behavior of inclusions in igneous magmas. *Journal of Geology*, **30**, 513–570.

BROWN, G. C. & FYFE, W. S. 1970. The production of granitic melts during ultrametamorphism. *Contributions to Mineralogy and Petrology*, **28**, 310–318.

BROWN, M., CANDELA, P. E., PECK, D. L., STEPHENS, W. E., WALKER, R. J. & ZEN, E-AN 1996. Proceedings of the 'Third Hutton Symposium on the Origin of Granites and Related Rocks'. *Transactions of the Royal Society of Edinburgh, Earth Sciences*, **87**, 1–361.

BROWN, P. E. 1988. Proceedings of the 'First Hutton Symposium on the Origin of Granites'. *Transactions of the Royal Society of Edinburgh, Earth Sciences*, **79**, 71–346.

—— & CHAPPELL, B. W. 1992. Proceedings of the 'Second Hutton Symposium on the Origin of Granites and Related Rocks'. *Transactions of the Royal Society of Edinburgh, Earth Sciences*, **83**, 1–508.

CHAPPELL, B. W. & WHITE, A. J. R. 1974. Two contrasting granite types. *Pacific Geology*, **8**, 173–174.

PATIÑO DOUCE, A. E. 1995. Experimental generation of hybrid silica melts by reaction of high-Al basalt with metamorphic rocks. *Journal of Geophysical Research*, **B100**, 623–639.

PITCHER, W. S. 1982. Granite type and tectonic environment. *In*: HSU, K. (ed.) *Mountain Building Processes*. Academic Press, London, 19–40.

—— 1997. *The nature and origin of granite*, 2nd ed. Chapman & Hall, London.

RAMBERG, H. 1981. *Gravity, deformation and the Earth's crust*, 2nd ed. Academic Press, London.

READ, H. H. 1958. *The Granite Controversy*. Thomas Murby & Co., London.

SANDER, B. 1930. *Gefugekunde der Gesteine*. Springer
Verlag, Berlin.

THOMPSON, A. B. 1982. Dehydration melting of pelitic
rocks and the generation of H_2O-undersaturated
granitic liquids. *American Journal of Science*, **282**,
1567–1595.

TUTTLE, O. F. & BOWEN, N. L. 1958. *Origin of granite
in the light of experimental studies in the system
$NaAlSi_3O_8–KAlSi_3O_8–SiO_2–H_2O$.* Geological
Society of America Memoirs, **74**.

Some time–space relationships for crustal melting and granitic intrusion at various depths

ALAN BRUCE THOMPSON

Dept. Erdwissenschaften, ETH Zurich, CH-8092, Switzerland (e-mail: alan@erdw.ethz.ch)

Abstract: Crustal melting occurs at the higher grades of metamorphism. Migmatites reflect crustal anatexis without necessary additional heat supply. Granites usually result from crustal melting resulting from 'hot' geotherms, or reflect an external heat supply.

The commonest mechanisms of crustal melting are: (i) decompression of crust thickened into mantle in convergent orogens; (ii) asthenosphere upwelled beneth the crust in extensional orogens; (iii) massive invasion or underplating of crust by mantle magma generated by mantle decompression. This paper considers some of the processes governing timescales and lengthscales of these various magmatic events with particular view to determining the intensity and duration of crustal melting, and relationships between the rates of melt extraction, ascent and emplacement.

Most intrusive granites and *in-situ* migmatites show geochemical characteristics of being generated by partial melting of crustal materials (e.g. Burnham 1997). *In-situ* granitic migmatites most likely represent small fractions of quite viscous melt, whereas intrusive granites represent injections of larger melt fractions of less-viscous felsic magma. Here we distinguish the migmatitic and granitic rocks mostly of metasedimentary origin (consisting of subequal amounts of plagioclase, alkali feldspar and quartz) from the felsic 'granitoids' (such as tonalite and granodiorite) produced by melting of amphibolitized basalt, or during the end-stages of fractionation of magmas generated by hydrous melting of fertile mantle peridotite.

Mid-crustal migmatites (*in-situ* melts of high-grade metamorphic rocks) require that heat was introduced or concentrated at these depths, whereas intrusive granites indicate that crustal melts from deeper in the crust were transported upwards. Mid-crustal heat sources include (i) concentrations of radiogenic elements (e.g. Chamberlain & Sonder 1990; Sandiford & Hand 1998), (ii) injected hot granites (Finger & Clemens 1995) or (iii) intrusion of mantle-derived basaltic magma (e.g. Shaw 1980; Spera 1980; Huppert & Sparks 1988*a, b*). In other cases of crustal melting the heat source is the mantle and the commonest mechanisms of crustal melting occur from (iv) thickening continental crust so that the buried root is subjected to higher temperatures by sideways heating from hotter mantle, (v) underplating of continental crust at the MOHO by basaltic magma or (vi) asthenosphere upwelled

beneath thinned crust. All cases involve addition of heat at a particular depth (isobaric heating), whereas the case of exhumation of thickened crust in addition can involve decompression through an appropriate solidus reaction (decompressional dehydration melting, Fig. 1).

It is useful here to view the behaviour of crustal melts in terms of sequential processes in the source depths, in the ascent regions and at the different emplacement levels in the crust. This paper considers some aspects of the timescales and lengthscales of these processes with a view to establishing which high temperature events could result in particular kinds of crustal melting at different depths and times in an orogenic history. It will be emphasized that quite different mechanisms of crustal melt generation and magmatic ascent operate at different crustal levels and at different times during orogeny. This reflects lithospheric layering of fertile rock composition and the changing distribution of heat sources and rheology (e.g. Vigneresse 1995). It is also very useful to examine the behaviour of crustal melt within the system in terms of volumetric flow rates of extraction from the source region (Q_W), ascent (Q_A) and emplacement (Q_E), using the nomenclature of Cruden (1998).

Some general considerations about crustal melting, *in-situ* migmatites and intrusive granites

The geotherm in stable continental crust is estimated to lie no higher than about 500 °C at

From: CASTRO, A., FERNÁNDEZ, C. & VIGNERESSE, J. L. (eds) *Understanding Granites: Integrating New and Classical Techniques.* Geological Society, London, Special Publications, **168**, 7–25. 1-86239-058-4/99/$15.00 © The Geological Society of London 1999.

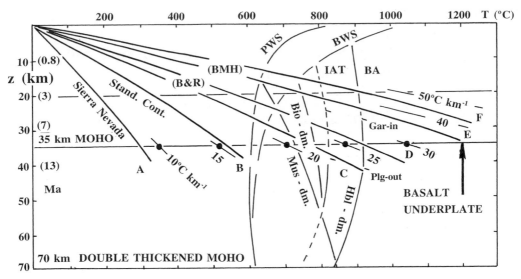

Fig. 1. Conductive geotherms for continental crust together with the location of melting reactions for various crustal rocks at H_2O-saturated conditions (PWS, pelite wet solidus; BWS, basalt wet solidus), and for dehydration melting (muscovite, Mus-dm; biotite, Bio-dm in pelite; for hornblende in basaltic amphibolite (BA) and for biotite and hornblende in Island Arc Tholeiite (IAT)—from summary by Thompson & Connolly (1995, fig. 7). The geotherms labelled A to F, calculated by Lachenbruch & Sass (1978, fig. 9-5), correspond to values of mantle (reduced) heat flow (Q_m) and surface heat flow (Q_s) of respectively (mW m^{-2}) A = 16.7, 37.6; B = 33.4, 54.4; C = 50.2, 71.1; D = 66.9, 87.8; E = 83.7, 104.6; F = 104.6, 121.3, computed for radiogenetic heat production in the top 10 km, A_0 = 2.1 µW m^{-3}, and thermal conductivity, K = 2.5 W m^{-1} K^{-1}. These authors (fig. 9.5) relate the calculated geotherms to temperature distributions they consider appropriate to A, Sierra Nevada; B, Standard Continent; C–D, Basin and Range (B&R); E–F, Battle Mountain High (BMH).

The thermal conduction time constant (Ma) for particular crustal thicknesses ($\tau = z^2/4\kappa$) is shown for a thermal diffusivity of $\kappa = 10^{-6}$ m^2 s^{-1} along the left axis. The light dashed lines through the solid circles at the 35 km Moho show simplified linear geothermal gradients (10 to 50 °C km^{-1}).

a 35 km MOHO (Fig. 1; gradient of about 15 °C km^{-1}). These temperatures are lower than the solidus temperatures for even H_2O-saturated minimum melting of two feldspar + quartz + mica rocks (Fig. 1, Table 1). The pelite wet solidus, PWS, lies at about 10 °C lower than the temperatures of the H_2O-saturated, or 'wet', granite solidus (e.g. Wyllie 1977). The solidi for these metasedimentary rocks all lie at lower temperatures than the basalt 'wet' solidus (BWS, e.g. Wyllie 1977; Burnham 1997), which are approximated by the melting of plagioclase + quartz + H_2O. Such H_2O-saturated solidi only apply when there is sufficient H_2O to saturate the melt with a free H_2O-fluid phase. For likely porosities of lower crustal rocks (<1%) and measured solubilities of H_2O in granitic liquids (>15 wt% at 10 kbar, Burnham 1997), such melt fractions are not likely to exceed a few percent (e.g. Thompson & Connolly 1995). H_2O-saturated solidi are not the reactions to be considered for crustal melting, but are quite appropriate to be considered for the granite solidus upon magmatic cooling, because H_2O is

continually concentrated into decreasing melt fractions. Appropriate for crustal melting solidi are the fluid-absent (dehydration-melting) reactions where hydrous minerals deliver their dehydration H_2O directly to a partial melt, without the appearance of a free H_2O-fluid phase. These reactions occur at temperatures higher than the H_2O-saturated solidus but much lower than the anhydrous (dry) solidi for feldspar + quartz.

Crustal melting temperatures and geothermal gradients

Crustal anatexis requires anomalously high temperatures within continental crust through localization of some heat source, or by decompression of heated rock during crustal thinning, through a dehydration-melting solidus with positive dP/dT. A finite overstepping of these solidi is required for a sufficiently large melt fraction (>30 vol%, e.g. Wickham 1987) to form and to begin segregation. The dehydration melting reactions shown in Fig. 1 show quite

Table 1. *Physical quantities, abbreviations and representative values used*

Mantle (reduced) heat flow, Q_m 16.7–104.6 mW m^{-2}
Surface heat flow, Q_s 37.6–121.3 mW m^{-2}
Thermal conductivity, $K = 1.5$–2.5 W m^{-1} K^{-1}
Radiogenetic heat production, $A_o = 2.0$–2.5 μW m^{-3}
Thermal diffusivity, $\kappa = 1.0$–1.2×10^{-6} m^2 s^{-1}
Exhumation rate, 0.17–0.70 mm a^{-1} (35 mm a^{-1} = 35 km Ma^{-1})
Latent heat of melting, $L_{basalt} = 500$, $L_{crust} = 200$, $L_{andesite} = 220$ kJ kg^{-1}
Heat capacity ($Cp_{crust} = 1$ kJ kg^{-1}; $Cp_{basalt} = 1$ kJ kg^{-1})
T_{MOHO}, 760 °C
Incubation time, 20 Ma
($t_{sp} = d^2/\kappa = (5\ \mathrm{km})^2/10^{-6} = 0.86 \times 10^6$ years apart)
z_{MOHO}, 35 km
Whole lithosphere homogeneous thinning factor 100%, $\beta = 2$
Radiogenic granite slab ($A_{gr} = 4.1$ μW m^{-3}) emplaced initially at 18–22 km depth
τ_1 (years), time when a basalt sill reaches 60% crystallinity
τ_2 (years), time when a crustal melt layer itself becomes 50% crystalline
One-sided (half) spreading rate (**u**) mm a^{-1}
Intermediate viscosity range of magma 10^3–10^6 Pa s
H_2O fluxes through lower-crustal rocks 10^{-10}–10^{-12} m^3 m^{-2} s^{-1}
$\dot{\gamma}$, shear strain-rate 10^{-9}–10^{-17} s^{-1}
$\dot{\varepsilon}$ bulk strain-rate 10^{-9}–10^{-17} s^{-1}

distinct temperature requirements for different lower crustal rock compositions (e.g. at a 35 km MOHO the following beginning-of-melting temperatures are indicated: muscovite pelite ≈ 725 °C, biotite + plagioclase + quartz gneiss ≈ 800 °C, intermediate metavolcanic ≈ 825 °C, basaltic amphibolite ≈ 910 °C). The attainment of 25 vol% melt requires additional temperature increase (20–60 °C) depending on the shape of the melt fraction (f) versus temperature curve.

The steady-state geotherms discussed by Lachenbruch & Sass (1978, fig. 9-5), for Sierra Nevada (A) and Standard Continent (B) lie at temperatures lower than even PWS (Fig. 1), whereas geotherms for Basin and Range (B&R, C to D) and Battle Mountain High (BMH, E to F) require an additional heat source (such as a basaltic underplate at 1200 °C).

Optimum crustal melt production and depths of crustal melting

The conditions for optimum crustal melt production are governed by source fertility (which is understood as how close the crustal rock composition approaches the granite eutectic/minimum melt composition, e.g. Thompson 1996), the amount of H_2O at the melting site, the ambient local temperature, and the nature (intensity, temperature and location) of the heat supply. Layered rocks of different fertility may generate quite variable amounts of melt and likewise local rheology can exert quite strong controls on melt segregation and mobility. For the case of crustal melting at a particular depth the anatectic source regions can sometimes be identified by introduced or proximate heat sources (e.g. Brown 1994; Brown et al. 1995). If these melts can segregate and ascend, their ability to cause further crustal melting by 'contact melting around hot granites' decreases progressively with higher emplacement level reflecting the temperature difference to the ambient shallower crustal rocks.

Models of decompression dehydration-melting of thickened crust show generation of partial melt mainly in the lower part just above the deepened MOHO. Such deep migmatites are consistent with the higher temperatures required by dehydration melting of micas in metasediments (Fig. 1). It has already been noted that migmatites at shallower depths could indicate lower temperature solidus conditions (if free H_2O were available) (PWS in Fig. 1, but then only a few percent melt would result, e.g. Thompson & Connolly 1995) and such migmatites could never become granites. Shallow level migmatites thus indicate that a mid-crustal heat source was present (because of the large temperature increase required to reach the dehydration-melting solidi), or that the migmatite terrains were tectonically emplaced into the mid crust.

Migmatites at different crustal levels

Many migmatites remain *in-situ* and injected migmatites do not appear to have risen far from their sites of generation (e.g. Mehnert 1968). Low temperature (610–650 °C) *in-situ* mid-crustal migmatites may well be evidence for H_2O-saturated melting (e.g. Clemens 1990; Thompson 1990), but require local sources of H_2O (such as nearby dehydration or a mid-crustal aquifer).

Much higher temperatures are required to induce dehydration melting in metasedimentary schists and gneisses when no free H_2O is present compared to the H_2O-saturated cases (e.g. at 10 kbar ≈ 35 km, biotite + plagioclase + quartz requires about 810 °C compared to 620 °C when wet). Often high temperature (T)–low pressure (P) metamorphic terrains may contain exposed migmatites that generally appear to reflect regional scale contact metamorphism induced around intruded gabbroic mantle magmas (e.g. Barton & Hanson 1989; Barton *et al.* 1991). If some migmatites have originated by contact melting around basalt they should show spatial relations to abundant mafic intrusions and show contact metamorphic effects from subsequent magma pulses (e.g. McLellan 1989).

Depth of emplacement of granites

In their survey of depth of emplacement of plutons, Barton *et al.* (1991, p. 800) concluded that granitic magmas intruded down to depths of about 5 km were emplaced into opening cavities (by displacement of the free boundary of the Earths surface?), whereas plutons at depths of 8–12 km intruded beneath a rigid rock barrier. In general, a mid-crustal brittle cap of quartz-dominated rheology down to 8 km would limit diapiric granitoid ascent (at 300 °C, a gradient of $300/8 = 37.5$ °C km^{-1}; e.g. Bergantz 1991, p. 25), but would permit dyke ascent through brittle fractures. Thus the depth of intrusion beneath an impermeable layer is controlled partly by amount of magma, and its buoyancy, but also by the rate of opening of a cavity compared to rate of freezing of the injected magma. The room problem is avoided in transtensional settings (Glazner 1991; Hutton 1992).

We will return later to the problems of determining which aspects of emplacement or ascent are related to the granitic magma and which aspects are related to the surroundings. It is most useful next to consider the processes in the various geological situations where crustal melting can occur. One main problem to be considered is how the ambient geotherm is raised above the relevant solidus temperature of crustal melt rocks to a point where the granitic melt fraction is sufficient to be extracted. The base of the continental crust is the optimum level to attain the high temperatures for crustal melting. The first order controls on the process of crustal melting are the ambient temperature relative to the rocks' solidus, the temperature of, and the conductive length to, the heat supply.

Extent of crustal melting in thickened crust and in the roots of mountains

Thickening of the continental crust causes perturbation of the geotherm during the formation of mountain roots. External heating of buried continental crust by the adjacent mantle and internal heating from decay of concentrated radiogenic K, U and Th causes the buried rocks to heat during their exhumation (e.g. England & Thompson 1984). Several modelling studies have a bearing on determining the extent of crustal melting in space and time. Models for regional metamorphism in continental collision orogens show that crustal decompressional dehydration-melting is possible in felsic and pelitic lithologies simply from thermal relaxation of the standard geotherm perturbed by thickening during formation of a mountain root. The melting occurs only in the base of thickened crustal root within 10 km above thickened Moho (see Thompson & Connolly 1995). Up to about 25 vol% crustal melting is thus possible in muscovite pelites and biotite gneisses without incursion of additional mantle heat into the crustal root. Mantle involvement in crustal melting is more obvious for the case of underplating of continental crust at the Moho by basaltic magma.

Anatexis of metapelites for thermal relaxation of standard geotherms in thickened continental crust

The crustal collision Case (a) in Fig. 2 for thrust thickening from 35 to 70 km perturbs a 'standard' geotherm by instantaneous doubling of 35 km crust through a saw-tooth thrusted geotherm ($K = 2.25$ W m^{-1} K^{-1}; $Q_m = 30$ mW m^{-2}; $A_o = 2.0$ μW m^{-3}; e.g. England & Thompson 1984; central case in their fig. 3e). After thickening then occurs a 20 Ma incubation time-lag followed by linear erosional thinning, at the rate of 35 km/100 Ma (0.35 mm a^{-1}), until 120 Ma. The melting history results from evolution of geotherms through the solidus reactions summarized in Fig. 1 (see Thompson & Connolly 1995). Melt fractions at each crustal depth were obtained from the various experimental determinations of melt fraction with increasing T at 10 kbar (summarized by Gardien *et al.* 1995) by scaling parallel to the illustrated P–T slopes of the melting reactions, shown in Fig. 1. The contours of 5 wt% melt increase (Fig. 2a), relative to the H_2O-saturated pelite solidus

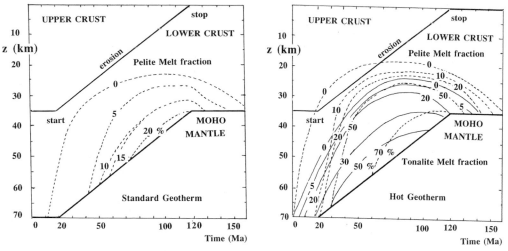

Fig. 2. Depth (km)–time (Ma) diagram showing melt production by dehydration melting of pelite and tonalite in continental collision orogens during exhumation of 35 km crust which has been doubly thickened. A 20 Ma incubation time-lag is followed by linear erosional thinning, at the rate of 35 km over 100 Ma (=0.35 mm a^{-1}) for thrust thickening of two different geotherms; (**a**) standard geotherm, $K = 2.25$ W m^{-1} K^{-1}, $Q_m = 30$ mW m^{-2}; $A_o = 2.0$ μW m^{-3} (e.g. Thompson & Connolly 1995, p. 15570, fig. 5); (**b**) hot geotherm, $K = 1.5$ W m^{-1} K^{-1}; $Q_m = 40$ mW m^{-2}; $A_o = 2.33$ μW m^{-3} (e.g. De Yoreo et al. 1989). In both of these models the melt was retained in situ (even as long as 150 Ma, i.e. 30 Ma after erosional uplift stopped at 120 Ma). In (a) the contours of 5 wt% melt increase relative to the H$_2$O-saturated pelite solidus (PWS from Fig. 1, labelled 0%). In (b) the outer melt contours are for pelite (dashed lines) and inner contours for amphibolite (solid lines, relative to BWS from Fig. 1, labelled 0%). Melt fractions were obtained, by scaling parallel to P–T slopes of melting reactions, from various experimental results of melt fraction (f) v. T at 10 kbar (summarized by Gardien et al. 1995).

(PWS from Fig. 1, labelled 0%), and the overlap of 10 and 15% melt fractions results from the dehydration-melting steps in the solidus curves (see also Bergantz & Dawes 1994). This history leads to a maximum melt generation of less than 25% in 'fertile' metapelite lithologies just above the buried Moho. These modest melt fractions are likely to remain close to their layers of origin as migmatites (e.g. Wickham 1987) and thus have been retained in-situ in the illustration in Fig. 2a. An important conclusion here is that while dehydration-melting of metasediments (muscovite pelites to biotite gneisses) is possible from perturbation of the standard geotherm, it is not possible for amphibolites to undergo fluid-absent melting.

A further major control on the degree of melting is the rate of thinning of thickened crust. The rate of 0.35 mm a^{-1} produces the effects illustrated in Fig. 2a, i.e. a maximum temperature of about 860 °C for a sample beginning exhumation from 70 km. Faster exhumation results in lower temperatures along P–T–t paths. Thus if this rate is doubled (c. 0.70 mm a^{-1}), or halved (c. 0.17 mm a^{-1}), then maximum temperatures of 810 °C, 910 °C, respectively are reached, rather than 860 °C, for a 70 km sample. These latter

higher temperatures would result only in beginning of melting in amphibolite, whereas higher melt fractions in metapelites would be evidence for very slow exhumation rates.

Anatexis of metapelites and metavolcanics for 'hot' geotherms

Much greater melt fractions (50–60 vol% in Fig. 2b) will be generated with 'hotter' initial geotherms ($K = 1.5$ W m^{-1} K^{-1}; $Q_m = 40$ mW m^{-2}; $A_o = 2.33$ μW m^{-3}; e.g. De Yoreo et al. 1989). Such larger amounts of melt will exhibit greater tendency for magma extraction. In Fig. 2b the outer melt contours are for pelite (dashed lines, relative to PWS) and inner contours for amphibolite (solid lines, relative to BWS from Fig. 1). Moreover, hotter geotherms would result in partial melting throughout metapelites in the lower 35 km of the thickened column and into the upper part. The degree of melting becomes even more prominent (see England & Thompson 1984, fig. 3g & h) for lower values of thermal conductivity (more basaltic rocks in the lower crust, e.g. Furlong et al. 1991, p. 443). In this case (Fig. 2b) the crustal melting source is longer lived, of the order of 100 Ma, as governed by the time-lag of the buried root and the duration of the exhumation. Such high melt fractions would not remain *in-situ*, although they have been constrained to do so for the illustration of Fig. 2b to be able to relate the amount of melt production to heat supply. Because the melt fraction contours intersect at changing angles through the exhumation history shown in Fig. 2b, the melt proportion in the two lithologies at any horizon is distinctive of a particular crustal depth and particular time. Melt extraction from meta-pelite layers (as indicated by proportion of restite to migmatite) will always be more efficient than from amphibolite assuming that both would be characterized by similar 'critical melt fractions'.

Anatexis of fertile lithologies for 'hot' geotherms with a highly radioactive mid-crustal layer

Patiño Douce *et al.* (1990) combined a 'hot' geotherm model for a layered crust with $Q_s = 60$ mW m^{-2} and $Q_m = 28$ mW m^{-2}, with a highly radioactive layer located between 12 and 25 km with $A_o = 2.5$ µW m^{-3}. This configura-tion resulted in initial T_{MOHO} of 760 °C at 38 km. These values generate even more upper crustal melting than the case illustrated in Fig. 2b, partly because of the shallower depth to z_{MOHO} of 54.5 km at 30 Ma, and partly because of 20% more heat generation than in the cases shown in Fig. 2.

Very high amounts of U and Th, physically concentrated from the sedimentary deposition stage, have been considered capable of inducing *in-situ* crustal melting (e.g. Chamberlain & Sonder 1990; Sandiford *et al.* 1991). Both of these mechanisms increase the heat supply and will result in an increase of temperature and thus degree of anatexis. However, such 'hotter than standard' geotherms require an addition of mantle heat (either as basaltic magma, or as uplifted asthenosphere), or mechanisms of con-centrating radiogenic elements exactly at the right place and time during the orogenic cycle.

Space and time limitations on crustal melting in thickened crust without addition of mantle magma

From the several investigations of decompression melting of crustal rocks (Fig. 1) it has been shown that up to 25% melt can be generated from thermal relaxation of standard geotherms (Fig. 2a) with an incubation time of 20 Ma. Increasing mantle heat flow (Q_m) and radiogenic

heat production (A_o), or decreasing thermal con-ductivity (K) within expected ranges (Table 1), will induce extensive dehydration-melting of metapelites and permit substantial melting of amphibolites for hot initial geotherms (Fig. 2b; see De Yoreo *et al.* 1989; Patiño Douce *et al.* 1990; Thompson & Connolly 1995).

Such hot initial geotherms are themselves abnormal as they could imply that the lower continental crust is permanently partly molten. That this does not appear to be the case from seismic evidence (e.g. Mooney & Meisner 1992) indicates either that the lower crust is not normally metapelitic, or that hot crust cannot be thickened to 70 km (R. Weinberg pers. comm. 1998), or that such 'hot geotherm' models are too simplistic. It should be noted that a hot initial geotherm could occur during a period of heating immediately prior to crustal thickening. Geolo-gically this would be equivalent to thickening of already extended and heated crust (e.g. P. H. Thompson 1989).

A major factor contributing to the attainment of highest temperatures along the decompression *P–T*-path is the time lag between burial and the beginning of exhumation (England & Thompson 1986; Zen 1988). The duration of the time lag reflects either the timescale of delamination of dense (eclogitic) crustal roots (Richardson & England 1979; Houseman *et al.* 1981), or the restoring forces of the mantle opposing the incursive thickened crustal root. Modelling by Schulmann *et al.* (1999) estimates the duration of the restoring events to be about 10 Ma after continental collision.

Another major aspect of such melting models involving decompression of thickened crust, is that even if partial melt is generated throughout the lower crust these melts remain *in situ* as mig-matites deeper than 25 km. Thus for an erosion-controlled exhumation of thickened crust, these melts if they remained *in situ* would never be exposed in the erosion surface nor even to mid-crustal levels. Thus to appear at higher levels in the crustal column and indeed to be visible at all during an exhumation history, such *in-situ* migmatites must be excavated tectonically, to-gether with their restites, from the source region to some shallower crustal level. Some further addition of heat is needed for crustal melts to be able to segregate from their restite and ascend to shallower crustal levels to appear as intrusive granites. The additional mantle heat sources reflect either lithospheric extension with conco-mitant upraising of the hotter asthenosphere, or delamination of the lower lithosphere to be replaced by hot asthenosphere.

Crust and mantle melting consequent upon lithospheric extension

Pure-shear lithospheric extension (uniform horizontal stretching) results in asthenospheric upwelling (e.g. McKenzie 1978). Thermal relaxation of instantaneously thinned lithosphere according to the formulations of Jarvis & McKenzie (1980) follows exponential decay with a time constant of about 60 Ma. Thus crustal melting and metamorphism within this time range may be caused by the extension.

The numerical experiments of lithosphere extension by Loosveld (1989) encompass quite a range of geological scenarios that lead to various degrees of crustal melting (Fig. 3a). The original work was directed to understanding the role of the mantle asthenosphere in high T–low P metamorphic terrains. I have integrated the results of his numerical experiments for the

purpose of determining lengthscales and time-scales of melting, by simply evaluating the time interval during which a given depth exceeds a specific temperature. The crustal and lithosphere thicknesses are differently effected. For cases A and E in Fig. 3a, the crust is thinned to 17 km; for B, C, D and F the crust remains at 35 km. Cases A, B, C show various degrees of lithosphere extension, whereas D, E and F show different detachment and excision. For case E detachment at 20 km has excised the lower crust, permitting upper crust to directly overlie the mantle. The results of the calculations are similar for A and F, B and D, C and E, and have been shown as 700, 800 and 900 °C isotherms as paired results in Fig. 3a (solid lines, dashed lines and dotted lines, respectively).

Cases A and F produce the highest temperatures (Fig. 3) at the shallowest depths, but no crustal melting (crustal thickness = 17 km) for

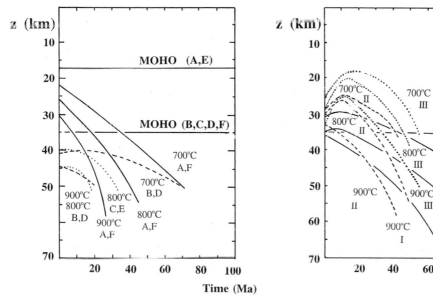

Fig. 3. Depth–time plots of three isotherms (700 °C, 800 °C, 900 °C) marking steps of crustal melting by extension or delamination, integrated from the extensional numerical experiments of Loosveld (1989, figs 6 and 7). The results of the calculations are similar for A and F, B and D, C and E, and have been shown as 700, 800 and 900 °C isotherms as paired results in (a) (solid lines, dashed lines and dotted lines respectively): **(a) A**, whole lithosphere homogeneously instantaneously thinned by 100% ($\beta = 2$; McKenzie 1978); **B, C**, 100%, 300% homogeneous thinning of lithosphere; crust at 35 km; **D**, simple shear thinning of mantle lithosphere (horizontal extension is 100 km) with a 12° detachment at the base of the crust (throw 21 km); **E**, as D but detachment in mid crust (17 km); **F**, complete delamination of mantle lithosphere and replacement by hot asthenosphere ($T_t = 0 = 1350$ °C; Houseman *et al.* 1981). Other parameters: pre-thinning thickness crust (=35 km); lithosphere (=100 km). $K = 3$ W K^{-1} m^{-1}, $\kappa = 1.2 \times 10^{-6}$ m^2 s^{-1}; $A_0 = 2.5$ μW m^{-3} over upper 10 km. A radiogenic granite slab ($A_{gr} = 4.1$ μW m^{-3}) was emplaced initially at 18 to 22 km depth. **(b)** Case I (solid lines), 'cold mode' delamination, where the deeper temperature boundary ($T_a = 1350$ °C at 37.5 km) is instantaneously reset after delamination (Bird and Baumgardner 1981); case II (dashed lines), 'hot mode' delamination, where asthenospheric convection continues for 1 Ma; case III (dotted lines), 'hot mode' delamination, where asthenospheric convection continues for 20 Ma ($T_a = 1300$ °C). Other parameters: T_a at 150 km = 1350 °C; $K = 2$ W m^{-1} K^{-1}, $\kappa = 10^{-6}$ m^2 s^{-1}; $A_0 = 2.5$ μW m^{-3} over upper 15 km.

classical case-A 'McKenzie-type' extension (McKenzie, 1978, p. 22). In contrast, case F (classical delamination beneath 35 km thick crust, Houseman *et al.* 1981) produces minor lower crustal melting of amphibolites (at 900 °C for 10 Ma) and extensive lower to mid-crustal melting of biotite gneisses at about (800 °C for 20 Ma), and muscovite schists (at about 700 °C for 35 Ma).

Case D produced similar results to case B, as does case E to case C. However, in all cases crustal melting ($T > 700$ °C) does not occur in the crust of standard thickness (z_{MOHO} at 35 km). H_2O-saturated melting ($T_{\text{solidus}} \approx 620$ °C at 10 kbar ≈ 35 km) could occur for short times (<20 Ma) for modes B, C, D, E (Fig. 3a), but would require availability of free H_2O fluid exactly at this time in space. Thus only the single extreme case of full-scale delamination where lithosphere is replaced by asthenosphere at 35 km (case F) could result in extensive dehydration-melting of metasediment, but only minor amounts of basaltic amphibolite melting.

Delamination of mantle lithosphere

In the extreme cases of mantle lithosphere attenuation (e.g. Loosveld & Etheridge 1990), or complete excision by detachment of dense eclogite (e.g. Houseman *et al.* 1981), hot asthenospheric mantle directly underlies lower crust (35 km thick in Fig. 3b). The examples in Fig. 3b are extended from case F in Fig. 3a.

Three experiments by Loosveld (1989, fig. 7) have been evaluated in terms of depth-time evolution in Fig. 3b. For 'cold-mode' delamination (Fig. 3b, solid lines, case I), the deeper temperature boundary ($T_a = 1350$ °C) is instantaneously reset after delamination at 37.5 km (Bird & Baumgardner 1981). 'Hot mode' delamination, where asthenospheric convection continues in the upwelled asthenosphere for 1 Ma (Fig. 3b, dashed lines, case II, $T_a = 1300$ °C at 35 km) or 20 Ma (Fig. 3b, dotted lines, case III, $T_a = 1350$ °C at 35 km). For the isotherms illustrated, 'cold mode' delamination (case I) results in melt production for 5 to 6 km above the 35 km Moho lasting up to about 40 Ma in muscovite (700–750 °C), but hardly at all in biotite lithologies (800–850 °C). 'Hot mode' delamination produces widespread lower crustal melting for case III (20 Ma convection, Fig. 3b) even of 900 °C amphibolites for 35 Ma. Amphibolites would just melt (at 900 °C) to 10 Ma for case II (1 Ma convection, Fig. 3b) and melting of biotite and muscovite would be extensive.

Thus crustal melting would be a usual consequence of delamination of mantle lithosphere,

but rarely a consequence of lithospheric extension unless basaltic melts advectively transfer the heat. Decompression melting of the mantle during lithospheric extension or delamination is likely. Thus basaltic magmas will in some cases underplate the lithosphere and in other cases intrude it. Such ponded basaltic magma can induce widespread melting in fertile lithologies of the lower crust following some period of preheating during which the ambient geotherm is increased by the earlier phases of underplating.

Basaltic underplating and crustal melting

'When ascending mantle magma reaches the base of the lithosphere it is likely to form pools of magma' (Turcotte 1987, p. 69). The calculations by Wells (1980) of the effects of crustal thickening by magmatic-accretion resulted in very high relative temperatures at all crustal levels. The thickness of the magmatic underplate depends upon the rate of extension in rift zones, and ablative subduction beneath island arcs.

Calculations of the lengthscales and timescales of underplating, such as those of Waters (1990, p. 250), indicate that for intrusion of basaltic magma at 1200 °C (latent heat, $L_{\text{basalt}} = 500$ kJ kg^{-1}) into crustal depths of 25–35 km, enormous thickness of underplate are required. To achieve 800 °C at 20 km depth, initial basaltic intrusion thicknesses of 10, 25 and 60 km underplated at depths of 25, 30, 35 km are required.

'The emplacement and crystallisation of individual basaltic sills, and the generation of felsic crustal melts above the sills are short-lived events compared to the overall duration of the crustal-scale orogenic event' (Waters 1990, p. 250). The longer the time gap between episodes of underplating, the more the thermal effect of individual intrusions is dissipated by conduction (see Hanson & Barton 1989). Thus, a crustal melting episode lasting 30 Ma requires an underplate of more than double the thickness needed at 5 Ma. Thinner underplating is required if the magmatic accretion is contemporaneous with extension. The difficulty of inducing much crustal melting at shallower levels, because of the low ambient gradients lies behind the extremely large amount of basaltic underplate required to cause Namaqualand granulite formation (Waters 1990), or extensive crustal melting (e.g. Wickham 1987; Brown *et al.* 1995).

Timescales for basaltic underplating

Timescales for basaltic underplating obtained by Loosveld & Etheridge (1990, p. 264) were calculated for exponential decay of mantle heat flow

($Q_m = 35\,\mathrm{mW\,m^{-2}}$) following underplating of basaltic magma. The thermal time constant used by them ($\tau_c = d^2/\pi^2\kappa$) means that for thermal diffusivity (κ) $= 10^{-6}\,\mathrm{m^2\,s^{-1}}$, for $d = 25$–60 km, then $\tau_c = 2$–12 Ma, where d is the half-width of the basaltic sill and π^2 is a scaling factor.

Volumes and rates of basalt underplating

Crustal melting near the Moho, or at any other depth to where basaltic sills can intrude, is quite efficient if the intrusions are emplaced within a few half widths and within several d^2/κ years of each other (Hanson & Barton 1989, p. 10 369). The efficiency of this process in natural situations can be examined if we consider ranges of magma flux. For example the total volume of the Sierra Nevada plutons to a depth of 10 km is about $10^6\,\mathrm{km^3}$ over an area of about $10^5\,\mathrm{km^2}$, and the duration of the overall emplacement of the order of 10^8 years (Shaw 1980). This gives a rate of about $10^{-2}\,\mathrm{km^3\,a^{-1}}$. Hanson & Barton (1989, p. 10 369) have calculated required magma fluxes to maintain emplacement relative to the dimensionless time of $\kappa t/d^2 = 1$. Thus for plutons of half width ($d = 5$ km) a magma flux of c. $200\,\mathrm{km^3\,km^{-2}}$ per 10^6 years is required over a time interval of ($t_{sp} = d^2/\kappa) = (5\,\mathrm{km})^2/10^{-6} = 0.86 \times 10^6$ years apart. While the length and timescales for basaltic magmatism reflect the processes of mantle melting, the emplacement depth and quantity of injected basaltic magma can be a localized heat source for crustal melting.

Basaltic magma sills as heat sources for crustal melting

The Moho and upper/lower crust lithology changes present density and viscosity barriers for entrapment of basaltic magma (e.g. Turcotte 1987). The ponding of basalt at these depths either requires that the magma can raise the roof rocks, or that horizontal space is provided by opening of extensional cavities. Because of the limited amount of magma pressure available, that required to lift the roof rocks leads to favoured depths of emplacement of only a few (2–3) km (Johnson & Pollard 1973; Corry 1988; Hogan et al. 1998). Deeper intrusions would thus indicate local but deep pull-apart and the presence of a local impermeable rheological cap (rigid lid). This magma can induce widespread melting in fertile lithologies of the lower crust after some period of preheating.

Various estimates have been made of mantle derived magma fluxes through the lithosphere. Observations of the Kilauea Iki basaltic eruptions on Hawaii during 1959 1960 gave flux rates of 50–$150\,\mathrm{m^3\,s^{-1}}$ (Williams & McBirney 1979, p. 232). Fracture propagation calculations by Turcotte (1987, p. 73) gave basaltic magmatic volume fluxes of $96\,\mathrm{m^3\,s^{-1}}$. Critical magma velocities of $>10^{-3}\,\mathrm{m^3\,s^{-1}}$ up fractures are required to avoid cooling by wall rocks (Marsh 1978; Spera 1980). If magma flux rates are always smaller in convergent regimes beneath island arcs, compared to extensional settings (continental rifts or mid-oceanic ridges), then crustal rheology rather than source events are then rate controlling.

Natural boundaries for depth of intrusion of mantle magma

Mid-crustal lithological changes, major low-angle faults, and the Moho serve as potential density discontinuities where basaltic magma can become trapped (e.g. Turcotte 1987; Glazner & Ussler 1988). These cases require space to be made for the invading magma—especially if the scenario envisaged is to lead to noticeable crustal melting. Basalt migration to higher crustal levels is thus favoured in brittle crust early in an orogenic history, i.e. before brittle–ductile layering contrasts are strongly developed and before extensive crustal melting prevents upward migration of granitic magma by inhibiting fracture. Hildreth (1981, p. 10183) and Huppert & Sparks (1988a, p. 620) noted that layers of later basaltic melt will find difficulties in intruding through partially melted crustal layers.

Degree of crustal melting by basaltic sills at different crustal levels

The major effect on the degree of crustal melting by injecting basaltic sills is the temperature of magma and of country rock. This effect can easily be seen in Fig. 4 which shows the relative masses of basalt at 1200 °C injected into the middle crust (at 20 km) that are needed to cause crustal melting starting from various initial geothermal gradients. This simplified format permits immediate scaling to any crustal depth and can easily be applied to sills of different thickness reflecting different magma flux rates (e.g. Shaw 1980).

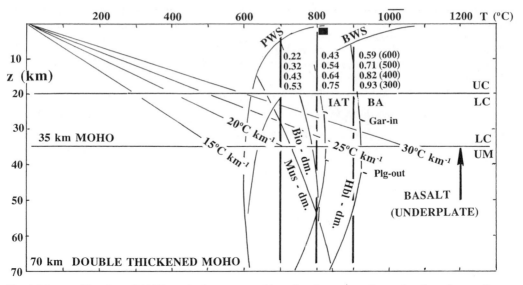

Fig. 4. Masses of basalt at 1200 °C required to cause melting of a given mass of crustal rock, at intermediate crustal levels. The simplified linear geotherms from Fig. 1 are shown relative to the experimentally-determined solidus curves. The calculations shown for 20 km depth (an approximate upper crust lower crust boundary, UC/ LC) are given as mass ratios of basalt/crust with the initial ambient temperatures shown in brackets (300–600 °C). The ratios are related to the isotherms at 700 °C, 800 °C and 900 °C, proxying for the dehydration melting reactions for muscovite (Mus-dm), biotite (Bio-dm) and hornblende (Hbl-dm). This format (linear geotherms and isothermal solidi) permits easy comparison at any crustal depth. Values of heat capacity ($Cp_{crust} = 1$ kJ kg^{-1}; $Cp_{basalt} = 1$ kJ kg^{-1}) and latent heat ($L_{basalt} = 500$ kJ kg^{-1}; $L_{crust} = 200$ kJ kg^{-1}) were used.

Convective versus conductive time scales for cooling of basaltic sills and adjacent granite partial melt layers

Hot liquid basalt, from mantle melting, emplaced (at 1200 °C) beneath fertile continental crust at initial Moho temperatures exceeding 500 °C will induce crustal melting (Lachenbruch & Sass 1978; Shaw 1980; Huppert & Sparks 1988a, b). Convection in the basaltic sill causes initial heat loss to be greater than for the case of conductive heat loss and consequently higher contact temperatures for a short time (e.g. see Hess 1960; Carslaw & Jaeger 1959, section 12.4 (eqn. 11); Jaeger 1964, p. 455; Spera 1980). Multiple pulses of basaltic injection are needed to induce extensive contact melting of crust.

For example, the calculations of Ghiorso (1991) for cooling of 10 km cubes of olivine tholeiite or boninite mantle magma show well the effects of convective versus conductive cooling. At a distance of 1 km from the contact of a 10 km cubic olivine tholeiite magma body (at 1220 °C), a maximum temperature of about 900 °C is attained after about 50×10^3 years for convective cooling, compared to about 750 °C after about 800×10^3 years (Ghiorso

1991, fig. 15.7) for pure conductive cooling. For the boninite magma (at 1375 °C) also of a 10 km cube, the convective model attains almost 1000 °C at 1 km from the contact in about 140×10^3 years, whereas the conductive model attains about 780 °C after about 800×10^3 years (Ghiorso 1991, fig. 15.8). Convection permits the same amount of heat to be lost faster than the case of conductive cooling—this is important for our purposes where the initial spurt of heat from the convecting basaltic magma may permit a particular higher temperature melting reaction to be crossed for a short time at a distance close to the contact.

Huppert & Sparks (1988a, p. 242) noted that if an episode of basaltic intrusion into the fertile continental crust occurs it results in rapid and transient generation of silicic magma on quite a short time scale (c. 10^2 years). A prerequisite for such crustal melting is that fertile lithologies must already be at sufficient temperature ($T_{MOHO} > 500$ °C) so that some melting as well as contact metamorphism will occur in response to the cooling and crystallization of the intruded basalt. As is discussed further below, this procedure is not likely to be efficient for high level intrusions where the ambient temperatures are much less (e.g. 200 °C at 10–15 km).

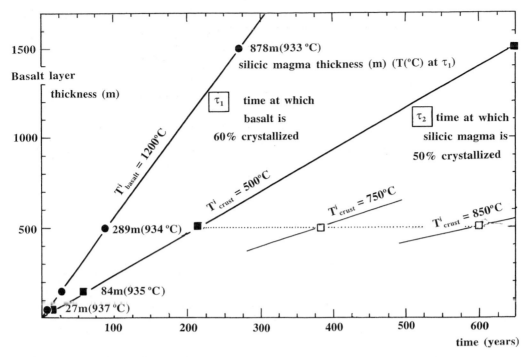

Fig. 5. Space–time relations for crustal melting (initially at 500 °C, heavy line in the middle) by basaltic sills (initially at 1200 °C). The time (τ_1, years), when the basalt sill reaches 60% crystallinity and convection stops, is shown (left heavy curve) for different basalt sill thickness (m). The thickness of the crustal melt layer (m) and temperature (°C) at τ_1 are noted next to the solid curve. Also plotted (from Huppert & Sparks 1988a, table 2, p. 613) is the time τ_2 when the crustal melt layer itself becomes 50% crystalline. The effect of higher initial country rock temperature (light lines, 750 °C, 850 °C) on prolonging the liquid history of the granite layer is shown by the postponement of the time to crystallize 50% (obtained from Huppert & Sparks 1988a, table 3, p. 614).

If total fusion of crustal layers were to occur then convection in this granite layer is possible. The basalt heat source itself stops convecting when the crystal content reaches about 65%. Huppert & Sparks (1988a, p. 240) have calculated time scales for crustal melting near basaltic sills that only last from a few years to a few hundred years.

The length scales and time scales for silicic magma generation by partial fusion of fertile crust from a convecting basaltic sill are shown in Fig. 5. This figure is constructed from results of calculations by Huppert & Sparks (1988a, tables 2 and 3). The values next to the heavy curve τ_1 are the thickness of the crustal melt layer (m) and its temperature (°C) at τ_1. In Fig. 5 the ratio of thicknesses of basalt sill ($T_{solidus} = 1200$ °C) to crustal melt ($T_{solidus} = 850$ °C) only varies from 1.85 to 1.71, clearly much greater than the examples illustrated for the conduction case in Fig. 4.

The ascent of mantle magma is controlled mainly by the rheology of the overlying lithosphere rather than by the rheology of the magma itself. Because of such rheological limits, magma invasion is favoured at shallower depth early in the stages of crustal extension. This occurs because the mantle magma acting as a heat source will induce crustal melting and these partially melted layers are less susceptible to fracture.

Rate determining steps for crustal melting

Obviously the very occurrence of migmatites and granites means that the geological circumstances leading to crustal melting were successful. The amount of crustal melting is mainly determined by the duration and magnitude of the heat sources and by how long a given rock stays at temperatures higher than its appropriate solidus. A special class of crustal melts will occur at 'wet spots' where the amount of melt produced is directly proportional to the amount of free H_2O, as well as temperature. The increased solubility of H_2O in granitic melts with increasing pressure would, for a given amount of free H_2O, favour greater extent of H_2O-saturated melting at shallower depth, whereas the positive dP/dT of the geotherm leads to increasing T with increasing depth. Thus the middle crust is the most likely location for any H_2O-saturated crustal melting to produce noticeable amounts of anatectites.

Fluid present melting can generate only a few percent melt at lower temperatures than dehydration (fluid-absent) melting of hydrous minerals (Clemens & Vielzeuf 1987; Thompson & Connolly 1995; Burnham 1997). Free H_2O could arrive along fractures or other higher permeability pathways to the potential migmatite source as ascending deep waters from lower crustal dehydration sources after passage through infertile crust (with average fluxes of 10^{-10} to 10^{-12} m^3 m^{-2} s^{-1}, e.g. Thompson 1997), or as descending surface fluids which would produce isotopically distinctive 'shallow migmatites' (e.g. Wickham 1987). Whether access of H_2O or rock temperature was controlling the amount of partial melting in natural migmatites, can be determined by comparing the results of geohygrometry (thermodynamic determination of activity of H_2O) with those of geothermometry relative to the experimentally determined $P-T$ conditions for H_2O-saturated or fluid-absent (dehydration-melting) reactions in thickened crustal roots.

Crustal dehydration-melting requires temperatures in excess of 750 °C for observable melting, 850 °C is required for substantial melting, and 950 °C for melt expulsion by compaction collapse of higher melt fractions. Experimentally determined melt fractions (f) with increasing T show steps related to dehydration-melting reactions (e.g. Gardien et al. 1995). Steps related to near-eutectic melting are high and narrow, while those reflecting peritectic melting are broader. To achieve a particular melt fraction requires different amounts of overstepping of a particular solidus, because these $f-T$ curves are far from linear (e.g. Bergantz & Dawes 1994).

In considering the length- and time-scales for crustal melting, it is important to determine which factors control the rates of melt generation (Q_G), and volumetric melt fluxes out of granitic source regions (Q_W), during magma ascent (Q_A), and to emplacement sites (Q_E).

It was emphasized above that the various mechanisms leading to crustal melting depend upon the intensity and duration of a series of geological processes. It is constructive to examine which factors lead to increased melt fraction and to determine which factors are likely to be rate-controlling in the various geological scenarios that can lead to crustal melting. These have been summarized in Table 2 with a view to permitting a diagnostic relationship between type of crustal

melting and the likely geological scenarios leading up to it.

Relative rates for melt generation, melt segregation and melt migration

The relative rates of melt generation (Q_G) and segregation (Q_S) determine whether migmatites separate from restite above some critical melt fraction. The relative rates of melt segregation (Q_S) and melt extraction (Q_W, Table 3) determine whether migmatites remain, or granites are formed leaving behind melt-drained restitic regions in the source. The rates of melt segregation (Q_S) and melt extraction (Q_W, Table 3) is also influenced through feedback processes by the rate of melt migration or ascent (Q_A).

Events in granitic source regions

For migmatites, melting is regulated by heat supply, H_2O availability, and small melt fractions are inhibited in their segregation and ascent. Once a large amount of granitic magma are formed (Table 2), magma evolution will be controlled by external factors in addition to the heat supply (e.g. rheology of surrounding rocks, Table 3).

The granitic magma ascent is inhibited by the amount of magma (sufficient for segregation and ascent up dykes or as diapirs; Clemens & Mawer 1992; Petford et al. 1993) and by increasing magma viscosity or resistant country-rock rheology when the granitic melt becomes trapped beneath a magmatard (Table 3; Parsons et al. 1992).

For felsic magmas a balance is achieved between heat-supply control on magma production and its ascent to the surface. For explosive felsic volcanics, fluid exsolution and bubble growth control their development rather than heat supply.

Segregation of granitic magma and accumulation beneath low permeability magmatards

It is important to determine when it is the properties of the magma versus the country rock that controls the rates of magma extraction (Q_W), ascent (Q_A) and emplacement (Q_E). This can be investigated in numerical experiments.

Stüwe et al. (1993) consider melt generation in the lower part of doubly thickened crust (with $T_{solidus} = 700$ °C and $T_{liquidus} = 1300$ °C) when the amount of melt produced reached a critical percentage (taken as 30 vol% following

Table 2. *Controls on crustal melting*

1. Scenario	2. Geological setting	3. Factors for high (f) melt fractions	4. Processes and rate controls
A. Melting without necessary additional heat supply			
1.1 Perturbation of geotherm during continental thickening	2.1 Thickened mountain roots	3.1 Longer time lag before exhumation	4.1 Migration of metamorphic H_2O, distribution of fertile lithologies
1.2 Decompression melting during exhumation of thickened mountain root	2.2 Thinning of thickened crust	3.2 Slower exhumation rate	4.2 P–T path relative to melt fraction contours
B. Melting resulting from 'hot' geotherms			
1.3 Increased concentration of radiogenic elements	2.3 Weathered granites; hydrothermal uranium deposit	3.3 Exhumation rate slow, cf. radiogenic decay rate	4.3 Physical concentration of U, Th and K, proximity to hydrothermal U + Th deposit
1.4 Increase in mantle heat supply	2.4 Continental rifts, extended back-arcs	3.4 Hot asthenosphere close to fertile lower crust	4.4 Upwelling of convecting asthenosphere more than conductive lithosphere
1.5 Thickening of recently thinned crust	2.5 Thickening of thinned crust	3.5 Rapid thickening of attenuated hot crust	4.5 Extension does not increase crustal T, but thickening hot geotherm will
C. External heat supply			
1.6 Extension of continental lithosphere	2.6 As 2.4	3.6 As 3.4	4.6 As 4.5
1.7 Delamination of continental lithosphere	2.7 Subducted-slab breakoff, extended back-arcs	3.7 As 3.4	4.7 Timescale for delamination 10–20 Ma, eclogitic but not felsic lower crust will delaminate
1.8 Contact melting adjacent to injected basaltic sills	2.8 As 2.4	3.8 Crust preheated to solidus by previous pulse	4.8 Short gap between successive magma pulses
1.9 Contact melting adjacent to 'hot' granites	2.9 Any of the above	3.9 As 3.8	4.9 Preheated crust and high H_2O

Table 3. *Relative volume fluxes of granitic melt during extraction (Q_W), ascent (Q_A) and emplacement (Q_E)*

$Q_E > Q_A$	$Q_E = Q_A$	$Q_E < Q_A$
flat tabular-sills	equidimensional intrusions	vertical sheets-dykes
cavity opening faster than vertical ascent	diapirs	explosive eruption
$Q_E < Q_W$	$Q_W = Q_E$	$Q_W > Q_E$
short lived migmatites	lopoliths	long lived migmatites *(in-situ)*
$Q_W > Q_A$	$Q_W = Q_A$	$Q_W < Q_A$
deep migmatite sills beneath magmatards	migmatite = granite	restite only (migmatite gone to form granite)
overlying ductile rheology control	magma rheology = crustal rheology	brittle rheology control

Nomenclature extended from Cruden (1998).

Wickham 1987). In their models the melt was instantaneously removed from the lower crust and injected as a 3 km thick sill at 12 km depth. This may provide a reasonable model for superimposed multiple events at shallow levels (see Sandiford *et al.* 1991) where episodic (cyclical) production and extraction of melt occurs. The wavelength of cyclicity (Stüwe *et al.* 1993, p. 831) depends more upon processes at the source region (T_{MOHO}, vertical height of the melted region, the critical segregation volume) than at the depth of emplacement (strain rate, isostatic and thermal evolution of the orogeny, temperature distribution before and during the emplacement). Their models indicate a wavelength of episodicities of the order of tens of Ma, even if the deformation events occur over 150 Ma.

Natural magmatards will occur if the melt segregation rate (Q_W) or the emplacement rate (Q_E) are faster than the magma ascent rate (Q_A, Table 3).

Comparing the rates of crustal melting with ascent rates

Several workers (e.g. Petford 1996) noted that the rate limiting steps in pluton formation are usually not melt ascent (Q_A) or emplacement rate (Q_E), but rather how fast and how much melt can be created (Q_G), segregated in (Q_S) and extracted from (Q_W) source regions. Emplacement of granitic magma into laccoliths (Petford 1996) or lopoliths (Cruden 1998) takes some 10–1000 years, and represents a balance between thermal and chemical controls on melting and the mechanical controls on the ascent and migration of the segregated melts. These relationships are formulated in Table 3.

Limits on the ascent and emplacement of crustal melts

Notwithstanding the multitude of difficulties in crustal melt production, segregation and accumulation, granitic magmas must be able to rise sufficiently fast to prevent their freezing at any particular depth between source and emplacement, by heat loss to the surroundings (heat death, Marsh 1978; Spera 1980). Comparative geobarometries from within plutons, from their aureoles and from the far-field regions can distinguish high-level from lower-crustal intrusives (e.g. Anderson 1996).

Depth of emplacement of crustal magmas

Application of geobarometry to several intrusions and their aureoles have shown emplacement depths of 2–20 km (e.g. Anderson 1996). The inferred depths in metamorphic aureoles distant from intrusions has confirmed that plutons are usually intruded into upper crustal rocks and therefore rarely reflect tectonic exhumation of lower crustal magma bodies intruded into higher metamorphic grade rocks (e.g. Ellis & Obata 1992). If migmatites or granitic intrusions are contained within a tectonic slice of lower crust then the migmatites plus country rocks would show geobarometric pressure perhaps > 10 kbar and a formation age

older than emplacement-time in the exposed crustal sequence.

A range in depths from 2 to 20 km indicates that for such high level intrusions there are no general rheological barriers to magma ascent operational everywhere in continental crust. For intrusions of more typically calc-alkaline magmas (generated by fractionation of hydrous mantle magmas or melting of amphibolites), their emplacement is controlled by the propagation of opening cavities (Hanson & Glazner 1995; Cruden 1998; Yoshinobu et al. 1998). It would be useful to compare ascent rates and emplacement rates of calc-alkaline magmas with migration rates deduced for granitic magmas.

Magma ascent rates in extensional settings

Sleep (1975) showed that basaltic magma chambers at mid-ocean ridges can be maintained at shallow depth (c. 5 km), when the one-sided (half) spreading rate (u) is greater than c. 1 cm a^{-1}. Hanson & Glazner (1995) suggest that similar extension rates will be required for shallow (5–10 km thick) silicic magma chambers in continental crust. The ambient crust needs to be heated by early magma pulses so that later magma injections also do not undergo rapid freezing. For extension of a 30 km crustal thickness, formation of andesitic magma chambers 1 km wide emplaced at 8 km at a half-spreading rate (u) of 1 cm a^{-1} takes 3×10^5 a (=300 ka, for $T_{magma} = 1000\,°C$; $T_{solidus} = 850\,°C$, $L_{andesite} = 220\,kJ\,kg^{-1}$). The times for the crust to reach 850 °C scales as u^2/κ (κ = thermal diffusivity = $10^{-6}\,m^2\,s^{-1}$), so that for $u = 4$ cm a^{-1} emplacement takes 50 ka. (Note that a magma flux of $10^{-3}\,km^3\,a^{-1}$ per 30 km horizontal length of crust leads to $u = 1/2(10^{-3} \times 30) = 1.5$ cm a^{-1} per km of arc length.)

The fact that granitic emplacement rates (Q_E) deduced as filling rates of extensional cavities can be very similar in magnitude to Q_E deduced by thermal conduction arguments (Fig. 6). This further indicates that rates of magma ascent (Q_A) and emplacement (Q_E) are usually delicately balanced both at mid-ocean ridges and in convergent-arc settings. Grossly disparate rates of injection compared to rate of space creation (Table 3) will determine whether magmatism was active ($Q_A > Q_E$) or passive following space generation ($Q_E > Q_A$).

Felsic magmatism in convergent arcs following extension at average tectonic rates provides room and thermal environment for repeated granite intrusion to shallower depth (Fig. 6). Long lived shallow silicic magma chambers would increase the degree of contact metamorphism

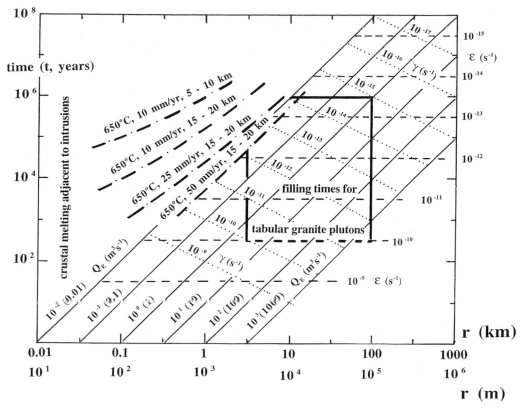

Fig. 6. Time–space nomogram showing the times (t, years) required to fill 3 km thick tabular plutons of various widths (=lengths, km) for a range of volumetric emplacement rates (Q_E m^3 m^{-2} s^{-1}). The corresponding shear and bulk strain rates ($\dot{\gamma}$ and $\dot{\varepsilon}$, s^{-1}) (calculated by Cruden 1998, p. 859) are indicated for his piston (dashed lines) and cantilever (dotted lines) models. The heavy lines outline the timespace region deduced by Cruden (1998) for common tabular granite plutons. Plots of 650 °C solidus isotherms for 5 × 10 km tabular intrusions (Yoshinobu *et al.* 1998, fig. 6) show overlap of conductive lengths only for high emplacement rates of spreading (*c.* 50 mm a^{-1}).

and enhance high T–low P metamorphism on a regional scale (Barton & Hanson 1989).

Ascent mechanisms of crustal melts

Rubin (1993) and Weinberg (1996) suggest that the ascent mechanisms change from dyke to diapir at an intermediate viscosity range of magma (10^3 to 10^6 Pa s). Diapirs make space by imposing flow on the surrounding crust. Weinberg & Podladchikov (1994) have shown that strain rate softening of the wall rocks may be an effective substitute for thermal softening. Thus the general lack of migmatites around most upper and mid-crustal plutons (Paterson *et al.* 1991) is a testimony to the easier deformability of crustal rocks, modelled by Westerley granite, rather than excessive magma temperatures.

Weinberg & Podladchikov (1994) have modelled granitoid diapirs rising through thermally graded crust with power-law rheology ($n = 2$). For an ambient geotherm of 30 °C km^{-1} (i.e. 750 °C at 25 km) spherical diapirs began rising with a velocity of 28 m a^{-1}, slowed down to *c.* 1 m a^{-1} near 20 km, and stopped near 15 km where $T_{\text{magmasphere}} = T_{\text{solidus}} = 650$ °C.

Relative rates of melt migration, ascent and emplacement

Regional compressive tectonic systems (such as convergent zones—magmatic arcs through to continental collision zones) would locally result in the building of flat-lying sills of intrusive magma, not of vertical dykes. The relative rates of melt migration (Q_W), ascent (Q_A) and emplacement (Q_E) may be usefully compared with a time–space nomogram, as shown in Fig. 6. This one follows the development of Cruden (1998) for the filling rates of tabular plutons.

The corresponding shear and bulk strain rates ($\dot{\gamma}$ and $\dot{\varepsilon}$, s^{-1}) (calculated by Cruden 1998, p. 859) are indicated for his piston (dashed lines) and cantilever (dotted lines) models. The heavy lines outline the time-space region deduced by Cruden (1998) for common tabular granite plutons of average thickness of 3 km. The region from $10^{2.5}$–10^6 years for plutons of width 5–100 km was obtained assuming that emplacement rate (Q_E) was balanced by granitic magma ascent rate (Q_A). Such a nomogram may be used to evaluate whether events in the magma source region, or during its migration, were responsible for particular tectonic occurrences. It is possible to superimpose timescales and lengthscales for a particular melting mechanism onto such a figure. For example, Yoshinobu *et al.* (1998) have calculated the conduction length scales for periodic melting at a line-source as a model for the thermal evolution of fault-controlled magma emplacement. Plots of their 650 °C solidus isotherms (fig. 6, hotter to the left to smaller length-scales) for 5×10 km tabular intrusions show overlap of conductive lengths only for high emplacement rates of spreading (c. 50 mm a^{-1}). Such calculations would be in keeping with faster emplacement rates (Q_E) than ascent rates (Q_A) in the origin of tabular plutons.

Some conclusions

The internal controls on crustal melting are (i) rock fertility and (ii) the amount of H_2O available at the solidus. The main external controls are related to the heat supply: (iii) magnitude, (iv) duration, (v) temperature of heat source and country rock and (vi) the conduction length into the country rock. For all cases of crustal melting the country rocks must be somewhat fertile. Only the C-type granites appear to reflect remelting of already melt-depleted source regions (e.g. Kilpatrick & Ellis 1992).

Migmatites reflect less than 25 vol% partial melting of metasedimentary source rocks usually by dehydration-melting of micas with a regulated moderate heat supply. Granites reflect moderate (c. 25–40 vol%) or high degrees (c. 40–60 vol%) of melting Moderate-degree granites reflect hot geotherms, radioactive layers or short distance to injected hot magmatic sills. High-degree granites require large volumes of preheated fertile crust with widespread access of external (mantle) heat sources.

The origin of migmatites is dominated by the internal controls in a weak heat supply. For the origin of moderate-degree granites the internal factors of fertility and amount of water are as important as the external factors of heat supply.

For the origin of high-degree granites, strong long-lived external heat sources are required (eclogitic delamination of thickened roots, or slab break-off, is likely).

A layered orogenic system with changing layers in extension and in compression could deliver melts upwards through dykes, but they would be rapidy quenched when they ponded as sills. Other convergent settings with oblique collision in transtension (e.g. magmatic back-arcs, Cordillera) would permit upwards access of mantle and crustal melts without a room problem.

The room problem is most probably dealt with in natural processes by a combination of extensional deformation (extension and transtension) coupled with evolving thermal structure: favouring firstly asthenospheric melting, secondly migration of these mantle melts, thirdly generation of crustal melts from ponded mantle melts, then fourthly ascent of crust melts. Some relationships between these quantities are explored in Table 3 here, and in Fig. 2 of Paterson & Miller (1998, p. 1347).

Ultimately the rate of space generation is governed by the crustal extension rate. Also even if crustal melt is produced locally in the convergent crustal column such migmatites could not rise far as granitic intrusions unless there is a significant transpressive component (Glazner 1991; Hutton 1992).

Thickened continental eclogitic roots can undergo delamination to be replaced by asthenosphere (Richardson & England 1979; Houseman *et al.* 1981). The timescale for lower crustal melting is here that for lower crustal decoupling, and the ascent of magma is rate-limited by crack propagation and related magma flow.

Paterson & Fowler (1993) remarked that more than one mechanism is required to explain the ascent and emplacement of shallow level plutons. Quite different mechanisms may dominate the ascent at different crustal depth ranges. For example, dyke ascent in layered crust favoured by colder geotherms and 'harder' rheologies, such as granulite compared to granite.

If ascent rate is faster than cavity opening rate, then magma pressure can cause fracture in extensional environments (Table 3). Passive extension governs balanced magma production rates.

Many discussions comparing ascent rates with emplacement rates, and most discussions disputing dykes versus diapirs use calc-alkaline and not granitic plutons for their examples. Much needed is a systematic formulation of granite-type related to intrusion form and size, intrusion depth, temperature of intrusion and rheology of overlying and underlying rocks, and tectonic history.

An understanding of the timing of granite intrusion and regional metamorphism in high temperature–low pressure terrains is needed to understand the role of crustal preheating in the origin of migmatites and granites.

With considerations such as these outlined here, it has become clear that the greatest difficulties in our evaluation of the various controlling steps is the lack of absolute geochronological limits on the duration of crustal melting and ascent and emplacement of that magma.

I thank R. Weinberg for a very helpful review; J. Clemens, J.-P. Burg and G. Bergantz for encouragement, and J.-L. Vigneresse for considerate editing. The work leading to this paper was supported by the ETH Zurich and the Swiss Nationalfonds.

References

ANDERSON, J. L. 1996. Status of thermobarometry in granitic batholiths. *Transactions of the Royal Society of Edinburgh: Earth Sciences*, **87**, 125 138.

BARTON, M. D. & HANSON, R. B. 1989. Magmatism and the development of low-pressure metamorphism: Implications from thermal modeling and the western United States. *Geological Society of America Bulletin*, **101**, 1051–1065.

——, STAUDE, J.-M., SNOW, E. A. & JOHNSON, D. A. 1991. Aureole Systematics. *In*: KERRICK, D. M. (ed.) *Contact metamorphism*. Mineralogical Society of America Reviews in Mineralogy, **26**, 723–847.

BERGANTZ, G. W. 1991. Physical and chemical characterization of plutons. *In*: KERRICK, D. M. (ed.) *Contact metamorphism*. Mineralogical Society of America Reviews in Mineralogy, **26**, 13–42.

—— & DAWES, R. 1994. Aspects of magma generation and ascent in continental lithosphere. *In*: RYAN, M. P. (ed.) *Magmatic Systems*. Acadamic Press, Inc., 291–317.

BIRD, P. & BAUMGARDNER, J. 1981. Steady propagation of delamination events. *Journal of Geophysical Research*, **86**, 4891–4903.

BROWN, M. 1994. The generation, segregation, ascent and emplacement of granite magma: The migmatite-to-crustally-derived granite connection in thickened orogens. *Earth-Science Reviews*, **36**, 83–130.

——, AVERKIN, Y. A., McLELLAN, E. L. & SAWYER, E. W. 1995. Melt segregation in migmatites. *Journal of Geophysical Research*, **100**, 15 655–15 679.

BURNHAM, C. W. 1997. Magmas and hydrothermal fluids. *In*: BARNES, H. L. (ed.) *Geochemistry of Hydrothermal Ore Deposits*, 3rd edn. John Wiley, New York, 63–123.

CARSLAW, H. S. & JAEGER, J. C. 1959. *Conduction of Heat in Solids*, 2nd ed. Oxford University Press, New York.

CHAMBERLAIN, C. P. & SONDER, L. J. 1990. Heat-producing elements and the thermal and baric patterns of metamorphic belts. *Science*, **250**, 763–769.

CLEMENS, J. D. 1990. The Granulite–Granite connexion. *In*: VIELZEUF, D. & VIDAL, P. (eds) *Granulites and Crustal Evolution*. NATO ASI Series, **311**, 25–36.

—— & MAWER, C. K. 1992. Granitic magma transport by fracture propagation. *Tectonophysics*, **204**, 339–360.

—— & VIELZEUF, D. 1987. Constraints on melting and magma production in the crust. *Earth and Planetary Science Letters*, **86**, 287–306.

CORRY, C. E. 1988. *Laccoliths; Mechanics of emplacement and growth*. Geological Society of America Special Paper, **220**.

CRUDEN, A. R. 1998. On the emplacement of tabular granites. *Journal of the Geological Society, London*, **155**, 853–862.

DE YOREO, J. J., LUX, D. R. & GUIDOTTI, C. V. 1989. The role of crustal anatexis and magma migration in the thermal evolution of regions of thickened continental crust. *In*: DALY, J. S., CLIFF, R. A. & YARDLEY, B. W. D. (eds) *Evolution of metamorphic Belts*. Geological Society, London, Special Publications, **43**, 187–202.

ELLIS, D. J. & OBATA, M. 1992. Migmatite and melt segregation at Cooma, New South Wales. *Transactions of the Royal Society of Edinburgh: Earth Sciences*, **83**, 95–106.

ENGLAND, P. C. & THOMPSON, A. B. 1984. Pressure-temperature-time paths of regional metamorphism: Part I: Heat-transfer during the evolution of regions of thickened continental crust. *Journal of Petrology*, **25**, 894–928.

—— & —— 1986. Some thermal and tectonic models for crustal melting in continental collision zones. *In*: COWARD, M. P. & RIES, A. C. (eds) *Collision Tectonics*. Geological Society, London, Special Publications, **19**, 83 94.

FINGER, F. & CLEMENS, J. D. 1995. Migmatization and "secondary" granitic magmas: effects of emplacement of "primary" granitoids in Southern Bohemia, Austria. *Contributions to Mineralogy and Petrology*, **120**, 311–326.

FURLONG, K. P., HANSON, R. B. & BOWERS, J. R. 1991. Modeling thermal regimes. *In*: KERRICK, D. M. (ed.) *Contact metamorphism*. Mineralogical Society of America Reviews in Mineralogy, **26**, 437–506.

GARDIEN, V., THOMPSON, A. B., GRUJIC, D. & ULMER, P. (1995) Experimental melting of biotite + plagioclase + quartz ± muscovite assemblages and implications for crustal melting. *Journal of Geophysical Research*, **100**, B8, 15 581–15 591.

GHIORSO, M. S. 1991. Temperatures in and around cooling magma bodies. *In*: PERCHUK, L. L. (ed.) *Progress in metamorphic and magmatic petrology*. Cambridge University Press, 387–417.

24 A. B. THOMPSON

GLAZNER, A. F. 1991. Plutonism, oblique subsidence, and continental growth. *Geology*, **19**, 784–786.

—— & USSLER III, W. 1988. Trapping of magma at midcrustal density discontinuities. *Geophysical Research Letters*, **15**, 673–675.

HANSON, R. B. & BARTON, M. D. 1989. Thermal development of low-pressure metamorphic belts: Results from two-dimensional numerical models. *Journal of Geophysical Research*, **94**, 10 363–10 377.

—— & GLAZNER, A. F. 1995. Thermal requirements for extensional emplacement of granitoids. *Geology*, **23**, 213–216.

HESS, H. H. 1960. *Stillwater igneous complex, Montana*. Geological Society of America Memoirs, **80**.

HILDRETH, W. 1981. Gradients in silicic magma chambers: Implications for lithospheric magmatism. *Journal of Geophysical Research*, **86**, 10 153–10 192.

HOGAN, J. P., PRICE, J. D. & GILBERT, M. C. 1998. Magma traps and driving presssure: consequences for pluton shape and emplacement in an extensional regime. *Journal of Structural Geology*, **20**, 1155–1168.

HOUSEMAN, G. A., MCKENZIE, D. P. & MOLNAR, P. 1981. Convective instability of a thickened boundary layer and its relevance for the thermal evolution of continental convergent belts. *Journal of Geophysical Research*, **86**, 6115–6132.

HUPPERT, H. E. & SPARKS, R. S. J. 1988a. The generation of granitic magmas by intrusion of basalt into continental crust. *Journal of Petrology*, **29**, 599–624.

—— & —— 1988b. The fluid dynamics of crustal melting by injection of basaltic sills. *Transactions of the Royal Society of Edinburgh: Earth Sciences*, **79**, 237–243.

HUTTON, D. H. W. 1992. Granite sheeted complexes— Evidence for the dyking ascent mechanism. *Transactions of the Royal Society of Edinburgh: Earth Sciences*, **83**, 377–382.

JAEGER, J. C. 1964. Thermal effects of intrusions. *Reviews of Geophysics*, **2**, 443–466.

JARVIS, G. T. & MCKENZIE, D. P. 1980. The development of sedimentary basins with finite extension rates. *Earth and Planetary Science Letters*, **48**, 42–52.

JOHNSON, A. M. & POLLARD, D. D. 1973. Mechanics of growth of some laccolithic intrusions in the Henry Mountains, Utah, I: Field observations, Gilberts model, physical properties and flow of the magma. *Tectonophysics*, **18**, 261–309.

KILPATRICK, J. A. & ELLIS, D. J. 1992. C-tape magmas: igneous charnockites and their extrusive equivalents. *Transactions of the Royal Society of Edinburgh: Earth Sciences*, **83**, 155–164.

LACHENBRUCH, A. H. & SASS, J. H. 1978. *Models of an extending lithosphere and heat flow in the Basin and Range province*. Geological Society of America Memoirs, **152**, 209–250.

LOOSVELD, R. J. H. 1989. The synchronism of crustal thickening and low-pressure facies metamorphism in the Mount Isa Inlier, Australia. 2. Fast convective thinning of mantle lithosphere during crustal thickening. *Tectonophysics*, **165**, 191–218.

—— & ETHERIDGE, M. A. 1990. A model for low-pressure facies metamorphism during crustal thickening. *Journal of Metamorphic Geology*, **8**, 257–267.

MARSH, B. D. 1978. On the cooling of ascending andesitic magma. *Philosophical Transactions of the Royal Society of London*, **A288**, 611–625.

MCKENZIE, D. P. 1978. Some remarks on the development of sedimentary basins. *Earth and Planetary Science Letters*, **40**, 25–32.

MCLELLAN, E. L. 1989. Sequential formation of subsolidus and anatectic migmatites in response to thermal evolution, eastern Scotland. *Journal of Geology*, **97**, 165–182.

MEHNERT, K. R. 1968. *Migmatites and the origin of granitic rocks*. Elsevier Science, New York.

MOONEY, W. D. & MEISSNER, R. 1992. Multi-genetic origin of crustal reflectivity: A review of seismic reflection profiling of the continental lower crust and Moho. *In*: FOUNTAIN, D. M., ARCULUS, R. J. & KAY, R. W. (eds) *Continental Lower Crust*. Elsevier, New York, 45–80.

PARSONS, T., THOMPSON, G. A. & SLEEP, N. H. 1992. Host rock rheology controls on the emplacement of tabular intrusions: Implications for underplating of extending crust. *Tectonics*, **11**, 1348–1356.

PATERSON, S. R. & FOWLER, K., JR. 1993a. Re-examining pluton emplacement processes. *Journal of Structural Geology*, **15**, 191–206.

—— & MILLER, R. B. 1998. Mid-crustal magmatic sheets in the Cascades Mountains, Washington, implications for magma ascent. *Journal of Structural Geology*, **20**, 1345–1363.

——, VERNON, R. H. & FOWLER, K. 1991. Aureole tectonics. *In*: KERRICK, D. M. (ed.) *Contact metamorphism*. Mineralogical Society of America Reviews in Mineralogy, **26**, 673–722.

PATIÑO-DOUCE, A. E., HUMPHREYS, E. D. & JOHNSTON, A. D. 1990. Anatexis and metamorphism in tectonically thickened continental crust exemplified by the Sevier hinterland, western North America. *Earth and Planetary Science Letters*, **97**, 290–315.

PETFORD, N. 1996. Dykes or diapirs? *Transactions of the Royal Society of Edinburgh: Earth Sciences*, **87**, 105–114.

——, KERR, R. C. & LISTER, J. R. 1993. Dike transport of granitoid magmas. *Geology*, **21**, 845–848.

RICHARDSON, S. W. & ENGLAND, P. C. 1979. Metamorphic consequences of crustal eclogite production in overthrust orogenic zones. *Earth and Planetary Sciences Letters*, **42**, 183–190.

RUBIN, A. 1993. Dykes vs diapirs in viscoelastic rock. *Earth and Planetary Science Letters*, **117**, 653–670.

SANDIFORD, M. & HAND, M. 1998. Australian Proterozoic high-temperature, low-pressure metamorphism in the conductive limit. *In*: TRELOAR, P. J. & O'BRIEN, P. J. (eds) *What Drives Metamorphism and Metamorphic Reactions?*

Geological Society, London, Special Publications, **138**, 109–120.

——, STÜWE, K. & POWELL, R. 1991. Mechanical consequences of granite emplacement during high T-low P metamorphism and the origin of 'anticlockwise' PT paths. *Earth and Planetary Science Letters*, **107**, 164–172.

SCHULMANN, K., THOMPSON, A. B. & JEZEK, J. 1999. Rapid Vertical Tectonics. *Terra Nova Abstracts*, symposium G-03.

SHAW, H. R. 1980. The fracture mechanism of magma transport from the mantle to the surface. *In*: HARGRAVES, R. B. (ed.) *Physics of Magmatic Processes*. Princeton University Press, Princeton, N.J., 201–264.

SLEEP, N. H. 1975. Formation of oceanic crust: Some thermal constraints. *Journal of Geophysical Research*, **80**, 4037–4042.

SPERA, F. J. 1980. Aspects of magma transport. *In*: HARGRAVES, R. B. (ed.) *Physics of Magmatic Processes*. Princeton University Press, 263–323.

STÜWE, K., SANDIFORD, M. & POWELL, R. 1993. Episodic metamorphism and deformation in low-pressure, high-temperature terrains. *Geology*, **21**, 829–832.

THOMPSON, A. B. 1990. Heat, fluids, and melting in the granulite facies. *In*: VIELZEUF, D. & VIDAL, P. (eds) *Granulites and Crustal Evolution*. NATO ASI Series, **311**, 37–57.

—— 1996. Fertility of crustal rocks during anatexis. *Transactions of the Royal Society of Edinburgh: Earth Sciences*, **87**, 1–10.

—— 1997. Flow and focusing of metamorphic fluids. *In*: JAMTVEIT, B. & YARDLEY, B. W. D. (eds) *Fluid flow and transport in rocks*. Chapman & Hall, 297–314.

—— & CONNOLLY, J. A. D. 1995. Melting of the continental crust: Some thermal and petrological constraints on anatexis in continental collision zones and other tectonic settings. *Journal of Geophysical Research*, **100**, 15 565–15 579.

THOMPSON, P. H. 1989. Moderate overthickening of thinned sialic crust and the origin of granitic magmatism and regional metamorphism. *Geology*, **17**, 520–523.

TURCOTTE, D. L. 1987. Physics of magma segregation processes. *In*: MYSEN, B. O. (ed.) *Magmatic Processes: Physicochemical Principles*. The Geochemical Society, Special Publications, **1**, 69–74.

VIGNERESSE, J. L. 1995. Crustal reime of deformation and ascent of granitic magma. *Tectonophysics*, **249**, 187–202.

WATERS, D. J. 1990. Thermal history and tectonic setting of Namaqualand granulites. *In*: VIELZEUF, D. & VIDAL, P. (eds) *Granulites and Crustal Evolution*. NATO ASI Series, **311**, 243–256.

WEINBERG, R. F. 1996. Ascent mechanism of felsic magmas: news and views. *Transactions of the Royal Society of Edinburgh: Earth Sciences*, **87**, 95–103.

—— & PODLADCHIKOV, Y. Y. 1994. Diapiric ascent of magmas through power law crust and mantle. *Journal of Geophysical Research*, **99**, 9543–9560.

WELLS, P. R. A. 1980. Thermal models for the magmatic accretion and subsequent metamorphism of continental crust. *Earth and Planetary Science Letters*, **46**, 2532–265.

WICKHAM, S. M. 1987. The segregation and emplacement of granitic magmas. *Journal of the Geological Society, London*, **144**, 281–297.

WILLIAMS, H. & McBIRNEY, A. R. 1979. *Volcanology*. Freeman-Cooper.

WYLLIE, P. J. 1977. Crustal anatexis. *Tectonophysics*, **43**, 41–71.

YOSHINOBU, A. S., OKAYA, D. A. & PATERSON, S. R. 1998. Modeling the thermal evolution of fault-controlled magma emplacement models: implications for the solidification of granitoid plutons. *Journal of Structural Geology*, **20**, 1205–1218.

ZEN, E. 1988. Thermal modelling of stepwise anatexis in a thrust-thickened sialic crust. *Transactions of the Royal Society of Edinburgh: Earth Sciences*, **79**, 223–235.

Granitic melt viscosities

DONALD B. DINGWELL

Bayerisches Geoinstitut, Universitaet Bayreuth, D-95440 Bayreuth, Germany
(e-mail: don.dingwell@uni-bayreuth.de)

Abstract: The viscosities of granitic melts play a crucial role in controlling the kinetics and dynamics of magma transport including generation, segregation, ascent, differentiation and emplacement. No general theory of liquid viscosity has been successfully applied to date to the description of granitic melt viscosities. The approach in the earth sciences so far has been empirical and experiments designed to measure the viscosity of granitic melts have made considerable progress in improving our picture of the composition-, temperature- and pressure-dependence of granitic melt viscosities. This chapter summarizes the recent progress in parameterizing the viscosity of granitic melts with respect to the most important compositional variables such as alkali/aluminium ratio, silica content, water content, anorthite content. The pressure-dependence of granitic melt viscosities is very slight. The temperature-dependence, in contrast, is a very strong and non-linear function of the reciprocal absolute temperature (i.e. non-Arrhenian). Indirect determinations of the viscosity of granitic melts have also contributed substantial information on the composition-, pressure- and temperature-dependence of melt viscosity. These measurements are particularly useful where direct viscosity determinations are difficult or impossible to perform. The outlook for a more complete picture of granitic melt viscosities in the next years is excellent. The combination of several factors, including conceptual advances in the quantitative understanding of relaxation in silicate melts, experimental advances in the measurement of viscosity of silicate melts, and the identification in the experimental studies to date of the areas of investigation where the effort should be concentrated in the future, give reason to expect a more or less complete picture, fully adequate for the purposes of modelling the dynamics of granitic magmas, in the near future.

Granitic magmatism has long been recognized to play a central role in the internal evolution of the continental crust. In the second half of this century a large research effort has been directed at understanding the equilibrium controls on the stability of phases in granitic magmas in order to constrain the pressures and temperatures of emplacement and crystallization of granitic rocks within the crust. Less effort has been expended on the exact description of the physical state of granitic magmas during their segregation, ascent and emplacement. This is despite the fact that almost for all aspects of magma segregation, transport, differentiation and emplacement or eruption, viscosity is one of the most important properties of a magma.

In recent years considerable debate has arisen over the physical and mechanistic nature of granitic magma transport within the crust. Unfortunately, the experimental data required for sophisticated testing of transport scenarios via numerical simulations are lacking. The situation is however steadily improving and in the past 5–10 years a renewal of interest in the experimental investigation of the physical properties has occurred. A great deal of progress has been made in the physical description of the rheology of granitic melts and a variety of new techniques and strategies have been developed along the way. It would seem appropriate and timely to summarize those advances here in the context of understanding granites via an integration of new and classical techniques.

Concepts and experiments

The composition and temperature dependence of viscosity is one of the most remarkable aspects of silicate melts. Viscosities of silicate melts range over more that ten log units as a function of either composition or temperature. Viscosity is closely related to the fundamental response time of the melt to displacements from its equilibrium state, the relaxation time of the structure. The study of viscosity of silicate melts is part of a

From: Castro, A., Fernández, C. & Vigneresse, J. L. (eds) *Understanding Granites: Integrating New and Classical Techniques.* Geological Society, London, Special Publications, **168**, 27–38. 1-86239-058-4/99/$15.00 © The Geological Society of London 1999.

much broader field of rheology, the study of deformation.

The viscosity of a Newtonian material is defined as the ratio of stress to strain rate

$$\eta = \sigma/\dot{\varepsilon}. \qquad (1)$$

In contrast, the elastic modulus of a Hookean elastic material is defined as the ratio of stress to strain

$$M = \sigma/\varepsilon. \qquad (2)$$

Whereas the elastic constants are almost universally reported in the SI unit of pressure, the Pascal (Pa), viscosity data are commonly reported in both the SI unit, the Pascal-second (Pa s) and the c.g.s. unit the poise (P). One poise is 0.1 Pa s and data are often reported as deci-Pascal-seconds (dPa s) as well. Major errors of interpretation have resulted from incorrect conversion in the literature. Viscosity composition relationships are often presented as the temperatures at which differing compositions have the same viscosity. These lines of constant viscosity are known as 'isokoms'.

Melt deformation can be entirely described in terms of two components, a volume and a shear component. The combination of volume and shear viscosity yields a longitudinal viscosity

$$\eta_l = \eta_v + (4/3)\eta_s \qquad (3)$$

which is relevant for the propagation of longitudinal seismic waves in melts. Data for longitudinal and shear viscosities illustrate that volume and shear viscosities are similar at very low strains and moderate strain rates (Dingwell & Webb 1990).

The ratio of Newtonian viscosity to Hookean elastic modulus yields a quantity in units of time. This ratio

$$\tau = \eta/M \qquad (4)$$

is the Maxwell relaxation time (Maxwell 1867). The near constancy of M, in contrast with the extreme variations observed in η, means that the onset of non-Newtonian rheology and the non-relaxed response of the melt phase, can be predicted by the knowledge of the Newtonian viscosity. The purely viscous response of a silicate melt is termed 'liquid' behaviour, whereas the purely elastic response is termed 'glassy' behaviour. The term silicate melt is used to describe molten silicates, independent of their rheological behaviour. The change from liquid to glassy behaviour describes a region of mixed liquid-like and solid (or glassy-) like behaviour, a region of 'viscoelastic' response.

Viscosity experiments can be subdivided into two types in several ways. Firstly, one can distinguish between those in which the strain rate is controlled and the resultant stress is measured (e.g. concentric cylinder) and those in which the stress is controlled and the resultant strain rate is measured (fibre elongation). Secondly, we can also separate those methods which involve only shear stresses (e.g. concentric cylinder) from those which involve a combination of shear and compressive or dilational stress (e.g. beam bending or fibre elongation). Further, for controlled strain experiments, distinction can be made between the application of a step function of strain and the continuous application of strain (e.g. stress relaxation versus steady-state flow). These distinctions are analysed in more detail by Dingwell et al. (1993) and Dingwell (1995).

The most widely used example of controlled strain rate and measured stress is the rotational (Couette) method (Ryan & Blevins 1987). The strain rate is imposed on an annulus of liquid filling the gap between an inner cylinder (or spindle) and the outer cylinder (or cup) which usually takes the form of a cylindrical crucible. This method is commonly used for viscosity measurements in the range of 10^0–10^5 dPa s. Shear viscosities in the range 10^9–10^{14} Pa s can be determined using fibre elongation techniques. Glass fibres with a diameter 0.1–0.3 mm and lengths 10–18 mm are commonly used with strain-rates typically in the range 10^{-8}–10^{-4} s^{-1} and tensile stresses up to c. 10^8 Pa. In this geometry, the observed viscosity η_{elong} is the elongational viscosity and is related to the shear viscosity η_s by

$$\eta_{\text{elong}} = 9\eta_v\eta_s/(3\eta_v + \eta_s) \qquad (5)$$

where η_v is the volume viscosity. In a viscous sense, volume strain is limited in magnitude and thus the volume viscosity of a melt approaches an infinite value with increasing time, such that Eqn 5 becomes

$$\eta_{\text{elong}} = 3\eta_s \qquad (6)$$

for times greater than the relaxation time of the melt. Equation 6 is known as Trouton's rule.

Absolute shear viscosities in the range 10^9–10^{11} Pa s can be determined using micro-penetration techniques. This involves determining the rate at which an indenter under a fixed load moves into the melt surface. For the case of a spherical indenter (e.g. Dingwell *et al.* 1992) the absolute shear viscosity is determined from

$$\eta_s = 0.1875 Pt / r^{0.5} \alpha^{1.5} \qquad (7)$$

(Pocklington 1940; Tobolsky & Taylor 1963) for the radius of the sphere r (m), the applied force P (N), indent distance α (m), and time t (s) ($t = 0$ and $\alpha = 0$ upon application of the force).

Cylindrical melt geometries can be determined by axial cylinder compression. Absolute viscosities in the range 10^4–10^8 Pa s (Gent 1960; Fontana 1970) and 10^7–10^{11} Pa s (Bagdassarov & Dingwell 1992) have been determined. For a cylindrical specimen of any thickness, the shear viscosity can be determined by

$$\eta_s = \frac{2\pi \, mgh^5}{3V \frac{\delta h}{\delta t}(\pi h^3 + V)} \qquad (8)$$

for the applied mass (kg), the acceleration due to gravity g, the volume V (m³) of the material, the height h (m) of the cylinder and time t (s) (Gent 1960; Fontana 1970), for the case in which the surface area of contact between the melt and the parallel plates remains constant and the cylinder bulges with increasing deformation. This is the 'no-slip condition'. For the case in which the surface area between the cylinder and the plate increases with deformation and the cylinder does not bulge, the viscosity is

$$\eta_s = \frac{mgh^2}{3V \frac{\delta h}{\delta t}}. \qquad (9)$$

This is the perfect slip condition.

The counterbalanced sphere and falling sphere viscometers are based on Stokes law

$$\eta = \frac{2\Delta\rho g r^2}{9v} \qquad (10)$$

where $\Delta\rho$ is the density difference between the falling sphere and the melt, r is the sphere radius and v is the terminal velocity of the sphere. Falling sphere viscometry has been employed at 1 atm and at high pressures (e.g. Shaw 1963; Fujii & Kushiro 1977; Ryan & Blevins 1987). A correction is normally applied to account for the non-infinite capsule dimensions and the viscous drag thereby exerted on the sphere by the capsule walls. Equation 10 thus becomes

$$\eta = 2\Delta\rho g r^2 /$$
$$9v \left[1 - 2.04 \frac{r}{r_c} + 2.09 \left(\frac{r}{r_c}\right)^3 - 0.95 \left(\frac{r}{r_c}\right)^5 \right] \qquad (11)$$

where r is the sphere radius and r_c is the capsule radius. The falling sphere method may be used for the simultaneous determination of density and viscosity. Very high pressure measurements of viscosity have been made by Kanzaki *et al.* (1987) who imaged the falling sphere in real time using a synchrotron radiation source and a MAC 80 superpress. This method extends the lower limit of measurable viscosity using the falling sphere method at high pressure to 10 dPa s.

Background of investigation

Experimental investigations of silicate melt viscosity were initiated early in this century to address the needs of a variety of applied fields. The two traditional materials applications were in the fields of glass and ceramics and metal extraction via slag separation. The data bases for and roles of rheology in these two fields have been summarized by Zarzicki (1991) and Turkdogan (1983), respectively. The glass and ceramics literature deals with melts whose properties are relevant at very high viscosity, with the textures of ceramics and the properties of glasses being engineered just above the glass transition temperature (see below). Thus in this field, the techniques employed and the phenomena observed are of particular relevance for the experimental investigation of granitic melt rheology.

It became clear in the first half of this century that the viscosities of melts whose chemical composition was similar to that of granites in terms of silica content and alkali/aluminium ratio were very high at the temperatures of interest to granitic magmatism. Kozu & Kani (1935) performed concentric cylinder viscometry on molten feldspars, and Bowen (1934) had already discussed the significance of such high viscosities for basic versus silicic melt transport, a discussion as relevant today as it was then (although the conclusions might differ).

Saucier and coworkers (Saucier 1952; Sabatier & Saucier 1955; Sabatier 1956; Saucier & Saplevitch 1958) began a study of the viscosity of acidic melts which employed different experimental designs including the first determination of the viscosity of a water-bearing rhyolite under pressure. The high temperature experiments of Saucier & Saplevich (1958) employed a tangential loading, and a loading-unloading procedure to obtain viscosities in the range of 10^9 to 10^{12} dPa s. Saucier (1952) describes the internally heated pressure vessel used for the determination of viscosity of a Retinite melt under 1100 °C and 180 bars. He explored several experimental designs for viscosity determination, including the concentric cylinder and falling sphere methods. He obtained data for a melt with 1.25% water. Sabatier (1956) deformed a cube of the same melt under a constant load at water pressures of 500 and 1000 bars.

In the 1960s a surge in experimental petrology led to the first attempts to quantify the influence of water on the viscosity of granitic melts. The compaction rate of hydrous volcanic ash was investigated by Friedman *et al.* (1963) who obtained data in the high viscosity range (10^9–10^{13} Pa s) for water contents of a few wt%. The first direct study of hydrous granitic melt viscosity using falling sphere viscometry was also reported in that year (Shaw 1963). In this study, the falling sphere or Stokesian settling method was employed to obtain viscosity data for water-rich (4–6 wt% H_2O) rhyolitic melts at 800 °C and 2 kbar in the viscosity range 10^3–10^5 Pa s.

After this brief sortie into experimental viscometry of granitic melts, attention in the 1970s turned to the parameterization of melt viscosities using empirical calculational schemes derived from the existing viscosity data base. Bottinga & Weill (1972) proposed a model based on anhydrous superliquidus viscosity data from simple silicate melt systems which appeared to account fairly well for the compositional dependence of andesitic to basaltic melt viscosities in the low viscosity, superliquidus region. Shaw (1972) incorporated the limited experimental data on hydrous granitic melt viscosities and thereby expanded the Bottinga & Weill (1972) data base to obtain a model capable of predicting the superliquidus viscosity of hydrous multicomponent melts. Both of these methods have enjoyed widespread application in the years since.

Interest in the experimental determination of melt viscosity picked up again in the 1970s due to the efforts of Kushiro and coworkers at the Geophysical Laboratory, who developed the falling sphere method for application at higher pressures (up to 25 kbar; Kushiro 1976, 1978*a, b*; Kushiro *et al.* 1976). The observation that the viscosity of silica-rich melts, such as albite and jadeite, decreases with increasing pressure was unexpected, and generated a considerable degree of debate about the physico-chemical and geophysical significance of this trend, but for purposes of application to granite petrogenesis the pressure-dependence is small enough to be neglected.

Recent advances

The influence of water on the viscosity of granitic melts was investigated in a series of early studies using differing geometries under elevated pressures in the 1950s and 1960s (Friedman *et al.* 1963; Shaw 1963; and references therein). On the basis of those data, and on the synthesis of anhydrous melt viscosities by Bottinga & Weill (1972), Shaw (1972) presented a model for the calculation of the viscosity of multicomponent melt viscosities which incorporated the effects of water known up to that time. The model has been very widely employed and appears to work satisfactorily for the low viscosity range where the assumption of an Arrhenian temperature dependence can be accepted:

$$\log \eta = a + b/T. \tag{12}$$

Silicate melts exhibit, in general, significant departures from an Arrhenian temperature dependence (Richet 1984). Coping with this non-Arrhenian behaviour turns out to be the greatest challenge facing the experimental investigation and theoretical modelling of melt viscosity. The pressure dependence of melt viscosity can, in contrast, be neglected for crustal pressures (Dingwell 1998, and references therein). The behaviour at issue here can be described with the help of Fig. 1 where the temperature dependence of the viscosities of hydrous calcalkaline granitic melts are illustrated for the low viscosity and high viscosity ranges.

The low viscosity range data have been obtained using falling sphere techniques (Schulze *et al.* 1996) and the high viscosity range data have been obtained using the micropenetration technique (Dingwell *et al.* 1996*a*). Although both studies were conducted on very similar compositions, it is clear from Fig. 1 that the slopes of the curves corresponding to their temperature dependences are quite different. The high viscosity range data exhibit much higher slopes than the low viscosity range data. The origin of this apparent discrepancy becomes obvious when both types of data are plotted together on a reciprocal absolute temperature (Arrhenian) plot. Figure 2 demonstrates this for the case of dry peralkaline granitic melts (Dorfman *et al.* 1996) where not only high and low viscosity data are presented, but also data obtained in the intermediate viscosity range as well. The curvature of the viscosity temperature relationships can be seen to increase with added Na_2O to a calcalkaline base. Thus peralkaline granitic melts have more non-Arrhenian viscosities than calcalkaline granitic melts. In fact this general type of behaviour is seen for the addition of all components to a metaluminous granitic melt (Hess *et al.* 1995, 1996). It appears to be a fundamental consequence of the depolymerization of

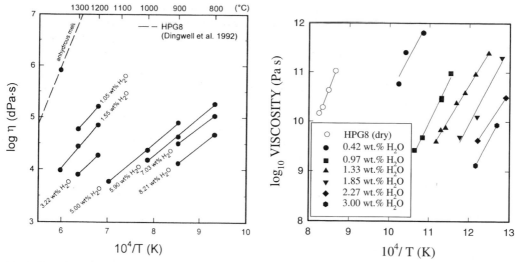

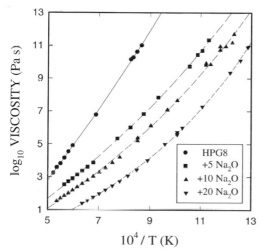

Fig. 1. Viscosity–temperature data for granitic melts with added water in the (**a**) low viscosity range and (**b**) high viscosity range. The apparent discrepancy in the temperature-dependence of the viscosity results from the significantly non-Arrhenian nature of the viscosity in such melts. This is the greatest experimental challenge to the description of Newtonian melt viscosities in granitic systems. Reproduced with permission from Schulze *et al.* (1996) and Dingwell *et al.* (1996*a*, *b*).

Fig. 2. Comparison of viscosity data from the intermediate viscosity region, derived from the centrifuge autoclave method together with data from concentric cylinder methods at lower viscosity and micropenetration methods at higher viscosity. Reproduced with permission from Dorfman *et al.* (1996).

the initially 'tectosilicate' melt network as network modifying cations are added (see however Toplis *et al.* 1997*a*, *b*). Non-Arrhenian behaviour can be described by the so-called Tamann–Vogel–Fulcher (TVF) equation

$$\log \eta = a + b/(T - c) \qquad (13)$$

where c is a constant. Despite knowledge of the likelihood of non-Arrhenian viscosity temperature relationships for hydrous granitic melts, no departure from the simple and inaccurate Arrhenian approximation was possible for several years. In fact, in the intervening 25 years since Shaw (1972), a number of investigations of the viscosities of hydrous granitic melts have been performed, and in the past few years, several new models have appeared for the calculation of hydrous melt viscosities. Three of these models (Baker 1996; Scaillet *et al.* 1996; Schulze *et al.* 1996) are based on Arrhenian temperature dependences and thus are restricted to the high temperature, low viscosity region.

This lack of progress on the non-Arrhenian effect was due to the nearly complete lack of data on the viscosities of water-bearing melts in the high viscosity range. Two things occurred in the early 1990s to alter this situation. Firstly, Richet *et al.* (1996) observed that dilatometric measurements of viscosity on water-bearing melts could be metastably performed at 1 bar pressure. This was followed by a flood of data at low but significant water contents at 1 bar pressure and high temperatures (Richet *et al.* 1996; Dingwell *et al.* 1996*a*, *b*, 1998*a*, *b*). This improvement, important as it was, still left the critical high viscosity region of water-rich granitic melts, where the greatest influence of non-Arrhenian behaviour is expected, uninvestigated. Thus the second improvement in the 1990s was the quantification of glass transition data in terms of melt viscosities (Dingwell & Webb 1990; Dingwell 1995) combined with the high temperature-high pressure direct observation of the glass transition temperature using spectroscopic and volume determinations of the relaxation process (e.g. Romano *et al.* 1994; Nowak & Behrens 1995; Shen & Keppler 1995). Dingwell (1998) combined these two aspects of melt relaxation and produced a consistent set of glass transition temperature estimates for water-rich granitic melts shown here in Fig. 3.

These data points provided the essential viscosity data for the high viscosity, water-rich region of the melt viscosity curves presented below for the Hess & Dingwell (1996) model. Combining such data together with the falling sphere determinations of relatively low viscosities (e.g. Schulze *et al.* 1996) and the micropenetration determinations of relatively high viscosities (e.g. Dingwell *et al.* 1996*a*), Hess & Dingwell (1996) were able to generate their non-Arrhenian model, the first of its kind, based on the total available data set up to fall 1996. These authors were able to set the pressure dependence of the viscosity of hydrous granitic liquids equal to zero for the pressure range up to 10 kbar and achieve a much improved fit to the entire available data set by the use of the first application of a non-Arrhenian model to these viscosity data over a very wide viscosity range via the relationship

$$\log \eta =$$
$$\frac{\{[-3.545 + 0\overset{.}{8}33 \ln(w)] + [9601 - 2368 \ln(w)]\}}{\{T - [195.7 + 32.25 \ln(w)]\}} \quad (14)$$

where η is in Pa s, w is water content in wt%, and T is in K.

The general formulation of the viscosity relationship is much simpler than that of Baker (1996) who incorporated structural modelling and yet the Hess & Dingwell (1996) reproduces the complete available experimental data set significantly better that either the Schulze *et al.* (1996), Baker (1996) or the Scaillet *et al.* (1996) models. That it is also a significant improvement on the Shaw (1972) model is demonstrated in Fig. 4 where the measured versus predicted viscosities for hydrous granitic melts are presented for all four models in comparison.

The general trends in viscosity as a function of temperature and water content are presented via output from the Hess & Dingwell (1996) model in Fig. 5. Clear and significant non-Arrhenian behaviour is observed for water-bearing granitic melts. The nonlinear decrease in isothermal viscosity with water addition is a strong function of temperature, being most extreme at low

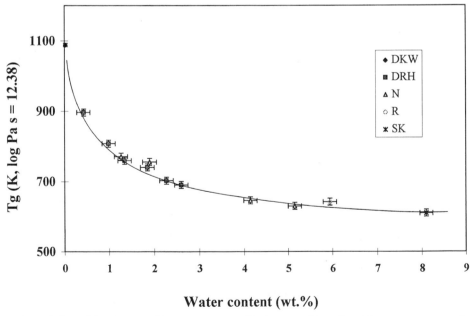

Fig. 3. The variation of the glass transition temperature of hydrous granitic melts with water content as obtained by volume, shear stress and structural relaxation studies. Reproduced with permission of Elsevier Science from Dingwell (1998) where original data sources are listed.

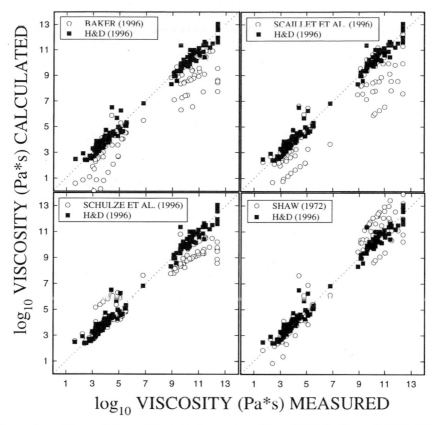

Fig. 4. Comparison of the predictions of the Arrhenian models of Shaw (1972), Scaillet *et al.* (1996), Schulze *et al.* (1996) and Baker (1996) with the non-Arrhenian model of Hess & Dingwell (1996). The significant deviations from the measured data for the Arrhenian models are, as expected, in the high viscosity range. Reproduced with permission from Hess & Dingwell (1996).

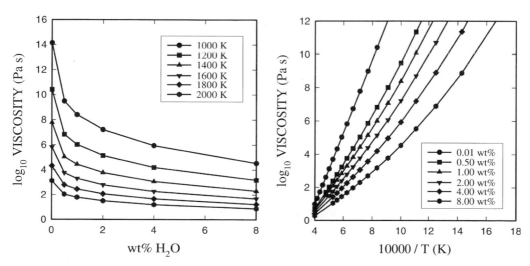

Fig. 5. The temperature and water content dependence of the shear viscosity of haplogranitic melts. This parameterization was compiled using spectroscopic, volume and shear stress relaxation data. Reproduced with permission from Hess & Dingwell (1996).

temperatures and high viscosities. This general trend qualitatively confirms the observation of Shaw (1963) but the degree of nonlinearity observed using the new model is much greater. This is most apparent at low water contents where the dependence of viscosity on water content is very great.

Further challenges

The viscosity of granitic melts of near metaluminous composition is very sensitive to the stoichiometry of the melt. Small excesses of alkali oxides, generating peralkaline compositions, strongly decrease the melt viscosity (see below). As a result, it is conceivable that the nonlinearity observed in Fig. 5 for synthetic melt compositions might not hold for the more complex multicomponent melt compositions found in nature. In order to test the applicability of the Hess & Dingwell (1996) model in natural systems, Stevenson et al. (1998) conducted a series of viscosity determinations on natural samples with minor water contents, with added water, and with water removed by remelting in air. Taking care to control other possible influences on viscosity such as variations in

oxidation state and other volatile (F, Cl) contents they obtained a set of viscosity data for several hydrous multicomponent melts that can be compared to those of the Hess & Dingwell (1996) model. The resulting good agreement suggests that the degree of nonlinearity indicated in that model for low water contents is an accurate reflection of the situation in natural granitic melt compositions (Fig. 6).

As noted above, significant variation of the alkali/aluminium ratio of granitic melts might have an important effect on the viscosity of the dry granitic melt and also on the intensity of the viscosity decrease with the addition of water. In order to investigate this possibility, the viscosities of peralkaline and peraluminous derivatives of the metaluminous melt compositions (both water-free and water-bearing) have been subjected to viscometry (Dingwell et al. 1998a, b). The results of these studies are illustrated together in Fig. 7 as isokom temperatures. The peraluminous melts exhibit similar viscosities, at all water contents, as the metaluminous melts. The peralkaline melts, in contrast, exhibit far lower viscosities than the metaluminous and peraluminous melts, at all water contents. Despite the relatively low viscosities of the

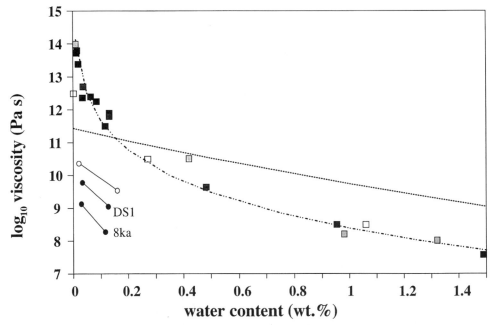

Fig. 6. Log viscosity (at $10^4/T = 10$) versus water content. Peralkaline rhyolites (filled circles) and peraluminous macusanite (open circles) plot off the trend defined by calc-alkaline rhyolites (black squares), BL6 samples (dark grey squares), HPG-8 (light grey squares) from Dingwell et al. (1996a), and the calc-alkaline andesite (open squares) from Richet et al. (1996). The Hess & Dingwell (1996) calculation scheme defined by the dashed curve closely fits the experimental data for the calc-alkaline compositions, whereas the Shaw (1972) method (dotted line) poorly approximates experimental data.

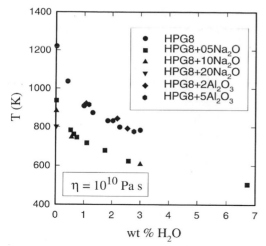

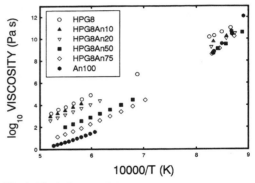

Fig. 8. The influence of the anorthite component on the viscosity–temperature relationship of calcalkaline haplogranitic melts. Note the strong viscosity decrease in the low viscosity range in contrast to the relative invariance in viscosity in the high viscosity range.

Fig. 7. Comparison of the influences of water on the viscosity of metaluminous (HPG8, Dingwell *et al.* 1996*a*), peraluminous (HPG8Al02, HPG8Al05, Dingwell *et al.* 1998*b*) and peralkaline (HPG8N5, HPG8N10, Dingwell *et al.* 1998*a*) granitic melts at the 10^{11} Pa s isokom. Note the extraordinarily low temperatures corresponding to the glass transition for approximately 3K/s cooling rate in the water-rich peralkaline melts. Reproduced with permission of Elsevier Science from Dingwell *et al.* (1998*a*).

peralkaline melts in the water-free state, the addition of water maintains its strong viscosity-reducing role in these compositions. The resulting viscosities for water-rich peralkaline melts are very low indeed.

Work in progress on the influences of variation in the proportions of normative albite, ortho-clase and quartz on granitic melt viscosities is revealing relatively small variations within the range of common granitic (*sensu lato*) rock compositions (Dingwell *et al.* in prep.). The influence of the anorthite component, on the other hand, shows quite a substantial decrease in melt viscosity in the low viscosity range but only a minor influence in the high viscosity range (Dingwell *et al.* in prep.). It is planned to incorporate such data in the Hess & Dingwell (1996)-based model in the near future. With the exception of fluorine (see references in Dingwell *et al.* 1996*b*, Fig. 8), further anionic components of granitic silicate melts have not been well-investigated to date. White & Montana (1990) demonstrated that CO_2 may be as effective as water, on a mole percent basis in reducing melt viscosity. Recently, however, a comparative study of the influences of F and Cl in synthetic alkali silicate melts has indicated that the influence of Cl on melt viscosity is very minor

in comparison to that of F (Dingwell & Hess 1998) and preliminary indications are that Br and I also play but a minor role (Dingwell, unpubl. data).

In specialized granitic melts and in pegmatitic systems, the compositions of the granite-derived water-rich melts can be highly enriched in a host of chemical components that are normally only present at trace element levels. Clearly, mapping out the viscosities of this multicomponent space of hydrous specialized melt chemistry is a major task because the effects of combining individual components on the viscosity of the resulting granitic melt are likely to be nonlinear. The task of dealing with this challenge has been initiated by studying the viscosities of granitic melts to which individual oxides have been added in varying amounts (Dingwell *et al.* 1996*b* and references therein). The individual influences of a large number of components on the viscosity of a metaluminous melt are compared in Fig. 9, where the effects that can be seen range quite widely.

Outlook

The outlook for improving the description of granitic melt viscosities in the next years is excellent. Several groups are intensively active in the area of melt viscosities at this time, a situation which has not been the case for the geosciences for decades. As a result we can expect that, if current activities continue in their intensity, fully generalizable models for the description of granitic melt viscosities as a function of composition, temperature and press-ure will be available within five years.

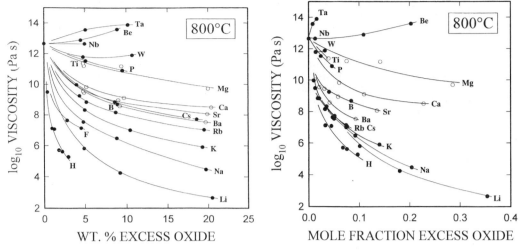

Fig. 9. The effects of various added components on the viscosity of a haplogranitic melt compared at 800 °C and 1 bar. See Dingwell *et al.* (1996*b*) for original references.

The critical experimental goals for the next years include: (1) precise definition of the pressure dependence of the viscosity of dry and hydrous granitic melts in order to refine this aspect of the modelling and comparison of data; (2) sufficient experimental determinations of the temperature dependence of hydrous melt viscosities such that the dependence of the degree of deviation of the viscosity–temperature relationship from the Arrhenian relationship can be simply related to composition; (3) evaluation of the importance of non-linear mixing effects of minor components in granitic melts on the variation of viscosity.

The critical modelling goals for the next years include: (1) the development of an algorithm for the description of melt viscosity which can cope with the sensitivity of melt viscosity to the alkali/aluminium ratio and therefore incorporate calcalkaline and peralkaline melt compositions in a single model; (2) the development of a quantitative link between water speciation and melt viscosity in multicomponent melts or, as an equally important alternative, an adequate understanding of the reasons why such a relationship does not exist; (3) the construction of a model for the non-Arrhenian temperature dependence of the viscosity that is generalizable for multicomponent melts.

The work of the Bayreuth group discussed here has been generously supported by the Deutsche Forschungsgemeinschaft (Di 431), the Alexander-von-Humboldt-Stiftung and the European Community. Careful reviews by J.-L. Vigneresse and an anonymous referee are gratefully acknowledged.

References

BAGDASSAROV, N. & DINGWELL, D. B. 1992. A rheological investigation of vesicular rhyolite. *Journal of Volcanological and Geothermal Research*, **50**, 307–322.

BAKER, D. R. 1996. Granitic melt viscosities: Empirical and configurational entropy models for their calculation. *American Mineralogist*, **81**, 126–134.

BOTTINGA, Y. & WEILL, D. F. 1972. The viscosity of magmatic liquids: a model for calculations. *American Journal of Science*, **272**, 438–475.

BOWEN, N. L. 1934. Viscosity data for silicate melts. *Transactions of the American Geophysical Union*, **15**, 249–255.

DINGWELL, D. B. 1995. Relaxation in silicate melts: some applications in petrology. *In*: STEBBINS, J. F., DINGWELL, D. B. & McMILLAN, P. W. (eds) *Structure and Dynamics of silicate melts*. Mineralogical Society of America Reviews in Mineralogy, **32**, 21–66.

—— 1998. Melt viscosity and diffusion under elevated pressures. *In*: HEMLEY, R. & MAO, D. (eds) *Ultrahigh Pressure Mineralogy*. Mineralogical Society of America Reviews in Mineralogy, **37**, 397–424.

—— & HESS, K.-U. 1998. Melt viscosities in the system Na-Fe-Si-O-F-Cl: contrasting effects of F and Cl in alkaline melts. *American Mineralogist*, **83**, 1016–1021.

—— & WEBB, S. L. 1990. Relaxation in silicate melts. *European Journal of Mineralogy*, **2**, 427–449.

——, BAGDASSAROV, N. S., BUSSOD, G. Y. & WEBB, S. L. 1993. Magma rheology. *In*: LUTH, R. W. (ed.) *Experiments at High Pressure and Applications to the Earth's Mantle*. Mineralogical Association of Canada Short Course Handbook, **21**, 131–196.

——, HESS, K.-U. & KNOCHE, R. 1996*b*. Granite and granitic pegmatite melts: volumes and viscosities. *Transactions of the Royal Society of Edinburgh: Earth Sciences*, **87**, 65–72.

——, —— & ROMANO, C. 1998*a*. Extremely fluid behaviour of hydrous peralkaline rhyolites. *Earth and Planetary Science Letters*, **158**, 31–38.

——, —— & —— 1998*b*. Viscosity data for hydrous peraluminous granitic melts: incorporation in a metaluminous model. *American Mineralogist*, **83**, 236–239.

——, KNOCHE, R., WEBB, S. L. & PICHAVANT, M. 1992. The effect of B_2O_3 on the viscosity of haplogranitic melts. *American Mineralogist*, **77**, 457–461.

——, ROMANO, C. & HESS, K.-U. 1996*a*. The effect of H_2O on the viscosity of a haplogranitic melt under P-T-X conditions relevant to silicic volcanism. *Contributions to Mineralogy and Petrology*, **124**, 19–28.

DORFMAN, A., HESS, K.-U. & DINGWELL, D. B. 1996. Centrifuge-assisted falling sphere viscometry. *European Journal of Mineralogy*, **8**, 507–514.

FONTANA, E. H. 1970. A versatile parallel-plate viscosimeter for glass viscosity measurements to 1000 °C. *American Ceramic Society Bulletin*, **49**, 594–597.

FRIEDMAN, I., LONG, W. & SMITH, R. W. 1963. Viscosity and water content of rhyolitic glass. *Journal of Geophysical Research*, **68**, 6523–6535.

FUJII, T. & KUSHIRO, I. 1977. Density, viscosity and compressibility of basaltic liquid at high pressures. *Carnegie Institution of Washington Yearbook*, **76**, 419–424.

GENT, A. N. 1960. Theory of the parallel plate viscosimeter. *British Journal of Applied Physics*, **11**, 85–88.

HESS, K.-U. & DINGWELL, D. B. 1996. Viscosities of hydrous leucogranitic melts: a non-Arrhenian model. *American Mineralogist*, **81**, 1297–1300.

——, —— & WEBB, S. L. 1995. The influence of excess alkalis on the viscosity of a haplogranitic melt. *American Mineralogist*, **80**, 297–304.

——, —— & —— 1996. The influence of alkaline earth oxides on the viscosity of granitic melts: systematics of non-Arrhenian behaviour. *European Journal of Mineralogy*, **8**, 371–381.

KANZAKI, M., KURITA, K., FUJII, T., KATO, T., SHIMOMURA, O. & AKIMOTO, S. 1987. A new technique to measure the viscosity and density of silicate melts at high pressure. *In*: MANGHNANI, M. & SYONO, Y. (eds) *High Pressure Research in Mineral Physics: The Akimoto Volume*. American Geophysical Union, Geophysical Monographs, **39**, 195–200.

KOZU, S. & KANI, K. 1935. Viscosity measurements of the ternary system diopside-albite-anorthite at high temperatures. *Imperial Academy of Japan Proceedings*, **11**, 383–385.

KUSHIRO, I. 1976. Changes in viscosity and structure of melt of $NaAlSi_2O_6$ composition at high pressures. *Journal of Geophysical Research*, **81**, 6347–6350.

—— 1978*a*. Density and viscosity of hydrous calk-alkaline andesite magma at high pressures.

Carnegie Institution of Washington Yearbook, **77**, 675–677.

—— 1978*b*. Viscosity and structural changes of albite ($NaAlSi_3O_8$) melt at high pressures. *Earth and Planetary Science Letters*, **41**, 87–90.

——, YODER, H. S. & MYSEN, B. O. 1976. Viscosities of basalt and andesite melts at high pressures. *Journal of Geophysical Research*, **81**, 6351–6356.

MAXWELL, J. C. 1867. On the dynamical theory of gases. *Philosophical Transactions of the Royal Society of London*, **A157**, 49–88.

NOWAK, M. & BEHRENS, H. 1995. The speciation of water in haplogranitic glasses and melts determined by in situ near infrared spectroscopy. *Geochimica et Cosmochimica Acta*, **59**, 3445–3450.

POCKLINGTON, H. C. 1940. Rough measurement of high viscosities. *Proceedings of the Cambridge Philosophical Society*, **36**, 507–508.

RICHET, P. 1984. Viscosity and configurational entropy of silicate melts. *Geochimica et Cosmochimica Acta*, **48**, 471–484.

——, LEJEUNE, A.-M., HOLTZ, F. & ROUX, J. 1996. Water and the viscosity of andesite melts. *Chemical Geology*, **128**, 185–197.

ROMANO, C., DINGWELL, D. B. & STERNER, S. M. 1994. Kinetics of quenching of hydrous feldspathic melts: quantification using synthetic fluid inclusions. *American Mineralogist*, **79**, 1125–1134.

RYAN, M. & BLEVINS, J. Y. K. 1987. *The viscosity of synthetic and natural silicate melts and glasses at high temperatures and at 1 bar (10^5 Pa) pressure and at higher pressures*. United States Geological Survey Bulletin, **17864**.

SABATIER, G. 1956. Influence de la teneur en eau sur la viscosité d'une rétinite, verre ayant la composition chimique d'un granite. *Comptes Rendus de l'Académie des Sciences de Paris*, **242**, 1340–1342.

—— & SAUCIER, H. 1955. Quelques expériences sur la déformation des roches éruptives acides à haute température. *Comptes Rendus de l'Académie des Sciences de Paris*, **241**, 1145–1147.

SAUCIER, H. 1952. Quelques expériences sur la viscosité à haute température des verres ayant la composition d'un granite. *Meeting Société Française de Minéralogie*, **75**, 1–45.

—— & SAPLEVITCH, H. 1958. Nouvelles expériences sur la déformation des roches vitreuses à haute température. *Comptes Rendus de l'Académie des Sciences de Paris*, **247**, 1214–1217.

SCAILLET, B., HOLTZ, F., PICHAVANT, M. & SCHMIDT, M. 1996. The viscosity of Himalayan leucogranites: implication for mechanisms of granitic magma ascent. *Journal of Geophysical Research*, **101**, 27691–27699.

SCHULZE, F., BEHRENS, H. & HOLTZ, F. 1996. The influence of water on the viscosity of a haplogranitic melt. *American Mineralogist*, **81**, 1155–1165.

SHAW, H. R. 1963. Obsidian-H_2O viscosities at 1000 and 2000 bars in the temperature range 700 to 900 °C. *Journal of Geophysical Research*, **68**, 6337–6343.

38   D. B. DINGWELL

—— 1972. Viscosities of magmatic silicate liquids: an
     empirical method of prediction. *American Journal
     of Science*, **272**, 870–889.
SHEN, A. & KEPPLER, H. 1995. Infrared spectroscopy of
     hydrous silicate melts to 1000 °C and 10 kbars:
     direct observation of water speciation in a
     diamond anvil cell. *American Mineralogist*, **80**,
     1335–1338.
SIMMONS, J. H., OCHOA, R. & SIMMONS, K. D. 1988.
     Non-Newtonian viscous flow in soda-lime-silica
     glass at forming and annealing temperatures.
     Journal of Non-Crystalline Solids, **105**, 313–322.
STEVENSON, R., BAGDASSAROV, N., DINGWELL, D. B. &
     ROMANO, C. 1998. The influence of trace amounts
     of water on obsidian viscosity. *Bulletin of
     Volcanology*, **60**, 89–97.
TOBOLSKY, A. V. & TAYLOR, R. B. 1963. Viscoelastic
     properties of a simple organic glass. *Journal of
     Physical Chemistry*, **67**, 2439–2442.

TOPLIS, M., DINGWELL, D. B. & LENCI, T. 1997.
     Peraluminous viscosity maxima in Na_2O-Al_2O_3-
     SiO_2 liquids: Triclusters in tectosilicate melts?
     Geochimica   et   Cosmochimica   Acta, **61**,
     2601–2612.
TOPLIS, M., DINGWELL, D. B., HESS, K.-U. & LENCI, T.
     1997. Viscosity, fragility, and configurational
     entropy of melts along the join SiO_2-$NaAlSiO_4$.
     American Mineralogist, **82**, 979–990.
TURKDOGAN, E. T. 1983. *Physicochemical properties of
     molten slags and glasses*. The Metals Society.
WHITE, B. S. & MONTANA, A. 1990. The effect of H_2O
     and CO_2 on the viscosity of sanidine liquid at high
     pressures. *Journal of Geophysical Research*, **95**,
     15 683–15 693.
ZARZICKI, J. 1991. *Glasses and the vitreous state*.
     Cambridge University Press.

Geophysical imaging of the shape of granitic intrusions at depth: a review

L. AMÉGLIO[1,2] & J. L. VIGNERESSE[2,3]

[1]LMTG, UMR CNRS no. 5563, 38 rue des Trente-Six Ponts, 31400 Toulouse, France
Present address: Rhodes University, Department of Geology, PO Box 94,
Grahamstown 6140, South Africa (e-mail: ameglio@rock.ru.ac.za)
[2]CREGU, UMR CNRS no. 7566 G2R, BP 239, 54506 Vandoeuvre-lès-Nancy cedex, France
(e-mail: jean-louis.vigneresse@g2r.u-nancy.fr)
[3]ENS Géologie Nancy, BP 40, 54501 Vandoeuvre-lès-Nancy cedex, France

Abstract: This review deals with direct evidence (field statements and geochemistry) and indirect observations (modelling experiments, analogical models, and geophysics) on granite plutons to model their shape at depth. 3D modelling of granite plutons can be achieved using geophysical tools. Amongst these tools, heat-flow and heat-generation data used earlier to estimate the thickness of granitic plutons appear inadequate. Electrical methods are strongly influenced by near-surface heterogeneities and temperature, which minimize their effectiveness at depth. Magnetic surveys provide information on contacts between pluton and country rocks, since magnetic halos are appropriate to delineate surface contours, but the technique lacks the resolution to reveal deep boundaries. Anisotropy of magnetic susceptibility is particularly well adapted to determine the internal structures of plutons. Seismic profiles at usual frequencies (30–80 Hz) define the layered structure of the floor of several bodies but fail to show the rock-type variations and their internal fabrics, because of their low impedance contrasts. Conversely, high-resolution seismic reflection profiles reveal fault structures in granites but the pluton's floor remains transparent. Gravity measurements have been widely applied in granites and owing to the 3D inversion of data, the shape of the pluton at depth and depth of its floor may be derived with confidence from density contrasts.

Granitic intrusions, which are a major constituent of the crustal basement, have been widely investigated for centuries. The specific problem of their accommodation in the Earth's crust has been addressed by increasingly numerous publications dealing mainly with geochemical, petrological and structural studies. Geochemistry and petrology have yielded a wealth of data concerning crystallization properties of silicate melts and also the age, source regions, and/or wall-rock contamination of magmas (see Miller *et al.* 1988). Structural studies, applied to a large number of plutons throughout many orogenic belts of varying age, has shown several degrees of interactions between magma and major tectonic faults or shear zones (Hutton 1997). With time, ideas and paradigms to resolve the so-called 'space problem' have evolved from consideration of *in-situ* assimilation or batch melting, emplacement in fold hinges and faults, stoping, forceful intrusion, diapirism, *in-situ* ballooning, emplacement through dykes and sills, and pluton growth by imbricated sheets or growing cavities (see Pitcher 1993). This has led to 3D models to image magma emplacement, and consequently constrain the proposed emplacement mechanisms of granites according to tectonic regimes (see for example: McCaffrey 1992; Jacques & Reavy 1994; Koukoukevlas *et al.* 1996). However, for the overwhelming majority of granitic plutons, outcrops provide only subhorizontal sections restricted to sub-surface observations (< 100 m). Moreover, proposed theoretical concepts consider movements and deformation without estimating original forces occurring at depth, and frequently they cannot resolve conflicting hypotheses derived from the same data set. For example, in the Sidobre granite pluton (Massif Central, France), geochemical and structural data led earlier to interpretation of this massif as a diapir rooted at 7 km (Borrel 1978). Recently, Darrozes *et al.* (1994) have proposed that the pluton was emplaced in an extensional tectonic regime with a supposed sill shape being

From: CASTRO, A., FERNÁNDEZ, C. & VIGNERESSE, J. L. (eds) *Understanding Granites: Integrating New and Classical Techniques*. Geological Society, London, Special Publications, **168**, 39–54. 1-86239-058-4/99/$15.00 © The Geological Society of London 1999.

confirmed by a detailed gravimetric study (Améglio *et al.* 1994) indicating that the massif was shallowly rooted (<3 km). The modes of granite emplacement and their internal dynamics are still not fully established.

It appears obvious that modelling of the present shape of plutons at depth could limit the plausible geological interpretations. To infer deeper geological features, geophysical data are required. Until recently this kind of approach to granites has been neglected, representing only some 5% of the total research activity on granites since 1994 (Clarke 1996). Geophysics applied to granitic plutons is the main concern of this contribution. The first part will consider results obtained at outcrop level using various mapping methods, and experiments in the laboratory. The second section aims to review the advances in geophysical studies to infer the morphology of the granitic plutons at depth.

Shape of granite plutons: an overview

The genesis of granitic rocks is the result of four main steps: generation, segregation, ascent and emplacement of granitic magma. This makes up a concept formalized by Petford *et al.* (1997) in which the granitic pluton (the final product) contains within it information that summarizes each step in its formation. These particulars are, in random order: rheological properties of silicate melts, internal magmatic structures, 3D shape, regional tectonic settings and the surrounding rocks. Each can be interpreted from direct observations (morphological notes in the field, fabric-structure measurements, petrography and geochemistry) and/or indirect observations (modelling experiments, analogue models, and geophysical surveys). We discuss below the advances in each kind of observation, and their application in inferring the present geometry of the granitic plutons at depth.

Direct observations

Field observations. Except in a few spectacular cases where the pluton shape at depth is revealed in the topography (Greenland, Bridgwater *et al.* 1974; Himalaya, Le Fort 1981), field data on granites are essentially restricted to surface observations. Laccolithic intrusions are widely exposed and Corry (1988) provides a comprehensive data base showing horizontal and vertical dimensions of these bodies, using estimated and measured data. Usually the shape at depth of granitic intrusions is derived by extrapolation using: (i) outcrops of contacts; (ii) magmatic structures; (iii) extrapolated

magmatic *P–T* conditions in the aureole of metamorphism; (iv) regional deformation of wall rocks.

The contact between a plutonic rock and its country rock can be summarized as being either discordant or concordant depending on whether or not contacts cut across the country-rock structures (Park 1983). On the one hand, discordant cross-cutting margins suggest a subjacent plutonic form with continuation to unknown depths and argue in favour of forceful intrusion. On the other hand, concordant margins have been taken to indicate laccolithic forms although these relations would also be consistent, but not necessarily (see Castro & Fernandez 1998), with the roof area of a diapir shape. For example, the Beinn an Dubhaich granite (Skye, Scotland) was earlier interpreted as a boss, then as a sheet intrusion, and finally, according to sub-surface (<50 m) borehole information, as a shallow funnel-shaped intrusion (see Raybould 1973). Twenty-three years later, local geophysical investigations pointed to this granite being a steep-sided stock extending down to about 1 km depth (Goulty *et al.* 1996). Finally, Geoffroy *et al.* (1997), based on internal magmatic structures, have proposed a ring-dyke complex at shallow depth. Clearly, purely geological evidence by itself appears too restrictive to define conclusively the three-dimensional shape of plutons.

Structural measurements through granitic plutons, especially the anisotropy of magnetic susceptibility (AMS) method (Borradaile 1988), provide significant constraints on magma flow (Borradaile & Henry 1997; Bouchez 1997). This leads to 3D extrapolated models of emplacement modes (Hutton *et al.* 1990; Bouillin *et al.* 1993; Leblanc *et al.* 1996; Anma 1997; Archanjo *et al.* 1998). Unfortunately, extrapolation to the third dimension is not always possible since the outcrop level is most of the time a sub-horizontal section through the bodies. It may also be ambiguous in the vertical dimension since vertical lineation and vertical foliation could reflect either upward flow, contractional deformation, or transpressional deformation (see Fossen *et al.* 1994). Another complication comes from different interpretations of the structural pattern according to various authors, and the main emplacement processes considered by them: diapirism (Cruden 1990), ballooning (Paterson & Fowler 1993), or permitted intrusions (Hutton 1997). In the Elberton granite (northeast Georgia, United States), AMS combined with palaeomagnetic data indicates possible post-emplacement tectonic rotation of the pluton (Ellwood *et al.* 1980). Consequently, the greatest care must be taken in the kinematic interpretation of raw structural data in terms of flow direction for the magma emplacement and deformation of wall-rocks. However, when correlated with other studies such as those dealing with gravity or geochemistry, a structural study by AMS remains the method best suited to infer the paths of magma emplacement (see next section).

In the case of a deeply eroded massif, a vertical section in the order of one or two kilometres may be available at surface. In the Bergell pluton (Central Alps, southern Switzerland and northern Italy), the main body intruded rocks of different metamorphic grade, offering at least a 10 km crustal depth profile, and its 3D shape is inferred as a 'ball' (Rosenberg et al. 1995, fig. 4) with a diameter of c. 9 km. Unfortunately, only one eighth of the total vertical aspect is really observed at the outcrop level, hence the shape proposed appears strongly subjective. In a wealth of Hercynian granitic plutons in the Pyrenees Range, Gleizes et al. (1997) claim that differences in the magmatic pattern throughout an individual pluton or between plutons may be explained by different outcrop levels revealing the metamorphic grade registered in wall-rocks. This led authors, in the case of the Mont-Louis–Andorra granitic pluton, to model in 3D the shape at depth of the floor (Bouchez & Gleizes 1995, fig. 8a). Unfortunately, the proposed extrapolation conflicts with Bouguer anomalies (Améglio 1998, fig. 76) respecting the feeder zones of the pluton.

Wall-rocks have also been investigated for their deformation to propose emplacement mode of granitic plutons. In the Bergell pluton (Rosenberg et al. 1995), local observations on the walls and the floor of the pluton provide useful information to constrain the force driving final emplacement. They show that the upper parts of the pluton were affected by ballooning, whereas the floor was folded during regional deformation. In southeastern Sweden, post-orogenic granites pass transitionally into augen gneisses that are generally folded conformably with surrounding rocks. Wikström (1984) suggests that, according to the 'balloon tectonic model' (Ramsay 1981), augen gneisses are caps or down-pointing flukes of the post-orogenic granites in a structure similar to one of Ramberg's (1981) centrifuged analogue experiments. The extrapolated shape in a cross-section of a 'beachcomber' (see Fig. 3; Wikström 1984, p. 413) appears, however, geologically unrealistic. Deformation criteria commonly used have been summarized by Paterson & Tobish (1988) for several granitic plutons according to their pre-, syn- or post-tectonic settings. It appears that no single and clear criterion can establish the timing of pluton emplacement, its relation to metamorphism, or its geometrical links with regional deformation. Moreover, strain/kinematic reconstructions in wall-rocks are difficult to interpret because they frequently include superimposed deformations induced by many chronologically later tectonic events which can overprint the structures. Hence, because of the inherent ambiguities in surrounding rocks, 3D extrapolations appear subjective. Nevertheless, internal fabrics in granites, which provide a snapshot of the deformation in the part of the crust where magma is emplaced, are of great use to infer tectonic models (see Paterson & Tobish 1988).

Petrography and geochemistry. Final crystallization pressures, and thus the depth of emplacement of granitic magmas, can be estimated using the experimental calibration of some geobarometers and geothermometers (Blundy & Holland 1990; Schmidt 1992; Powell & Holland 1994). However, the scatter in results (due to the water content, oxygen fugacity and activity of F and Cl elements in a magma), and their lack of concordance, indicate that reconstruction of the magmatic conditions from mineral composition is not very reliable. Moreover, multiple intrusive pulses occurring in many plutons complicate the interpretation of re-equilibration to late-magmatic and/or subsolidus conditions. In spite of these constraints, in the well-exposed La Gloria pluton in the Chilian Andes (with more than 2.5 km of relief, and in a pyramidal shape indicated by projecting the wall contacts), Cornejo & Mahood (1997) reconstructed magmatic gradients identified as differential cooling rates related to vertical position inside the pluton. However indicative of the broad conditions of emplacement geochemical information may be, it cannot be extrapolated to depth to infer the shape of the floor, and even worse, its depth since it cannot be expressed by a quantitative relation.

Whole-rock chemical analysis is usually used to constrain the age and source regions of magma. This can also be used to follow the magma flow with respect to the whole shape of the pluton and thus constrain the origin of petrographical and chemical zonation. The Fichtelgebirge granitic pluton (Germany) contains several facies showing a continuum in chemistry, attributed to progressive differentiation. The distribution of facies compared with the shape of the floor of the pluton shows that late-magmatic facies remain near the root zones while less differentiated facies are more distant (Hecht et al. 1997). In the Cabeza de Araya pluton (Extremadura, Spain), Vigneresse &

Bouchez (1997) outline a similar link between the positions of feeders and petrochemical units. Moreover, further refinements using structural patterns allow authors to propose a dynamic scenario of the granite emplacement process and to discuss the succession of magma pulses. These examples prove, unfortunately, that proposed 3D reconstructions are inadequate without evidence provided by other methods.

Indirect observations

Numerical modelling. Widely developed during the last 20 years as a result of advances in computer techniques, numerical modelling examines the flow of magma in dykes and focuses on the velocity of magma as a function of dyke width, in order to infer the volume of magma that can be displaced before crystallizing (Delaney & Pollard 1982). Under a constant magma pressure, a major result is that the dyke propagation correlates positively with the dyke length and correlates inversely with magma viscosity. Consequently, thin vertical dykes transport basalt to the upper crust, whereas equidimensional sills are commonly inferred to accommodate granite emplacement (Lister & Kerr 1991; Rubin 1995). Looking at the interaction between the stress field and magma rheology leads to similar conclusions: dykes are preferentially formed with low-viscosity magmas, while stocks result from high-viscosity magmas (Emerman & Marrett 1990). Numerical experiments have also been conducted by Guglielmo (1993) to quantify interactions between regional tectonic-related and local pluton-related strain fields. Three-dimensional simulation models examine the position, type and kinematic evolution of structures around plutons that expand in a non-coaxial tectonic environment. Models assuming an initial spherical shape for the pluton cannot offer an objective view of the 3D shapes of plutons obtained during experiments. However, they offer practical guidelines for field work to characterize deformations in wall-rocks.

Analogue experiments. Pioneering experiments from Ramberg's group led to diapiric schemes of magma emplacement, leading to the popularized (and unverified) inverted drop shape (Ramberg 1981; Cruden 1990; Weinberg 1997) in which the aspect ratio between horizontal and vertical axis is less than 1. Recently, diapirism has been examined above modelled subduction zones (Anma & Sokoutis 1997). It appears that buoyant air inclusions (analogue material of the magma according to the authors) in a more viscous fluid leave an inclined diapir tail near the trench axis where shear deformation is greatest. By contrast, away from the trench axis, where horizontal extension prevails, intrusions become tabular in shape. Unfortunately, these 'diapiric experiments' are conducted within a plastic medium with a viscosity contrast unrealistically low for upper-crustal settings. It is also interesting to note that the analogue experiments could not simulate diapir intrusion for more realistic viscosity contrasts, but did show the development of dikes (Ramberg 1981).

Other recent analogue experiments of granitic intrusions have been significantly improved by Vendeville & Jackson (1992), Roman-Berdiel *et al.* (1997) and Benn *et al.* (1998*a*), and contrast with the earlier diapiric models often conducted in homogeneous environments. In particular, they address the importance of rheological layering and the possible influence of fault activity which introduces inhomogeneities in the upper crust. From those experiments, it appears that: (i) the shape of the intrusion is greatly influenced by the deformation, and (ii) intrusions approximate the shape of laccoliths with an aspect ratio of about 4 to 5. However suggestive these experiments may be, they cannot adequately address the complexity of mechanisms operating in the crust (stress patterns, thermal anomalies, lithostatic load, internal magma pressures). Moreover, the granite emplacement time-span is often a hundred times greater, or more, than in the analogue experiments.

Geophysical data. In the following, we deal with the main processing methods used in geophysical data interpretation. Results of most of geophysical investigations of granitic bodies by authors and others will be examined in the next section. To interpret adequately the effect (or anomaly) induced by the object which may be detected at depth (granitic bodies), the choice of the type of geophysical survey is important since it can severely bias the results. Anomalies are often interpreted along profiles leading to 2D structures, or directly on maps with the aim of investigating the 3D shape of the buried bodies. In the former case, a strong assumption implies that the body is infinite in the direction (y) perpendicular to the profile. In a 3D interpretation, the body has no preferential shape in any of the three coordinates (x, y, z). Between those two end-members, 2.5D interpretation methods introduce numerical corrections to 2D formulas, which take account of the finite extension in the (y) direction. In practice the 2D methods are faster and consequently cheaper than the 3D ones. However, when dealing with geological structures, a map representation provides structural information that cannot be inferred from successive cross-sections.

Characterization of mass distribution, hence the pluton's shape and/or depth of its floor, is achieved by two main calculation processes, which are direct and inverse techniques (Vigneresse 1990). Direct or forward modelling assumes *a priori* a mass and computes the corresponding anomaly which in turn is compared to the measured anomaly. Then, the shape of the initial body is modified step-by-step until the computed and measured anomalies equate (Talwani & Ewing 1960). Some filtering operations (high- and low-pass filters) and field transformation encompassing first or second vertical derivatives, and upward or downward continuity, are also applied to measurements before calculation to extract one parameter carrying information on the source body (Bhattacharyya 1967; Telford *et al.* 1990). The inverse technique (Gottlieb & Duchatau 1996) allows the removal of (most of) the *a priori* information on the source body. Model parameters are computed directly from measurements. The process can be iterative with a starting model (Cordell & Henderson 1968) or purely matricial. In that case, a linear system of equations is solved to compute the respective densities of pieces of the structure (Lines & Treitel 1984). Other non-linear systems of equations deal with the shape of the bodies within which predetermined densities have been posted (Richard *et al.* 1984). Constrained systems of equations can be constructed which allow a stochastic approach to the mass distribution (Tarantola & Valette 1982). Geophysics which refers to physical properties (such as seismic velocity, rheological properties) of the crust (Schön 1996) remains the best method to investigate the shape of plutons at depth since it links geological observations to physical laws.

In summary the overwhelming majority of direct investigations provide observations along cross-sections and cannot be extended to depth with confidence. The general concept, supported by the pluton's relations to the country rocks, is that the shape is characterized by sub-vertical walls with a slight inward dip. This has led to the popular, and 'fallacious' idea (see Vigneresse 1995a, fig. 1) of the inverted tear-drop or diapiric shape. Based on petrographical and geochemical studies, magmatic structures, and wall-rock deformation, diapirism, ballooning, or permitted intrusions are all generally accepted. The difficulty lies in assessing the relative importance of each concept to explain how space is made during pluton emplacement. This conflict of opinion clearly demonstrates that available geological evidence cannot conclusively support any single hypothesis. Frequently some geometrical landmark provided by other methods is needed to secure interpretation. Indirect observations seem to be most suitable even if the information about the 3D shape is limited. Numerical modelling focuses only on the rheological characteristics of magma (velocity, viscosity, volume) and the path of its ascent in the crust as dykes. Following analogue experiments, it appears that the shapes of the intrusions are strongly controlled by the regional or local deformation at the time of the emplacement and this results in laccolith-shaped bodies. However, experiments are too simple to furnish a general understanding of the problems. Moreover, interpretations still tend to be general rather than specific. Assessment of deeper features is essential and geophysics is the best tool for this.

Major results of geophysical studies on granite plutons

Numerous papers and books have discussed or reviewed geophysical techniques applied to different rocks, and at several scales. In the following section only those elements that indicate the current shape of granitic intrusions at depth will be reviewed. Background descriptions of the principles of each geophysical method are available for example in Telford *et al.* (1990).

Geothermal data

Gamma-ray spectrometry is used to measure the radiogenic elements (also named heat-producers) within rock (mainly K, U and Th), and also heat-generation which reflects their global abundance. The remarkable uniformity of heat-generation for a given igneous unit, even over large areas, suggests that these parameters are diagnostic properties of igneous rocks. They are actually used by geologists to map individual units or to correlate rock types (Jaques *et al.* 1997). U and Th are magmatophile elements, and thus are amongst the first elements to be mobilized during melting of the lower crust. They are concentrated in the magma formed, and they are easily carried towards the upper crust through magma ascent and intrusion. As a result, the upper crust (with abundant granitoids and low-grade metamorphic rocks) is enriched in those incompatible elements, whereas the lower crust (granulitic facies) is generally depleted. Therefore U, Th and

K can be monitored as remote sensing elements in the study of the crustal structure (Wollenberg & Smith 1987).

Several studies have demonstrated that surface heat-flow is linearly correlated with surface heat-generation (Roy et al. 1968; Lachenbruch 1970). The slope of that linear relationship is called the thermal depth (D). It has the dimension of a length and is commonly related to the form of heat production distribution with depth. Since granitic rocks are commonly enriched in heat-producing elements, estimation of the thickness of granitic intrusions has been achieved using this approach (Tilling 1974). However, over a geothermal province, the linear relationship between heat-flow and heat-generation is not strictly followed because of crustal heterogene-ities, or because points cluster within an elongated pattern. This is demonstrated in the Tinemaha Granodiorite (Sierra Nevada batholith) by Sawka & Chappel (1988). These authors show that the thermal depth value for the granodiorite ($D = 2.2$ km) is significantly less than that of the batholith ($D = 10.1$ km). This result implies that plutons within a heat-flow province may not follow the same distribution law for heat-producing elements. Moreover, a magnetic and gravity survey of the same pluton (Oliver 1977) provides a depth estimate of c. 4.5–5.5 km. Finally, a reappraisal

of data and related arguments to estimate the depth of plutons from the linear relation between heat-flow and heat-generation has been conducted from a compilation of worldwide measurements (Vigneresse & Cuney 1991). No correlation is observed between the depth of the granites and the thermal depth (Fig. 1).

Electrical methods

Electrical properties appear theoretically useful in that they are highly sensitive to subtle changes within the crust. However, the interpretation of measurements is difficult due to the great number of parameters that can produce equal changes in the resistivities of rocks (Ward 1976). In granite, electrical properties appear to be controlled dominantly by the amount of free water and by temperature (Olhoeft 1981). Consequently, electrical methods are sensitive to conductors localized beneath the site of the measurements. They are well suited to mapping of late faults in granite. On the scale of a granitic massif (leucogranite of Saint Goussaud, France), electrical measurements can also be interpreted in terms of contrasted resistivities (Fig. 2a), and in terms of boundaries between granite and surrounding micaschists (Fig. 2b) (Vasseur et al. 1990). Note that electrical data present weak resolution at depth (less than 500 m). The granite

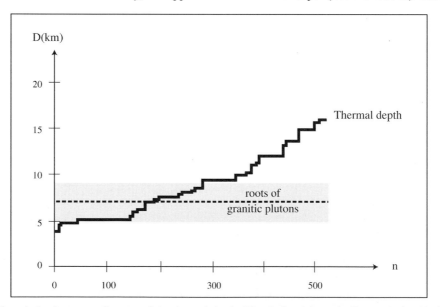

Fig. 1. Cumulative frequency diagram of the thermal depth (D) calculated from the linear relationship between heat-flow and heat-generation (after Vigneresse & Cuney 1991). Worldwide data without reference to the age of the region are weighted according to the number (*n*) of points to compute the correlation. Greyish zone corresponds to the downward extent of the granites deduced from gravity data. No correlation is observed between the depth of the granites and the thermal depth.

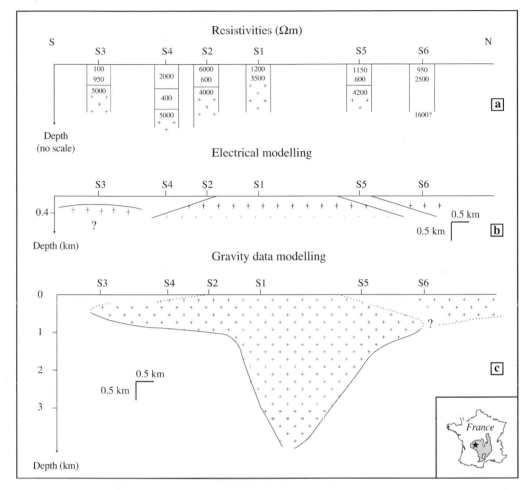

Fig. 2. Shape at depth of the Saint Goussaud leucogranite (Limousin, France) modelled using electrical and gravity data (after Jover 1986 and Vasseur *et al.* 1990). Scales of sections have been corrected to avoid the vertical exaggeration. S₁ to S₆ are electrical soundings. Crosses represent granite. (**a**) Geoelectric schematic cross-section indicating contrasted resistivities (in Ωm) at depth. (**b**) Geological interpretation of resistivities in terms of boundaries between granite and surrounding micaschists. The granite is dipping at low angles under its cover-rocks. Data present a weak resolution with depth (less than 500 m). (**c**) Cross-section through the granite showing its morphology with depth as inferred from gravity data. Contacts with wall-rock and floor are well defined by gravity, whereas the roof appears better constrained by electrical data.

is dipping at a low angle under its cover-rocks. Extrapolation to depth appears, however, to conflict with the shape modelled by gravity data (Fig. 2c) (Jover 1986). Magneto-telluric surveys present a better vertical resolution. Through the Carnmenellis batholith (Cornubian, England), magneto-telluric soundings (Beamish 1990) provide a 1D smooth inversion vertical resistivity log indicating an electrical base to the granite at 14 km, in agreement with regional gravity models.

However interesting these results are, they cannot lay claim to resolve successfully the whole

3D shape of plutons because: (i) parameters taken into account in electrical surveys are too varied to yield secure interpretations and (ii) they provide only limited information concerning either deep boundaries such as the floor (magneto-telluric soundings) or the roof (electrical profiles).

Magnetic methods

Magnetic data are particularly valuable in determining basement structures and lithological continuities under cover or in areas of poor

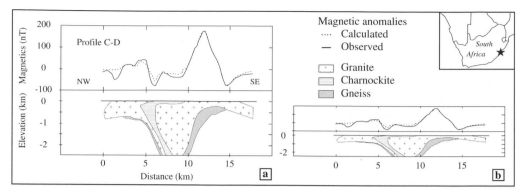

Fig. 3. 2D magnetic model across the Ntimbankulu granitic pluton (Kwatzula–Natal, South Africa) (after Maré & Thomas 1998). (**a**) Picture showing a magnetic high on the southeastern side of the magnetic profile, attributed to the effect of gneissic rocks and suggesting a ring-like feature which is smaller on the northwestern side. A mushroom-shape is proposed for the pluton on this scale. (**b**) The same section using 1 × 1 scale. This completely changes the perception of the shape of the body: the mushroom-shape geometry turns into a flattened cupola. The origin of the ring-like magnetic signature of this pluton is open to debate.

outcrop, since data comprise responses from magnetic minerals only (Jaques *et al.* 1997). The high sensitivity and small grid size of the resulting maps often yield mappable features too subtle to be observed in outcrop or hand specimen. In some granitic plutons in the Archean Abitibi greenstone belt (Quebec) petrographically similar intrusions may have either similar or different aeromagnetic expressions (Schwarz 1991). In granitic plutons a marginal magnetic halo may be related to an internal ring-dyke or to a contact-metamorphic aureole. It follows that qualitative and quantitative interpretation of probable pluton-caused anomalies is hazardous without ground information. In the Mesoproterozoic Ntimbankulu pluton (KwaZulu-Natal, South Africa), the ring dyke-like signature of the aeromagnetic data is induced, according to Maré & Thomas (1998), by remnants of quartz–feldspar gneiss trapped beneath the 'wings' of a mushroom-shaped body (Fig. 3). Unfortunately, the cross-section given (Fig. 3a) is not to scale. The correct 1 × 1 scale of the cartoon (Fig. 3b) reflects its true dimensions. The proposed mushroom-shape turns into a flattened cupola. This mistake and the lack of uniqueness of the proposed 2D models also introduce doubt about the proposed solution to explain the ring-dyke structure. In the Agaçören Granitoid (central Turkey), gabbroic rocks crop out in different localities and were interpreted as roof-pendants. The inferred structural position of these gabbros, combined with field investigations and aeromagnetic data (Kadioglu *et al.* 1998), prove that they have winding contacts with the host granitic body and that they display a gradual change in composition and texture, from

gabbro at the top and diorite at the foot, towards the contacts with granitoid. 2D magnetic and pseudogravimetric modelling suggest the existence of a shallow cone-shaped and deeply buried gabbroic body (Kadioglu *et al.* 1998). In the Mont-Louis–Andorra granitic pluton (Pyrenees), the magnetic susceptibility of the rock accurately reflects the modal content of ferromagnesian phases, and provides an efficient tool for regional mapping of petrological variations in granitoid plutons (Gleizes *et al.* 1993).

These examples demonstrate that magnetic surveys lack the resolution at depth to reveal the pluton's floor with confidence, but are particularly well adapted to delineating lateral extension of the granitic facies, or to localize the metamorphic aureole that surrounds granitic intrusions.

Seismic surveys

Study of the velocity of propagation of artificially induced waves provides cross-sectional images of the layers of rock beneath Earth's surface. Surveys include seismic refraction as well as seismic reflection measurements. In seismic refraction studies, velocities appropriate to granites are usually underlain at mid-crustal depth by higher velocities (Wenzel *et al.* 1987). Unfortunately, the refraction method does not have the resolution to determine whether the transition is abrupt or gradual downward into migmatites, gneiss or other rocks. By contrast, reflection seismic surveys are frequently designed to reveal such.

High-resolution seismic reflection over the leuconoritic-noritic Bjerkreim-Sokndal (Norway) layered intrusions provides a high-quality seismic image (Deemer & Hurich 1997) of the denser rock type cited as the source of lower crustal reflectivity. Reflections are produced by sub-horizontal boundaries between major (>100 m) lithological units of varying average modal composition, and small-scale modal layering (centimetres to several metres). A synformal structure, controlled by a detailed comparison with exposed geology, is proposed by these authors. These results show that the modest acoustic impedance contrast commonly observed within leuconoritic-noritic series is sufficient to produce reflections over continuous boundaries. Within the Lake District batholith (Cumbria, northern England), seismic reflection profiles reveal also a succession of reflector-poor and reflector-rich areas (Evans et al. 1993). This is interpreted as a main sheet-like body some 1100 m thick underlaid with poorly reflective tabular laccolithic granitic intrusions, interfingered with reflective wedges of country rock, to give the batholith a 'cedar-tree' profile (Corry 1988, fig. 8).

In granitic intrusions with a more homogeneous modal composition, impedance contrasts (that determine the reflection of seismic waves) are very poor because of the low density and gradual velocity changes within different granitic facies. This results in granites being objects transparent to common seismic frequencies ranging from 30 to 80 Hz (Fig. 4). Matthews (1987) proposed that granites can act as 'a window into the lower crust' beneath them. In Fig. 4 several reflectors that appear just beneath the pluton or from the bottom account for the inaccuracy of the proposed gravity model (shared area). These form a subhorizontal zone with an appreciable thickness of around 0.5 to 4 km. Other published examples suggest that this is often the case (Lynn et al. 1981; Jurdy & Phinney 1983; Goulty et al. 1992). Several explanations are possible and we summarize below the four alternatives which appear the most likely: (i) some silicic magma systems or epizonal batholiths are underplated by basaltic magma (Duffield et al. 1980); (ii) melting of crustal rocks by mantle basalt with the formation of cumulate layers (Taylor 1980); (iii) sub-batholithic migmatites (Lynn et al. 1981); (iv) tectonic contacts such as thrust zones (Jurdy & Phinney 1983). Actually, each proposal merits consideration. Note the first alternative previously exposed shows similarities with magma storage zones described as laterally

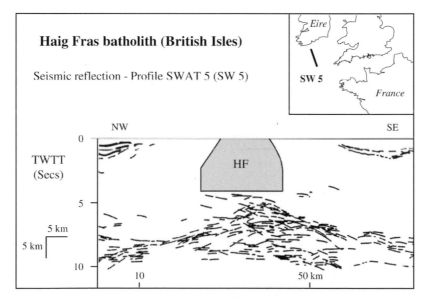

Fig. 4. Interpretation of a seismic reflection profile across the exposure of the Haig Fras (HF) batholith (British Isles) according to Matthews (1987). Black line in inset shows the location of the seismic profile. Vertical scale: 1 s TWTT approximately 3 km. The section shows reflectors visible under the granite, but it also shows the outline of a body (shaded area) modelled to fit the observed gravity anomaly (Edwards 1984). Granites appear to be devoid of reflectors. Reflections coming in earlier from beneath granites are commonly observed in other plutons. Basic layering at the bottom of the granitic pluton is an alternative explanation that contrasts with the concept of the pluton acting as 'a window into the lower crust', through which we see reflections from the lower crust with clarity.

elongated, planar and compartmentalized lenses connected by dykes below basaltic eruption sites (Wilson & Head 1988).

Seismic records in crystalline rocks are rare because most reflection work is motivated by petroleum exploration in sedimentary rocks. However, the results reviewed in this contribution indicate that seismic surveys would produce valuable information on the structure of igneous complexes. Within the Lac du Bonnet batholith (Canada), for example, high-frequency sources and receptors can be used to image numerous fractures (Mair & Green 1981). Concerning the contacts with surrounding rocks, these are always supposed to occur around the change in the reflection fabric in wall-rocks (Vejmelek & Smithson 1995). Data quality could readily be enhanced by using several crossing profiles, groups of several geophones, multi-fold coverage, and/or low frequencies (<30 Hz). Moreover, wide-angle reflection experiments would resolve the velocities within rocks beneath plutons. The main constraint remains the cost of seismic surveys.

Gravity investigations

The sign of the gravity anomaly over a granitic mass depends solely upon density contrast between that body and the country rocks into which it is intruded. Thus, the anomaly over a granite can be either negative or positive in sign. The fact that it is usually negative is attributable to the comparatively low density of granitic rocks. Former modelling (e.g. Bott & Masson-Smith 1959; Isaacson & Smithson 1976) of these anomalies was often realized to confirm subsurface geological observations. On account of this *a priori* approach, and considering the non-uniqueness of 2D gravity data interpretation depending on density contrasts used, granitic plutons were schematically represented as bosses with steep contacts extending at depth to 15 km or more, and without a floor. This led to their popular representation as an inverted 'drop of magma' with a long vertical axis as in the concept of diapirs. Processing of gravity data is now more refined (see next paragraph), and except for some unrealistic geological models providing the same shapes as previously proposed (Dindi & Swain 1988; Everaerts *et al.* 1996), results seem to suggest pluton shapes that are rather thin (<10 km) with inward-dipping contacts, or tabular or wedge-shaped (Hodge *et al.* 1982; Vigneresse 1988; Améglio *et al.* 1994; Benn *et al.* 1998*b*). As expected, McCaffrey & Petford (1997) show that laccoliths have a horizontal elongation (L) greater than their thickness (T), both being related by a power law $T = 0.12\ L^{0.88}$. Granitic plutons show a greater aspect ratio (L/T) and a lower power law coefficient of *c.* 0.80 (McCaffrey & Petford 1997). This leads to the conclusion that plutons are scale-invariant and with a self-affine elongation.

Within the last decade, gravity investigations over granitic plutons have provided a better understanding of the shape at depth of these bodies (Vigneresse 1990; Stettler *et al.* 1993; Améglio *et al.* 1994; Lyons *et al.* 1996; Behn *et al.* 1998; Dehls *et al.* 1998). Amongst the several methods used, the 3D iterative procedure of gravity data inversion of Cordell & Henderson (1968), adjusted by Vigneresse (1990), appears the best suited. Resolution of depth with a precision of 10–15%, depending on an adequate density contrast between granite and wall-rock, is indicated by indirect tests (Améglio *et al.* 1997). Thus, the bulk error on depth determination is kept quite small and compares favourably with geological observations. An example is given in Fig. 5 (see also Olivier *et al.* 1999): a detailed geometrical view can be obtained for the whole 'pipe' or 'lavatory pan' shape of the Aya granitic pluton (Basques Pyrenees). The funnel-shaped zone with an E–W-trending elliptical horizontal section, and a maximum depth of 3.5 km in the southern part of the pluton, determines the probable root of the pluton directly above basic facies. These deepest zones, provided they show vertical lineations and/or differentiated or late-magmatic facies in outcrop, are frequently interpreted as the feeder channels which fed the pluton (Vigneresse 1995*b*). The shape of the floor also constrains the magmatic pattern in the Sidobre pluton (Montagne Noire, France). In this pluton, lineations have very constant NNE–SSW trends and slight northward plunges in the northern half of the pluton, and southward plunges in the southern half of the pluton (Darrozes *et al.* 1994). The change of the lineation plunge occurs along an N75° E direction, in agreement with a general doming of the floor of the pluton along the same direction (Améglio 1998).

A review based on several years of gravity surveys allowed Améglio *et al.* (1997) to describe two main types of plutons. Flat-floored plutons are rather thin (a few kilometres), extend widely in every horizontal direction, and have several feeder zones. They are emplaced as sills in the upper crust, or within rather ductile environments in extensional tectonic contexts. These contrast with the thick (more than 10 kilometres) wedge-shaped plutons, extending largely in one direction along which a few root zones appear, which are V-shaped in transverse section, and

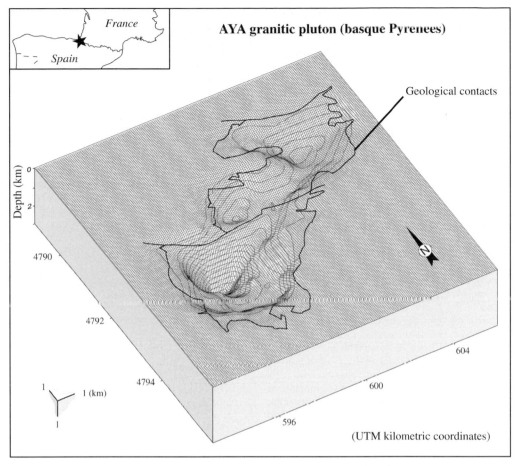

Fig. 5. 3D shape in view of the Aya granitic pluton (Basques Pyrenees) after Améglio (1998). The depth has been calculated using the inverse technique, specifically the 3D iterative procedure of Cordell & Henderson (1968), modified by Vigneresse (1990). Scale 1 × 1 × 1. Gravity reveals numerous details of the shape of the pluton.

which correspond to the infilling of dilatant volumes of the brittle crust during transcurrent tectonics. Note that this view, based on experience within the Variscan orogeny in France and Spain, does not necessarily represent all possible situations. However, a similar dichotomy in the shape of granitic bodies has also been proposed for several massifs from the western Bohemian massif (Trzebski *et al.* 1997). In that interpretation, the deeper and wedge-shaped massifs are also considered as having been emplaced during active tectonism which partly controlled the shape of their walls.

In the flat-lying South Mountain Batholith, Benn *et al.* (1998*b*) claim that magma was emplaced within a transpressive stress field. This is in conflict with the extensional tectonic regime commonly proposed for this flat shape (Vigneresse 1995*b*). Benn *et al.* (1998*b*) suggest that

crustal anisotropies may be the most important control on batholith assembly and on the final 3D shapes of plutons. Vigneresse *et al.* (1999) provide a more subtle and provocative explanation of the two-fold discrimination of the shape of granite plutons (see previous paragraph) as the result of a switch in the stress pattern caused by the emplacement of magma. This model contradicts the traditional notion of emplacement at a level of neutral buoyancy.

Conclusions

It appears that the explanation of how space is made during pluton emplacement cannot be reached without prior understanding of their present shapes at depth. Conclusions based only on direct observations, and purely on geological investigations, cannot be viewed with confidence.

To infer deeper geological features, geophysical data are essential. Geothermal data used earlier to determine the depth of the floor are not reliable. Electrical methods appear to be strongly influenced by near-surface heterogeneities which minimize their value in deep investigation. By contrast, magneto-telluric investigations seem to provide better results, but the precision at depth seems to be poor. Magnetic methods are well suited to providing information on the lateral extension of rocks, but they are not appropriate for determining the volumetric extension of granites because they lack resolution with depth. Seismic profiles at usual frequencies (30–80 Hz) fail to show the rock-type variations and *a fortiori* their internal fabrics, because of their low impedance contrasts. Conversely, being well adapted to horizontal structures, they image multi-reflector zones at the bottom of or just beneath several granites. This needs to be investigated further to assess its importance in the understanding of 'granite systems'. Interpretations of the shape of granitic plutons using all the earlier geophysical investigations still tend to be general rather than specific. Gravity, using 3D data inversion processes, appears best suited to inferring the whole morphology at depth of granitic intrusions. This can also benefit from structural data that record the last movements of the magma before complete crystallization. We claim that such integrated studies, applied routinely, should take place systematically before sophisticated studies are undertaken to image local phenomena.

Finally, in the upper and middle crust, the data on the pluton shape accumulated from various geophysical techniques clearly shows that granitic intrusions have a large aspect ratio. Their vertical extent is much lower than their lateral dimensions. This is in complete opposition to the concept of diapir-like intrusions, which popularly present the shape as an inverted 'magma drop'. In addition to the large aspect ratio, most plutons can be considered as flat bodies, the floors of which dip gently toward a restricted feeder zone (1.0–0.3 km in diameter). Conversely, some plutons, often associated with active shear zones, show steeply plunging walls and a wedge shape. Their floors are deeper than those of the preceding category of plutons and they do not show any evidence of a clearly localized feeder. This conclusion does not necessarily represent all possible situations. However, it points to features that may be general to most environments of pluton emplacement.

We are indebted to P. Ledru (BRGM Orléans, France) and J. P. Lefort (University of Rennes, France) for their constructive criticisms, and to H. Eales (Rhodes University, South Africa) for his comments and help in improving of English. J. Delhs and K. McCaffrey are warmly thanked for their critical reviews.

References

AMÉGLIO, L. 1998. *Gravimétrie et forme tridimentionnelle des plutons granitiques. Contributions aux études structurales.* PhD thesis, Paul Sabatier University, Toulouse, France.

——, VIGNERESSE, J. L. & BOUCHEZ, J. L. 1997. Granite pluton geometry and emplacement mode inferred from combined fabric and gravity data. *In*: BOUCHEZ, J. L., HUTTON, D. H. W. & STEPHENS, W. E. (eds) *Granite: From Segregation of Melt to Emplacement Fabrics.* Kluwer Academic Publishers, Dordrecht, 199–214.

——, ——, DARROZES, J. & BOUCHEZ, J. L. 1994. Forme du massif granitique du Sidobre (Montagne Noire, France): sensibilité de l'inversion des données gravimétriques au contraste de densité. *Comptes Rendus de l'Académie des Sciences, Paris,* **319**, 2, 1183–1190.

ANMA, R. 1997. Oblique diapirism of the Yakushima granite in the Ryukyu arc, Japan. *In*: BOUCHEZ, J. L., HUTTON, D. H. W. & STEPHENS, W. E. (eds) *Granite: From Segregation of Melt to Emplacement Fabrics.* Kluwer Academic Publishers, Dordrecht, 295–318.

—— & SOKOUTIS, D. 1997. Experimental pluton shapes and tracks above subduction zones. *In*: BOUCHEZ, J. L., HUTTON, D. H. W. & STEPHENS, W. E. (eds) *Granite: From Segregation of Melt to Emplacement Fabrics.* Kluwer Academic Publishers, Dordrecht, 319–334.

ARCHANJO, C. J., MACEDO, J. W. P., GALINDO, A. C. & ARAÚJO, M. G. S. 1998. Brasiliano crustal extension and emplacement fabrics of the mangerite-charnockite pluton of Umarizal, Nord-east Brazil. *Precambrian Research,* **87**, 1–2, 19–32.

BEAMISH, D. 1990. A deep geoelectric survey of the Carnmellis granite. *Geophysical Journal of the Interior,* **102**, 679–693.

BEHN, M. D., EUSDEN, J. D. & NOTTE III, J. A. 1998. A three-dimensional gravity model of the southern contact of the Sebago pluton, Maine. *Canadian Journal of Earth Science,* **35**, 649–656.

BENN, K., ODONNE, F. & SAINT BLANQUAT, M. 1998a. Pluton emplacement during transpression in brittle crust: New views from analogue experiments. *Geology,* **26**, 12, 1079–1082.

——, ROEST, W. R., ROCHETTE, P., EVANS, N. G. & PIGNOTTA, G. S. 1998b. Geophysical and structural signatures of syntectonic batholith construction: the South Mountain Batholith, Meguna Terrane, Nova Scotia. *Geophysical Journal International,* **136**, 144–158.

BHATTACHARYYA, B. K. 1967. Some general properties of potential fields in space and frequency domain: a review. *Geoexploration,* **5**, 127–143.

BLUNDY, J. D. & HOLLAND, J. B. 1990. Calcic amphibole equilibria and new amphibole-plagioclase geothermometer. *Contributions to Mineralogy and Petrology,* **104**, 208–224.

BORRADAILE, G. J. & HENRY, B. 1997. Tectonic applications of magnetic susceptibility and its anisotropy. *Earth-Science Reviews*, **42**, 49–93.

—— 1988. Magnetic susceptibility, petrofabrics and strain. *Tectonophysics*, **156**, 1–20.

BORREL, A. 1978. *Le massif granitique du Sidobre: pétrographie, structure, relation mise en place-cristallisation.* PhD thesis, Paul Sabatier University, Toulouse, France.

BOTT, M. H. P. & MASSON-SMITH, D. 1959. A gravity survey of the Criffel granodiorite and the new red sandstone deposits near Dumfries. *Proceedings of the Yorkshire Geological Society*, **32**, 3, 13, 317–332.

BOUCHEZ, J. L. 1997. Granite is never isotropic: an introduction to AMS studies of granitic rocks. *In*: BOUCHEZ, J. L., HUTTON, D. H. W. & STEPHENS, W. E. (eds) *Granite: From Segregation of Melt to Emplacement Fabrics.* Kluwer Academic Publishers, Dordrecht, 95–112.

—— & GLEIZES, G. 1995. Two-stage deformation of the Mont-Louis–Andorra granite pluton (Variscan Pyrenees) inferred from magnetic susceptibility anisotropy. *Journal of the Geological Society, London*, **152**, 669–679.

BOUILLIN, J. P., BOUCHEZ, J. L., LESPINASSE, P. & PÊCHER, A. 1993. Granite emplacement in an extensional setting: an AMS study of the magmatic structures of Monte Capanne (Elba, Italy). *Earth and Planetary Science Letters*, **118**, 263–279.

BRIDGWATER, D., SUTTON, J. & WATTERSON, J. 1974. Crustal downfolding associated with igneous activity. *Tectonophysics*, **21**, 1–2, 57–77.

CASTRO, A. & FERNANDEZ, C. 1998. Granite intrusion by externally induced growth and deformation of the magma reservoir, the example of the Plasenzuela pluton, Spain. *Journal of Structural Geology*, **20**, 9–10, 1219–1228.

CLARKE, D. B. 1996. Two centuries after Hutton's 'Theory of the Earth': the status of granite science. *Transactions of the Royal Society of Edinburgh*, **87**, 353–359.

CORDELL, L. & HENDERSON, R. G. 1968. Iterative three dimensional solution of gravity anomaly using a digital computer. *Geophysics*, **33**, 596–601.

CORNEJO, P. C. & MAHOOD, G. A. 1997. Seeing past effects of re-equilibration to reconstruct magmatic gradients in plutons: La Gloria Pluton, central Chilean Andes. *Contributions to Mineralogy and Petrology*, **127**, 159–175.

CORRY, C. E. 1988. *Laccoliths: mechanics of emplacement and growth.* Geological Society of America, Special papers, **220**.

CRUDEN, A. R. 1990. Flow and fabric development during the diapiric rise of magma. *Journal of Geology*, **98**, 681–698.

DARROZES, J., MOISY, M., OLIVIER, Ph., AMÉGLIO, L. & BOUCHEZ, J. L. 1994. Structure magmatique du granite du Sidobre (Tarn, France): de l'échelle du massif à celle de l'échantillon. *Comptes Rendus de l'Académie des Sciences, Paris*, **318**, 2, 243–250.

DEEMER, S. & HURICH, C. 1997. Seismic image of the basal portion of the Bjerkreim-Sokndal layered intrusion. *Geology*, **25**, 12, 1107–1110.

DEHLS, J. F., CRUDEN, A. R. & VIGNERESSE, J. L. 1998. Fracture control of late Archean pluton emplacement in the northern Slave Province, Canada. *Journal of Structural Geology*, **20**, 9–10, 1145–1154.

DELANEY, P. T. & POLLARD, D. D. 1982. Solidification of basaltic magma during flow in a dike. *American Journal of Science*, **282**, 856–885.

DINDI, E. W. & SWAIN, C. J. 1988. Joint three-dimensional inversion of gravity and magnetic data from Jombo Hill alkaline complex, Kenya. *Journal of the Geological Society, London*, **145**, 493–504.

DUFFIELD, W. A., BACON, C. R. & DALRYMPLE, G. B. 1980. Late Cenozoic volcanism, geochronology and structure of the Coso Range, Inyo County, California. *Journal of Geophysical Research*, **85**, 2381–2404.

EDWARDS, J. W. F. 1984. Interpretations of seismic and gravity surveys on the eastern part of the Cornubian platform. *In*: HUTTON, D. H. W. & SANDERSON, D. J. (eds) *Variscan Tectonics of the North Atlantic Region.* Geological Society, London, Special Publications, **xx**, 119–124.

ELLWOOD, B. B., WHITNEY, J. A. & WENNER, D. 1980. Age, paleomagnetism, and tectonic significance of the Elberton granite, Northeast Georgia Piedmont. *Journal of Geophysical Research*, **85**, 1481–1486.

EMERMAN, S. H. & MARRETT, R. 1990. Why dikes? *Geology*, **18**, 231–233.

EVANS, D. J., ROWLEY, W. J., CHADWICK, R. A. & MILLWARD, D. 1993. Seismic reflections from within the Lake District batholith, Cumbria, northern England. *Journal of the Geological Society, London*, **150**, 1043–1046.

EVERAERTS, M., POITEVIN, C., DE VOS, W. & STERPIN, M. 1996. Integrated geophysical/geological modelling of the western Brabant massif and structural implications. *Bulletin de la Société Géologique Belge*, **105**, 1–2, 41–59.

FOSSEN, H., TIKOFF, B. & TEYSSIER, C. 1994. Strain modeling of transpressional and transtensional deformation. *Norsk Geologiske Tidsskrift*, **74**, 134–145.

GEOFFROY, L., OLIVIER, Ph. & ROCHETTE, P. 1997. Structure of a hypovolcanic acid complex inferred from magnetic susceptibility anisotropy measurements: the Western Red Hills granites (Skye, Scotland, Thulean Igneous Province). *Bulletin of Volcanology*, **59**, 147–159.

GLEIZES, G., LEBLANC, D. & BOUCHEZ, J. L. 1997. Variscan granites of the Pyrenees revisited: their role as syntectonic markers of the orogen. *Terra Nova*, **9**, 38–41.

——, NÉDÉLEC, A., BOUCHEZ, J. L., AUTRAN, A. & ROCHETTE, P. 1993. Magnetic susceptibility of the Mont-Louis Andorra Ilmenite-type granite (Pyrenees): a new tool for the petrographic characterization and regional mapping of the zoned granite plutons. *Journal of Geophysical Research*, **98**, 3, 4317–4331.

GOTTLIEB, J. & DUCHATAU, P. 1996. *Parameter identification and inverse problems in hydrology, geology and ecology*. Kluwer Academic Publishers, Dordrecht.

GOULTY, N. R., DARTON, C. E., DENT, A. E. & RICHARDSON, K. R. 1996. Geophysical investigation of the Beinn an Dubhaich Granite, Skye. *Geological Magazine*, **133**, 2, 171–176.

——, LEGGETT, M., DOUGLAS, T. & EMELUS, C. H. 1992. Seismic reflection test on the granite of the Skye Tertiary igneous centre. *Geological Magazine*, **129**, 5, 633–636.

GUGLIELMO, G. 1993. Interference between pluton expansion and non-coaxial tectonic deformation: three-dimensional computer model and field implications. *Journal of Structural Geology*, **15**, 593–608.

HECHT, L., VIGNERESSE, J. L. & MORTEANI, G. 1997. Constraints on the origin of zonation of the granite complexes in the Fichtelgebirge (Germany and Czech Republic): evidence from a gravity and geochemical study. *Geologische Rundschau*, Suppl. **86**, 93–109.

HODGE, D. S., ABBEY, D. A., HARBIN, M. A., PATTERSON, J. L., RING, M. J. & SWEENEY, J. F. 1982. Gravity studies of subsurface mass distributions of granitic rocks in Maine and New Hampshire. *American Journal of Sciences*, **282**, 1289–1324.

HUTTON, D. H. W. 1997. Syntectonique granites and the principle of effective stress: a general solution to the space problem? *In*: BOUCHEZ, J. L., HUTTON, D. H. W. & STEPHENS, W. E. (eds) *Granite: From Segregation of Melt to Emplacement Fabrics*. Kluwer Academic Publishers, Dordrecht, 189–197.

——, DEMPSTER, T. J., BROWN, P. E. & BECKER, S. D. 1990. A new mechanism of granite emplacement: intrusion in active extensional shear zones. *Nature*, **343**, 452–455.

ISAACSON, L. B. & SMITHSON, S. B. 1976. Gravity anomalies and granite emplacement in west-central Colorado. *Geological Society of America Bulletin*, **87**, 22–28.

JACQUES, J. M. & REAVY, R. J. 1994. Caledonian plutonism and major lineaments in the SW Scottish Highlands. *Journal of the Geological Society, London*, **151**, 955–970.

JAQUES, A. L., WELLMAN, P., WHITAKER, A. & WYBORN, D. 1997. High-resolution geophysics in modern geological mapping. *Journal of Australian Geology and Geophysics*, **17**, 2, 59–173.

JOVER O. 1986. *Les massifs granitiques de Guéret et du Nord-Millevaches (Massifs Central français): analyse structurale et modèle de mise en place*. PhD thesis, Nantes University, France.

JURDY, D. M. & PHINNEY, R. A. 1983. Seismic imaging of the Elberton granite, Inner Piedmont, Georgia, using COCORP southern Appalachian data. *Journal of Geophysical Research*, **88**, 5865–5873.

KADIOGLU, Y. K., ATES, A. & GÜLEC, N. 1998. Structural interpretation of gabbroic rocks in Agacören Granitoid, central Turkey: field observations and aeromagnetic data. *Geological Magazine*, **135**, 2, 245–254.

KOUKOUKEVLAS, L., PE-PIPER, G. & PIPER, D. J. W. 1996. Pluton emplacement by wall rock thrusting, hanging-wall translation and extensional collapse: latest Devonian plutons of the Cobequid fault zone, Nova Scotia, Canada. *Geological Magazine*, **133**, 285–298.

LACHENBRUCH, A. H. 1970. Crustal temperature and heat production: implication of the linear heat flow relation. *Journal of Geophysical Research*, **75**, 3291–3300.

LEBLANC, D., GLEIZES, G., ROUX, L. & BOUCHEZ, J. L. 1996. Variscan dextral transpression in the French Pyrenees: new data from the Pic des Trois-Seigneurs granodiorite and its country rocks. *Tectonophysics*, **261**, 331–345.

LE FORT, P. 1981. Manaslu leucogranite: a collision signature of the Himalaya and a model for its genesis and emplacement. *Journal of Geophysical Research, Solid Earth*, **86**, 10 545–10 568.

LINES, L. R. & TREITEL, S. 1984. A review of least squares inversion and its application to geophysical problems. *Geophysical Prospecting*, **32**, 159–186.

LISTER, J. R. & KERR, R. C. 1991. Fluid-mechanical models of crack propagation and their application to magma transport in dykes. *Journal of Geophysical Research*, **96**, 10 049–10 077.

LYNN, H. B., HALE, L. D. & THOMPSON, A. 1981. Seismic reflections from the basal contacts of batholiths. *Journal of Geophysical Research*, **86**, 10 633–10 638.

LYONS, J. B., CAMPBELL, J. G. & ERIKSON, J. P. 1996. Gravity signatures and geometric configurations of some Oliveran plutons: Their relation to Acadian structures. *Geological Society of America Bulletin*, **108**, 7, 872–882.

MAIR, J. A. & GREEN, A. G. 1981. High-resolution seismic reflection profiles reveal fracture zones within a 'homogeneous' granite batholith. *Nature*, **294**, 439–442.

MARÉ, L. P. & THOMAS, J. 1998. Paleomagnetism and aeromagnetic modelling of the Mesoproterozoic Ntimbankulu Pluton, KwaZulu-Natal, South Africa: mushroom-shaped diapir. *Journal of African Earth Sciences*, **25**, 4, 519–537.

MATTHEWS, D. H. 1987. Can we see granites on seismic reflection profiles? *Annales Geophysicae*, **5B**, 4, 353–356.

MCCAFFREY, K. J. W. 1992. Igneous emplacement in a transpressive shear zone: Ox Mountains igneous complex. *Journal of the Geological Society, London*, **149**, 221–235.

—— & PETFORD, N. 1997. Are granitic intrusions scale invariant? *Journal of the Geological Society, London*, **154**, 1–4.

MILLER, C. F., WATSON, E. B. & HARRISON, T. M. 1988. Perspectives on the source, segregation, and transport of granitoid magmas. *Transactions of the Royal Society of Edinburgh*, **79**, 135–156.

OLHOEFT, G. R. 1981. Electrical properties of granite with implications for lower crust. *Journal of Geophysical Research*, **86**, 931–936.

OLIVER, H. W. 1977. Gravity and magnetic investigations of the Sierra Nevada batholith, California. *Geological Society of America Bulletin*, **88**, 445–461.

OLIVIER, P., AMÉGLIO, L., RICHEN, H. & VADEBOIN, F. 1999. Emplacement of the Aya Variscan granitic pluton (Basque Pyrenees) in a dextral transcurrent regime inferred from a combined magneto-structural and gravimetric study. *Journal of the Geological Society, London*, **156**, 991–1002.

PARK, R. G. 1983. *Foundations of structural geology*. Chapman & Hall, New York.

PATERSON, S. R. & FOWLER, T. K. 1993. Re-examining pluton emplacement processes. *Journal of Structural Geology*, **15**, 191–206.

—— & TOBISCH, O. T. 1988. Using pluton ages to date regional deformations: Problems with commonly used criteria. *Geology*, **16**, 1108–1111.

PETFORD, N., CLEMENS, J. C. & VIGNERESSE, J. L. 1997. Application of information theory to the formation of granitic rocks. *In*: BOUCHEZ, J. L., HUTTON, D. H. W. & STEPHENS, W. E. (eds) *Granite: From Segregation of Melt to Emplacement Fabrics*. Kluwer Academic Publishers, Dordrecht, 3–10.

PITCHER, W. S. 1993. *The nature and origin of granite*. Blackie, London.

POWELL, R. & HOLLAND, T. 1994. Optimal geothermometry and geobarometry. *American Mineralogist*, **79**, 120–133.

RAMBERG, H. 1981. *Gravity, deformation on the Earth's crust in theory, experiments and geological applications*. Academic Press, New York.

RAMSAY, J. G. 1981. Emplacement mechanics of the Chindamora Batholith, Zimbabwe. *Journal of Structural Geology*, **3**, 93.

RAYBOULD, J. G. 1973. The form of the Beinn an Dubhaich granite, Skye, Scotland. *Geological Magazine*, **110**, 4, 341–350.

RICHARD, V., BAYER, R. & CUER, M. 1984. An attempt to formulate well posed questions in gravity: application of linear inverse techniques to mining exploration. *Geophysics*, **49**, 1781–1793.

ROMAN BERDIEL, M. T., GAPAIS, D. & BRUN, J. P. 1997. Granite intrusion along strike-slip zones in experiment and nature. *American Journal of Science*, **297**, 651–678.

ROSENBERG, C. L., BERGER, A. & SCHMID, S. M. 1995. Observations from the floor of a granitoid pluton: inferences on the driving force of final emplacement. *Geology*, **23**, 5, 443–446.

ROY, R. F., BLACKWELL, D. D. & BIRCH, F. 1968. Heat generation of plutonic rocks and continental heat flow provinces. *Earth and Planetary Science Letters*, **5**, 1–12.

RUBIN, A. M. 1995. Getting dikes out of the source region. *Journal of Geophysical Research*, **100**, 5911–5929.

SAWKA, W. N. & CHAPPELL, B. W. 1988. Fractionation of uranium, thorium and rare earth elements in a vertically zoned granodiorite: implications for heat production distributions in the Sierra Nevada batholith, California, U.S.A. *Geochimica et Cosmochimica Acta*, **52**, 1131–1143.

SCHMIDT, M. W. 1992. Amphibole composition in tonalite as a function of pressure: an experimental calibration of the Al-in-hornblende barometer. *Contributions to Mineralogy and Petrology*, **110**, 304–310.

SCHÖN, J. H. 1996. Physical properties of rocks: fundamentals and principles of petrophysics. *In*: HELBIG, K. & TREITEL, S. (eds) *Seismic exploration*. Handbook of geophysical exploration, **18**. Pergamon Press.

SCHWARZ, E. J. 1991. Magnetic expressions of intrusions including magnetic aureoles. *Tectonophysics*, **192**, 191–200.

STETTLER, E. H., COETZEE, H. & ROGERS, H. J. J. 1993. The Schiel Alkaline Complex: geological setting and geophysical investigation. *South African Journal of Geology*, **96**, 3, 96–107.

TALWANI, M. & EWING, M. 1960. Rapid computation of gravitational attraction of three dimensional bodies of arbitrary shape. *Geophysics*, **25**, 203–225.

TARANTOLA, A. & VALETTE, B. 1982. Inverse problem = quest for information. *Journal of Geophysics*, **50**, 159–170.

TAYLOR, H. P. 1980. The effects of assimilation of country rocks by magmas on $^{18}O/^{16}O$ and $^{87}Sr/^{86}Sr$ systematics in igneous rocks. *Earth and Planetary Science Letters*, **47**, 243–254.

TELFORD, W. M., GELDART, L. P., SHERIFF, R. F. & KEYS, D. A. 1990. *Applied geophysics*. Cambridge University Press.

TILLING, R. I. 1974. Estimating the 'thickness' of the Boulder batholith, Montana, from heat-flow and heat-productivity data. *Geology*, **3**, 457–460.

TRZEBSKI, R., BEHR, H. J. & CONRAD, W. 1997. Subsurface distribution and tectonic setting of the late-Variscan granites in the northwestern Bohemian Massif. *Geologische Rundschau*, Suppl. **86**, 64–78.

VASSEUR, G., DUPIS, A., GALLART, J & ROBIN, G. 1990. Données géophysiques sur la structure du massif leucogranitique du Limousin. *Bulletin de la Société Géologique de France*, **8**, VI(6), 3–11.

VEJMELEK, L. & SMITHSON, S. B. 1995. Seismic reflection profiling in the Boulder batholith, Montana. *Geology*, **23**, 811–814.

VENDEVILLE, B. C. & JACKSON, M. P. A. 1992. The rise of diapirs during thin-skinned extension. *Marine and Petroleum Geology*, **9**, 331–353.

VIGNERESSE, J. L. 1988. Forme et volume des plutons granitiques. *Bulletin de la Société Géologique de France*, **8**, IV(6), 897–906.

—— 1990. Use and misuse of geophysical data to determine the shape at depth of granitic intrusions. *Geological Journal*, **25**, 248–260.

—— 1995a. Crustal regime of deformation and ascent of granitic magma. *Tectonophysics*, **249**, 187–202.

—— 1995b. Control of granite emplacement by regional deformation. *Tectonophysics*, **249**, 173–186.

—— & BOUCHEZ, J. L. 1997. Successive granitic magma batches during pluton emplacement: the case of Cabeza de Araya (Spain). *Journal of Petrology*, **38**, 12, 1767–1776.

—— & CUNEY, M. 1991. Are granites representative of the heat flow provinces? *In*: CERMAK, V. & RYBACH, L. (eds) *Exploration of the deep continental crust. Terrestrial heat flow and the Lithosphere structure*. Springer-Verlag, 86–110.

——, TIKOFF, B. & AMÉGLIO, L. 1999. Modification of the regional stress field by magma intrusion and formation of tabular granitic plutons. *Tectonophysics*, **302**, 203–224.

WARD, S. H. 1976. Special electromagnetics supplement. *Geophysics*, **41**, 1103–1255.

WEINBERG, R. F. 1997. Diapir-driven crustal convection: decompression melting, renewal of the magma source and the origin of nested plutons. *Tectonophysics*, **271**, 217–229.

WENZEL, F., SANDMEIER, K.-J. & WÄLDE, W. 1987. Properties of the lower crust from modeling refraction and reflection data. *Journal of Geophysical Research*, **92**, 11 575–11 583.

WIKSTRÖM, A. 1984. A possible relationship between augen gneisses and postorogenic granites in SE Sweden. *Journal of Structural Geology*, **6**, 409–415.

WILSON, L. & HEAD, J. W. 1988. Nature of local magma storage zones and geometry of conduit systems below basaltic eruption sites: Puu Oo Kilauea East Rift, Hawaii, example. *Journal of Geophysical Research*, **93**, 14 785–14 792.

WOLLENBERG, H. A. & SMITH, A. R. 1987. Radiogenic heat production of crustal rocks: an assessment based on geochemical data. *Geophysical Research Letters*, **14**, 295–298.

What do experiments tell us about the relative contributions of crust and mantle to the origin of granitic magmas?

ALBERTO E. PATIÑO DOUCE

Department of Geology, University of Georgia, Athens, GA 30602, USA
(e-mail: klingon@3rdrock.gly.uga.edu)

Abstract: The origin of different kinds of granitic rocks is examined within the framework of experimental studies of melting of metamorphic rocks, and of reaction between basaltic magmas and metamorphic rocks. Among the types of granitic rocks considered in this chapter, only peraluminous leucogranites represent pure crustal melts. They form by dehydration-melting of muscovite-rich metasediments, most likely during the fast adiabatic decompression that results from tectonic collapse of thickened intracontinental orogenic belts. All other granitic rocks discussed here represent hybrid magmas, formed by reaction of basaltic melts with metamorphic rocks of supracrustal origin. These hybrid rocks include Cordilleran granites, formed at or near convergent continental margins, strongly peraluminous 'S-type' granites, alumina-deficient 'A-type' granites, and rhyolites associated with continental flood basalts. The differences among these types of granites reflect differences both in their source materials and in the pressures at which mantle-crust interactions take place. In turn, these variables are correlated with the tectonic settings in which the magmas form. Hybrid mafic cumulates are also produced by mantle-crust interactions, simultaneously with the granitic melts. These cumulates range from orthopyroxene + plagioclase-rich assemblages at low pressure to clinopyroxene + garnet-rich assemblages at high pressure, and are known to be important constituents of the lower continental crust. With the exception of peraluminous leucogranites, generation of granitic magmas is almost always associated in space and time with growth, rather than just recycling, of the continental crust.

Igneous rocks of granitic composition *sensu lato*, i.e. spanning the range from quartz diorites to true granites and including both plutonic and volcanic rocks (throughout this chapter, I use the terms 'granite' and 'granitic rock' in this broad sense, unless otherwise specified), are the most distinctive components of the continental crust. Few geologists doubt today that crustal precursors are involved in the origin of most granitic rocks. However, granitic rocks encompass a wide diversity of compositions and tectonic associations, and these variables tend to be correlated. These long-known observations hint at different sources, conditions of formation, and relative contributions of mantle and crust in the origin of different kinds of granitic rocks. Such topics have been often probed using the trace element and isotopic compositions of granitic rocks. Less attention, however, has been paid to what major element compositions, and their causative agent which is phase equilibrium, can tell us about the origin of different kinds of granitic rocks, and

about the reasons for the common correlation between rock composition and tectonic environment.

A large data set of melt compositions generated experimentally from natural starting materials has become available over the last decade (see Appendix). This data set makes it possible to address the issue of the origin and diversity of granitic magmas from the point of view of major element compositions and phase equilibrium. This is the purpose of this chapter. Undoubtedly, any individual granitic rock is the product of a complex set of processes. The workings of some of these processes, most notably how chemical interactions between mantle and crust take place, are very incompletely understood. Nevertheless, an important conclusion that is reached here is that there are deep underlying similarities, not only in sources but also in conditions of formation, that explain why there are distinctive types of granitic rocks. Once the reasons for these similarities become clear, it is possible to

From: CASTRO, A., FERNÁNDEZ, C. & VIGNERESSE, J. L. (eds) *Understanding Granites:*
Integrating New and Classical Techniques. Geological Society, London, Special Publications,
168, 55–75. 1-86239-058-4/99/$15.00 © The Geological Society of London 1999.

gain a better understanding of why different types of granitic rocks tend to be associated with specific tectonic environments.

Major element compositions of experimentally produced crustal melts

Pervasive free aqueous fluids do not exist in the continental crust at depths greater than the uppermost few kilometers (Yardley & Valley 1997), except perhaps in those cases in which low-grade rocks are rapidly underthrust below higher-grade rocks. Therefore, generation of granitic magmas commonly takes place under fluid-absent conditions, by dehydration-melting reactions (e.g. Thompson 1982). In such reactions, hydrous minerals in metamorphic rocks break down incongruently, supplying chiefly water and feldspar components to a melt phase, and MgO, FeO, CaO and Al_2O_3 to refractory mineral phases. Only three abundant hydrous minerals persist to temperatures high enough to overlap with those at which water-undersaturated silicic melts form. These minerals—muscovite, biotite, and hornblende—are thus the only ones that commonly undergo dehydration-melting in the continental crust. Other hydrous minerals may be important sources for dehydration-melting at ultra-high pressures (e.g. zoisite in eclogites, Skjerlie & Patiño Douce 1998), but are not considered here.

The major element compositions of melts produced experimentally by dehydration melting of a wide range of metamorphic bulk compositions are reviewed in Fig. 1 (sources of data are given in the Appendix). Metamorphosed greywackes, mafic pelites and felsic pelites are all mica-rich sources, and are distinguished as follows: greywackes contain biotite + plagioclase and no aluminosilicate; mafic pelites contain biotite + aluminosilicate, with or without subordinate muscovite and/or plagioclase; muscovite is the dominant hydrous phase in felsic pelites, which may also contain plagioclase and subordinate biotite. The most obvious differences are those between amphibolite-derived melts and melts derived from mica-rich sources: the former are depleted in total alkalis, and enriched in lime, relative to the latter (Fig. 1a, d, f). Amphibolite-derived melts are most strongly depleted in K relative to melts of mica-rich sources, and are commonly trondhjemitic (e.g. Rapp *et al.* 1991, see also below).

The relative behaviors of alumina and mafic components differ notably between amphibole- and mica-derived melts. Melts of mica-rich sources have a relatively restricted range of total contents of Al_2O_3 + FeO + MgO + TiO_2, but a wide range of $Al_2O_3/(FeO + MgO + TiO_2)$ ratios. The opposite is true of amphibolite-derived melts (Fig. 1b). At crustal pressures, clinopyroxene + plagioclase are almost always present in the peritectic assemblage formed by incongruent dehydration-melting of amphibolites, and may explain the relatively constant ratio of alumina to ferromagnesian components by a buffering reaction such as the following one:

clinopyroxene +

$$[Al_2O_3]_{melt} \rightleftarrows plagioclase + [MgO, FeO]_{melt}.$$

This peritectic assemblage is rarely formed during melting of mica-rich sources, in which the clinopyroxene-in boundary is close to the plagioclase-out boundary (at $P \geqslant 15\,kbar$, see Patiño Douce & Beard 1995; Patiño Douce & McCarthy 1998). Buffering of alumina relative to ferromagnesian components is thus less likely in mica-derived melts. In contrast, quartz is almost always present in the melting residues of micaceous protoliths, but rarely in those of amphibolites. Quartz allows buffering of FeO and MgO melt contents by means of reactions such as

$$orthopyroxene \rightleftarrows quartz + [MgO, FeO]_{melt}.$$

This explains why melts derived from mica-rich sources tend to have lower, and less variable, contents of ferromagnesian components than amphibolite melts. Amphibolite-derived melts are on average less peraluminous than biotite-derived melts, but many of the former are nevertheless strongly peraluminous (Fig. 1e; see also Patiño Douce & Beard 1995).

There is considerable variability among melts derived from mica-rich sources. This variability reflects source composition. Melts of muscovite-rich sources (felsic metapelites) constitute the most distinctive group, depleted in FeO + MgO + TiO_2 relative to biotite-derived melts (Fig. 1a and b). Among biotite-derived melts, those derived from metagreywackes (sources that contain biotite + plagioclase but no aluminosilicate) are richer in CaO relative to both ferromagnesian components and alumina (Fig. 1c and d), than those derived from mafic metapelites (sources that contain biotite + aluminosilicate ± plagioclase).

Pressure of melting and infiltration of H_2O-rich fluids also affect phase relations and, hence, melt compositions. For melting of metagreywackes, the effect of pressure has been discussed in detail by Patiño Douce & Beard (1996), Patiño Douce (1996) and Patiño Douce & McCarthy (1998). Stability of plagioclase has a particularly

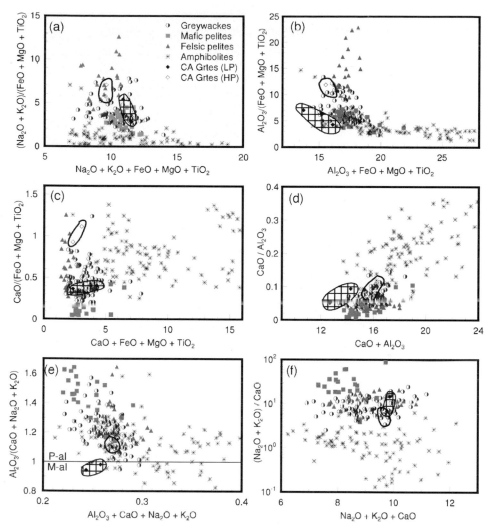

Fig. 1. Compositions of melts produced by experimental dehydration-melting of various lithologies (see appendix for sources of data). Melt compositions produced by dehydration-melting of calc alkaline granites ('CA Grtes') at 4 and 8 kbar ('LP' and 'HP', respectively) by Patiño Douce (1997) are also highlighted with closed curves (squared for 4 kbar and clear for 8 kbar). All values are in wt%, except for (e), which is in moles per 100 g or rock. Also in (e), P-al means peraluminous and M-al means metaluminous. Note that although the compositions of melts from different sources may overlap in some diagrams, every group of melts is distinctive in at least one set of major element compositional parameters. In most diagrams in this chapter, major element compositions are plotted as a ratio between two variables versus the sum of the same variables. These plots, which are conceptually similar to those used for radiogenic isotopes, bring out more clearly than traditional plots the differences and similarities among the various groups of granitic rocks. In particular, this way of plotting major element compositions yields tightly clustered trends, that are not as readily apparent if the same data are plotted in traditional compositional diagrams. Such trends are consistent with mixing or assimilation processes.

significant effect on melt compositions, and is in turn controlled to a large extent by incorporation of grossular component into garnet and diopside into clinopyroxene. In metagreywackes of intermediate Fe/Mg ratio (Mg number between approximately 40 and 60), the residual assemblages at $P < 10$ kbar are granulites dominated by orthopyroxene + plagioclase. Garnet becomes a residual phase at pressures between 10 and 12 kbar, and clinopyroxene at pressures of the order of 15 kbar. Orthopyroxene + plagioclase disappear from the residual assemblage at pressures between 15 and 20 kbar, so that at higher pressures the residues of dehydration-melting of metagreywackes are eclogites (garnet + Na-rich clinopyroxene). As the bulk composition becomes richer in Fe, and as long as oxygen fugacities remain at or below those of the QFM equilibrium, garnet is stabilized to lower pressures (down to approximately 5 kbar), and the stability fields of pyroxenes and plagioclase shrink. Under more oxidizing conditions, neither garnet nor pyroxenes are stable residual phases during melting of Fe-rich metagreywackes at low pressure. In this instance (for pressures of 5–10 kbar and oxygen fugacities above those of the QFM equilibrium) the residual assemblages formed by melting of Fe-rich metagreywackes are composed of plagioclase + magnetite (Patiño Douce & Beard 1996).

Free H_2O lowers the melting point of the assemblage plagioclase + quartz, but has a minimal effect on the stability of micas (Conrad et al. 1988; Patiño Douce 1996). Therefore, infiltration of aqueous fluids into upper amphibolite-grade metasediments causes them to partially melt at temperatures lower than those of the beginning of dehydration-melting. The relatively 'cool' and H_2O-rich melts formed in this way are noticeably enriched in Na_2O/K_2O compared to the hotter melts formed by dehydration melting of the same metasediments. The depression of the quartz + plagioclase solidus by the addition of H_2O becomes stronger with increasing pressure, whereas the thermal stability of micas expands with increasing pressure. In consequence, Na enrichment in melts derived from H_2O-fluxing of metasediments becomes stronger with increasing pressure (Patiño Douce 1996; Patiño Douce & Harris 1998). At pressures greater than 10 kbar and relatively low temperatures (of the order of 700 °C), H_2O-fluxed melting of K-rich metasediments yields trondhjemitic melts in equilibrium with mica- and garnet-rich residual assemblages. Trondhjemitic leucosomes in biotite-rich migmatites, and perhaps some intrusive trondhjemite plutons, may have this origin. Neither of these groups of rocks, however, are considered further in this chapter (but see Patiño Douce 1996; Patiño Douce & Harris 1998).

Dehydration-melting of calc-alkaline granitoid rocks that contain hornblende + quartz is particularly sensitive to pressure (Patiño Douce & Beard 1995; Patiño Douce 1997). At pressures below approximately 5 kbar, the peritectic assemblage formed by incongruent melting of hornblende + quartz is dominated by Ca-rich plagioclase (An_{40} to An_{80}) and orthopyroxene, whereas at higher pressures the solid products of this reaction consist predominantly of clinopyroxene, plus garnet at $P > 10$ kbar. This change in solid assemblage has strong effects on melt compositions (see Fig. 1). Melts generated by dehydration-melting of calc-alkaline precursors at low pressure ($\leqslant 5$ kbar) are distinctly depleted in Al and Ca relative both to their sources and to higher pressure melts (see also Patiño Douce 1997). Low-pressure melts are also metaluminous. All of these compositional features are lost as the depth of melting increases, and clinopyroxene ($\pm$ garnet) replace orthopyroxene + plagioclase in the residue (Fig. 1).

The major element compositions of anatectic melts are determined both by the compositions of their metamorphic precursors and by the conditions of melting. Melts derived from each group of quartzofeldspathic rocks, however, have distinct compositional ranges (Fig. 1). This fact is used here to trace the lineage of granitic rocks. In particular, I explore whether there are unique correlations between granite types, source composition, and conditions of formation, and whether input of mafic, mantle-derived material is evident in the major element compositions of granitic rocks.

Major element compositions and tectonic settings of silica-rich igneous rocks

This chapter focuses on six distinct groups of silicic igneous rocks. These groups are clearly defined on the basis of their major element compositions, and also of the tectonic environments in which they occur.

(i) Peraluminous leucogranites (PLGS): they generally form small intrusive bodies within medium- to high-grade orogenic metamorphic belts, related to episodes of tectonic crustal thickening. They include two-mica granites, muscovite-tourmaline granites, and muscovite-garnet granites.

(ii) Peraluminous S-type granites (PSGS): as originally defined by Chappel & White (1974), these are strongly peraluminous granitic rocks and their volcanic equivalents that are characterized by the presence of low-pressure, high-temperature mafic aluminous minerals, such as cordierite, spinel, or aluminous orthopyroxene. They characteristically occur as shallow batholiths with well developed contact aureoles (e.g. White & Chappell 1988), and their origin is not obviously related to episodes of crustal thickening.

(iii) Cordilleran peraluminous granites (CPGS): these are predominantly two-mica granites that are exemplified here by the Idaho Batholith (e.g. Hyndman 1983) and other two-mica granites of the Cordilleran Interior of

Western North America (Miller & Barton 1990). They tend to occur in large composite intrusions emplaced at high pressures (perhaps as much as 8–10 kbar, e.g. Zen & Hammarstrom 1984; Grover *et al.* 1992) in Cordilleran Interior settings such as that of Western North America. The mafic aluminous phases characteristic of PSGS are absent from Cordilleran peraluminous granites. They differ from the next group but, together with it, they make up Cordilleran granites.

(iv) Calc-alkaline granites (CAGS): this is the most characteristic group of Cordilleran granites. They are the voluminous orogenic granitic and rhyolitic rocks that are formed at convergent continental margins. They are metaluminous to weakly peraluminous rocks (in the sense of Miller 1985), that contain biotite, biotite $\pm$ hornblende, or, in strongly differentiated rocks, biotite $\pm$ muscovite (but the latter are rather scarce).

(v) Metaluminous, alkali-rich granites (MAGS): these are predominantly metaluminous 'A-type' granites and rhyolites (Collins *et al.* 1982; see also Patiño Douce 1997) and rapakivi granites (e.g. Haapala & Rämö 1992). They have notoriously low contents of CaO, Al_2O_3, and MgO, and always originate in regions of normal or less than normal crustal thickness (hence they are often called 'anorogenic granites', e.g. Anderson 1983). They commonly post-date major calc-alkaline magmatism. These Al-poor granites may occasionally be marginally peraluminous or marginally peralkaline. However, strongly peralkaline granites (e.g. riebeckite granites) and their volcanic equivalents (pantellerites, comendites, etc.), and strongly peraluminous topaz rhyolites, are altogether different groups, and are explicitly excluded from MAGS (see also Patiño Douce 1997).

(vi) Rhyolites associated with continental flood-basalt provinces (FBRS): these are distinctly metaluminous and Fe-rich rocks that share some characteristics with MAGS but constitute a recognizably different group.

This list is not comprehensive. Space limitations make it impossible to discuss here the origin of other important groups of silicic igneous rocks, most notably: strongly peralkaline granites, continental trondhjemites (most abundant in Archaean cratons), oceanic plagiogranites and topaz rhyolites.

In the following sections, compositional data taken from the literature (see the Appendix for sources) for rocks within each of these groups are compared to experimental data. In every case except CPGS, the data for each group come from multiple plutonic and/or volcanic associations

that are not genetically related and that may be widely separated in space and time. In spite of this lack of genetic association, the major element compositions of rocks within each group are distinctive and are clearly different from those of other groups. There is a close correspondence between major element compositions, which are controlled to a large extent by phase equilibria and, hence, by thermodynamic intensive variables, and the tectonic environment in which each group of rocks forms. This correspondence is unlikely to be coincidental. Rather, it must indicate some deep underlying uniformity in the processes and conditions responsible for the origin of silicic magmas within each group.

Natural granites v. experimental crustal melts

SiO_2 contents: are granites pure crustal melts?

Figure 2 shows the range of SiO_2 contents of melts produced experimentally from dehydration-melting of various kinds of micaceous metamorphic rocks, together with those of the six groups of silicic igneous rocks described above. Melts produced by incongruent dehydration-melting of mica + quartz are silica-rich, with SiO_2 contents $\geqslant 70$ wt%. This reflects the fact that mica is consumed before

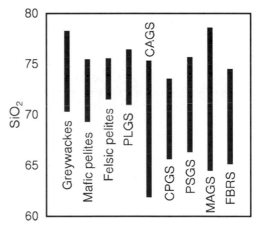

Fig. 2. Ranges of SiO_2 contents (in wt%) of three of the groups of experimental melts, compared to those of six groups of natural granites, as follows (see Appendix for sources of data). PLGS: peraluminous leucogranites. CAGS: calc-alkaline granites. CPGS: Cordilleran Interior peraluminous granites. PSGS: peraluminous S-type granites. MAGS: dominantly metaluminous 'A-type' granites. FBRS: rhyolites associated with flood basalts.

quartz during melting of most common meta-morphic rocks. Once all mica is consumed, the residual assemblage becomes notably refractory, as it is dominated by pyroxene and/or garnet and/or cordierite, ± plagioclase and quartz. Further melting of these refractory assemblages is very limited even at temperatures of the order of 1000–1100 °C (Patiño Douce & Johnston 1991; Patiño Douce & Beard 1995), which are approximately the highest temperatures recorded in the continental crust (e.g. Kilpatrick & Ellis 1992; Rudnick 1992; Downes 1993). Because melting essentially stops when micas are consumed, melt compositions do not undergo major further changes, and remain silica-rich to very high temperatures.

Among natural granites, only peraluminous leucogranites display a range of SiO_2 contents comparable to those of experimental melts (Fig. 2). All other groups include rocks with SiO_2 contents that range down to about 65 wt% or less. The lower cutoff value for each group is obviously an artifact of the particular data sets used in this study. The important point, however, is that suites of granitic rocks (and their volcanic equivalents) other than PLGS have SiO_2 contents that are significantly lower than those that can be obtained by melting of common quartzo-feldspathic rocks at geologically reasonable temperatures. The implication is that peralumi-nous leucogranites are the only group that is likely to represent pure anatectic crustal *melts*. Other processes, capable of yielding magmas with lower SiO_2 contents, must be involved in the origin of other groups of granitic rocks. For example, these less siliceous rocks may represent crystal-rich magmas that entrained a significant proportion of residual or peritectic phases, or they may be hybrid melts formed by chemical interaction of crustal rocks with mafic mantle-derived magmas. These two possibilities are not mutually exclusive, and at least some granitic rocks could be crystal-rich hybrid magmas.

Intrusion of, or underplating by, mafic magmas is almost certainly the source of heat in the origin of many granitic magmas (e.g. Huppert & Sparks 1988; Bergantz 1989). It seems unlikely, however, that interactions between the continental crust and invading mafic magmas be always limited to mere heat transfer. Chemical interactions can and do take place in many instances, as revealed by the isotopic compositions of silicic igneous rocks (e.g. Grove *et al.* 1982; Davidson *et al.* 1988; Hildreth & Moorbath 1988; Grunder 1995) and of mafic igneous cumulates that reside in the lower continental crust (e.g. Downes 1993; Rudnick & Fountain 1995). The nature of these chemical interactions is complex, and almost certainly involves assimilation, magma mixing, fractional crystallization and crystal accumu-lation (e.g. Hildreth & Moorbath 1988). It is not the purpose of this contribution to elucidate the details nor the relative importance of all of these processes, which, in any event, are likely to vary widely from one igneous suite to another. Rather, I will discuss what the major element compositions of granitic rocks reveal about the importance of basaltic influx in the origin of various kinds of granitic magmas and about the depth at which mantle–crust interactions take place.

Modelling mantle–crust interactions

Following Bowen (1922), Patiño Douce (1995) and McCarthy & Patiño Douce (1997) demon-strated experimentally that reaction of basaltic magmas with quartzofeldspathic rocks yields silicic magmas in equilibrium with mafic igneous cumulates that resemble rocks that are abundant in the lower continental crust. These experimen-tal results are used here to model the compo-sitional variation of natural silicic igneous rocks. Throughout this discussion, it is important to keep in mind that pressure exerts a strong influence on the compositions of the silicic magmas and of the complementary mafic cumu-lates, largely as a result of the transformation of orthopyroxene + plagioclase to clinopyrox-ene + garnet.

In some of the following diagrams (Figs 3, 4, 7 and 8), rock compositions taken from the literature are compared to the compositional ranges of crustal melts produced experimentally (from Fig. 1). With the exception of peralumi-nous leucogranites, all other groups of silicic igneous rocks include many samples that plot outside of all of the fields of crustal melts. Moreover, in many cases the data for each group of granitic rocks define fairly tight trends which, in these diagrams, suggest either mixing curves or crystal fractionation trends. Clearly, these trends are neither, because each group of granites includes samples from unrelated igneous com-plexes, separated in space and time. However, the very fact that such random collections of com-positional data, grouped together only on the basis of similarities in the tectonic settings of the rocks, define tightly clustered trends, suggests that the same general processes and materials are at work in the generation of all rocks within each group, regardless of their age and geographic location.

Also shown in these figures are calculated curves that model the compositions of melts that would be produced by reaction of a putative crustal melt end-member (generally chosen to correspond approximately to the most 'crustal-like' samples within each group, except for FBRS, see below) with a basaltic melt. These curves are not simple mixing lines between a silicic melt and a basaltic melt. Rather, they are reaction curves, calculated taking into account the mafic solid assemblages that are produced when crustal rocks react with basaltic melts (Patiño Douce 1995). The calculated curves include the effects of both the components that are added by the basalt and the components that are subtracted by crystallization of the solid refractory assemblage. These curves are obviously simplifications, because the crystallizing assemblages must change as the relative amounts of basalt and crustal rock change, and as temperature changes. However, they are useful models because they clearly show the relative effects of pressure and crustal composition on granite melt composition.

Assuming that the basaltic composition is always the same (a high-Al olivine tholeiite, see Patiño Douce 1995, for further details), the solid assemblage produced by the reaction is determined by the composition of the crustal assimilate and by pressure. Data from Patiño Douce (1995, 1997) were used to construct the modal and mineral compositions of the solid assemblages used to calculate the reaction curves. Different reaction curves are shown in all diagrams, for different pressures. The low pressure curves are constructed for $P \leqslant 5$ kbar, corresponding to interaction of basalts with crustal rocks in the uppermost 15–20 km of the continental crust. The high pressure curves, constructed for $P = 12–15$ kbar, model mantle-crust interactions at depths of 40–60 km, i.e. in the deepest portions of thickened continental crust. The area between both curves encompasses the range of depths at which mantle-crust interactions are most likely to take place.

In Figs 3 and 4, which deal with strongly peraluminous granites, it is assumed that the assimilate is a pelitic rock. The residual assemblage used to model assimilation at low pressure is a spinel norite, composed of plagioclase (An_{75}), aluminous orthopyroxene and spinel in proportions of 6:4:1, respectively. The high pressure assemblage is modeled as a garnet granulite, composed of garnet and plagioclase (An_{40}) in proportions of 4:1. In Figs 7 and 8, in which less aluminous granites are shown, it is assumed that the assimilate is a metagreywacke. The assemblage that models the reaction at low pressure is mangeritic, composed of plagioclase (An_{75}) and orthopyroxene in proportions of 2:1.

At high pressure, the model assemblage consists of clinopyroxene, garnet, plagioclase (An_{40}) and orthopyroxene in proportions of 10:5:4:1, respectively.

These assemblages are simplified models that of course cannot reproduce the whole range of mafic cumulates that are produced when basalts react with metamorphic rocks in nature. Moreover, each individual granitic rock within each group is almost certainly the product of many processes, including hybridization, magma mixing, fractional crystallization and crystal accumulation. The model reaction curves incorporate the chief effects of pressure and crustal assimilate composition on the assemblages that crystallize during mantle-crust interaction and, hence, on the compositions of hybrid silicic melts. The model curves therefore provide a useful framework within which to discuss the role of mafic magmas in the origin of the various groups of granitic rocks.

Strongly peraluminous granites

Two groups of strongly peraluminous granites, PLGS and PSGS, are often confused and lumped together in the literature. The fact is that, although both kinds of granites have metasedimentary precursors, they are distinctly different. Both groups are shown in Figs 3 and 4.

PLGS overlap with, or are very close to, the compositions of melts produced by dehydration-melting of muscovite schists (felsic metapelites). The compositional ranges spanned by PLGS and muscovite-derived melts are nearly the same. In contrast, melts generated by dehydration-melting of biotite-rich sources are enriched in ferromagnesian components compared to PLGS, and are distinctly different from the latter (Figs 3 and 4). Major element compositions (see also Patiño Douce & Harris 1998), in agreement with isotopic and trace element compositions (Le Fort et al. 1987; Nabelek et al. 1992; Inger & Harris 1993), thus indicate that peraluminous leucogranites are pure crustal melts, derived from dehydration-melting of muscovite-rich schists.

The fact that biotite dehydration-melting does not play a major role in the origin of PLGS is important and is probably a consequence of the different geometries of the muscovite and biotite dehydration-melting solidi. Muscovite breaks down along a solidus that is considerably flatter than that along which biotite breaks down, with the solidi for both micas crossing over at pressures perhaps of the order of 10–12 kbar (Fig. 5). The biotite dehydration-melting solidus is also nearly parallel to typical adiabatic decompression paths for thickened continental

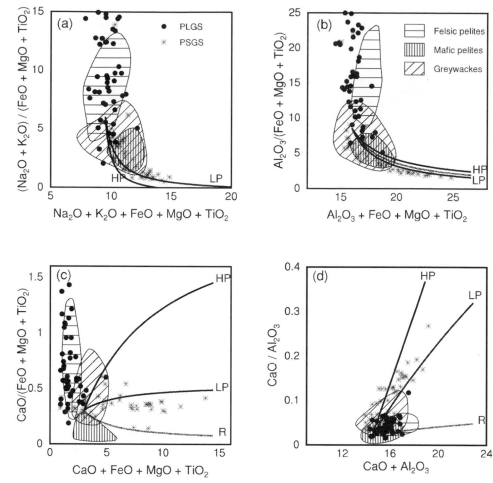

Fig. 3. Compositions of peraluminous leucogranites (circles) and S-type granites (asterisks), compared to melts produced by experimental dehydration-melting of various kinds of metasediments (fields drawn from data in Fig. 1). The thick solid lines are reaction curves that model the melt compositions that would be produced by hybridization of high-Al olivine tholeiite with metapelite (see also Patiño Douce 1995). At low pressure ('LP', $P \leqslant 5$ kbar) the reaction produces an assemblage dominated by plagioclase and aluminous orthopyroxene, whereas the dominant solid product of the reaction at high pressure ('HP', $P = 12$–15 kbar) is garnet, accompanied by subordinate plagioclase. The broken lines labeled 'R' show the magma compositions that would result from melt-restite mixing in a pelitic system with no addition of basaltic components (data from Vielzeuf & Holloway 1988; and Patiño Douce & Johnston 1991).

crust, whereas adiabatic decompression paths are likely to intersect the muscovite dehydration-melting solidus (Fig. 5). The implication is that, in the absence of basaltic intrusions, tectonic thickening of the continental crust may not generally produce temperatures high enough to induce dehydration-melting of metasediments. During adiabatic decompression attending rapid tectonic denudation of the thickened orogenic crust, peraluminous leucogranite magmas form by decompression-melting of muscovite schists (e.g. Harris & Massey 1994), perhaps supple-

mented by shear heating (e.g. Harrison *et al.* 1998). Metamorphic rocks in which biotite is the dominant hydrous mineral remain largely unmolten during the entire tectonic cycle, unless the crust is subject to additional heating by basaltic intrusion or underplating, or unless they are fluxed by H_2O-rich fluids (see also Patiño Douce & Harris 1998).

In contrast to PLGS, PSGS are not pure crustal melts. They are richer in lime and in ferromagnesian components (Fig. 3), and many also have lower silica contents (Fig. 2), compared

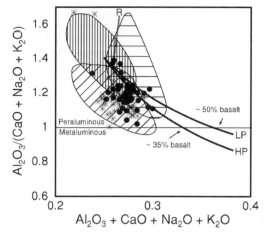

Fig. 4. Alumina saturation relationships (molar alumina over alkalis plus lime, abscissa shows moles per 100 g of rock) of peraluminous leucogranites and S-type granites, compared to those of experimental melts (symbols, fields and curves as in Fig. 3). Hybrid melts produced at low pressure with up to 50% of basaltic material in them are peraluminous (see also Patiño Douce, 1995). Incorporation of restite from melting of metapelites would displace magma compositions in a direction opposite to those of S-type granites.

to the melts that form by dehydration-melting of metasediments. Interaction of mafic magmas with crustal rocks (e.g. Elburg 1996b) and incorporation of restitic phases (White & Chappell 1977) have both been proposed as important processes in the origin of PSGS. These processes are not mutually exclusive and both could, in principle, account for the compositional characteristics of these granites. PSGS define notably coherent trends in the diagrams of alkalis v. ferromagnesian components and alumina v. ferromagnesian components (Fig. 3a and b). These trends are consistent with either basaltic input or restite mixing. Note, however, that high-pressure interaction of basalts with metapelites produces melts with $(Na_2O + K_2O)/(FeO + MgO + TiO_2)$ ratios that are too low compared to those of PSGS. The diagrams of lime v. ferromagnesian components and lime v. alumina (Fig. 3c and d) favor low-pressure hybridization of metapelites with basaltic melts as the most likely explanation for the origin of PSGS. The calculated low-pressure reaction curves match closely the compositional variation of PSGS. In contrast, high-pressure interaction of basalts with metapelites yields hybrid melts that are too rich in CaO compared to PSGS,

reflecting the breakdown of plagioclase and crystallization of almandine–pyrope garnet (see Patiño Douce 1995). Addition of pelitic restite leads to magma compositions that are depleted in CaO compared to PSGS.

An important consideration is whether hybrid magmas that contain a sizable mantle component can be as strongly peraluminous as S-type granites are. The experiments of Patiño Douce (1995) answered this question affirmatively, and those results are expanded in Fig. 4. At low pressure, the modeled hybrid melt compositions remain peraluminous with addition of up to 50% basalt. At high pressure, crystallization of garnet causes the melts to become metaluminous sooner. The strongly peraluminous compositions of S-type granites are thus not inconsistent with a hybrid origin. However, in this diagram (Fig. 4) PSGS do not follow any recognizable trend, and they tend to plot at lower values of alumina saturation indices (molar $Al_2O_3/(CaO + Na_2O + K_2O)$) than those of the model reaction curves. Incorporation of restitic phases formed by melting of a pure crustal pelitic source is not a likely explanation for this behavior, because it would shift granite compositions in the opposite direction, making them more peraluminous (Fig. 4). Incorporation of solid phases formed by hybridization (plagioclase + orthopyroxene) cannot be ruled out, but it would still fail to explain the absence of a recognizable trend in this diagram, which is at odds with the well-defined trends seen in Fig. 3.

Major element data show that S-type granites are neither pure crustal melts nor restite-rich magmas of pure crustal derivation. Rather, they are hybrid magmas, which may (but do not necessarily have to) contain some amount of entrained solid phases ('restite') of hybrid origin. Interaction of basaltic magmas with Al-rich metasedimentary rocks at relatively shallow levels, perhaps no deeper than 15–20 km (see also Patiño Douce 1995), is the best explanation for the origin of S-type granites. The solid assemblages formed by such interactions are dominated by aluminous orthopyroxene and Ca-rich plagioclase. Cordierite, which is a characteristic phase of PSGS, may either crystallize directly from Al-rich hybrid melts, or form by orthopyroxene-melt reactions during cooling and decompression (Castro et al. 1999).

An Al-rich metasedimentary component is undoubtedly required in order to generate the characteristic compositions of S-type magmas. However, there should be no doubt that basalts play a crucial role in the origin of S-type magmas, as sources of both heat and matter. The evidence furnished by experimental

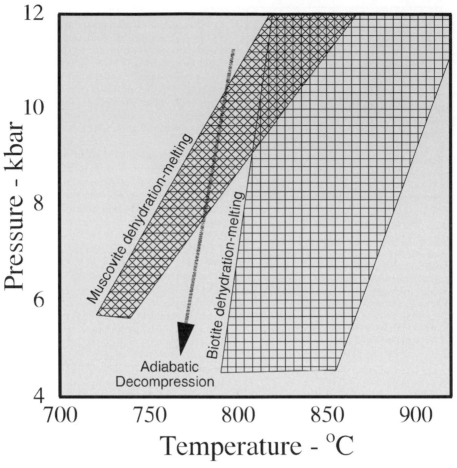

Fig. 5. Comparison of muscovite and biotite dehydration-melting ranges. The range for muscovite dehydration-melting is from Patiño Douce & Harris (1998), and that for biotite dehydration-melting is from Vielzeuf & Montel (1994), Patiño Douce & Beard (1995, 1996), and Vielzeuf & Clemens (1992). The arrow shows the typical slope of adiabatic decompression paths generated by tectonic collapse of thickened orogenic belts (after Harris & Massey 1994). Such paths are likely to intersect the solidi of muscovite schists, but not those of biotite schists. Figure modified after Patiño Douce & Harris (1998).

petrology to this effect is in full agreement with isotopic evidence (Gray 1984, 1990; Clarke *et al.* 1988; Collins 1996; Elburg 1996*a*; Castro *et al.* 1999), and with field evidence that demonstrates the presence of mafic magmas in S-type magma chambers (e.g. Elburg 1996*b*). The tectonic setting in which PSGS form is not clear. The fact that they are almost always shallow intrusions with well-developed contact aureoles and that their compositions require *shallow* interaction of basalts with Al-rich metasediments (because deep interactions lead to magmas more calcic and less peraluminous than S-type granites) suggest that they form where basalts intrude sections of mature metaclastic rocks of, at most,

normal thickness. This could happen after tectonic collapse of a thickened intracontinental orogen (i.e. of a collision zone), or as a result of lithospheric delamination at a continental margin with a thick cover of mature clastic sediments (e.g. Collins & Vernon 1994; Foster & Gray 1998).

The compositional and genetic differences between PLGS and PSGS are important. For historical reasons (Chappell & White 1974) the name 'S-type granites' should be applied only to PSGS. It would be preferable, however, to drop it from the geological nomenclature, because many other types of granites also include metasediments among their sources (see also below).

In particular, peraluminous leucogranites, even though they are pure melts of metasedimentary sources, should not be called 'S-type' granites, because their origin and compositions are radically different from those of S-type granites, as defined by Chappell & White (1974) in the Lachlan Fold Belt of SE Australia.

Cordilleran granites

Two distinct types of granitic rocks, CAGS and CPGS, are considered in this section. The vast majority of CAGS are metaluminous to weakly peraluminous, biotite and biotite $\pm$ hornblende granites. In contrast, peraluminous two-mica granites predominate among CPGS.

Because of the tectonic setting in which they form, at or near active continental margins, the possibility should be considered that Cordilleran granites are the products of melting of metamorphosed mafic igneous rocks, i.e. basaltic amphibolites. Figure 6 shows that this is not generally the case. Cordilleran granites tend to be more alkali-rich than melts derived from basaltic amphibolites, even at very low degrees of melting (Wolf & Wyllie 1991). The greatest discrepancy is in potassium content (Fig. 6c). Dehydration-melting of basaltic amphibolites is a likely explanation for the origin of trondhjemitic rocks (see, for example, Wyllie et al. 1996), but cannot account for the origin of Cordilleran granites.

Data from the partial melting experiments of Patiño Douce & Beard (1995) on a quartz amphibolite (SQA) are shown separately from those of other amphibolite starting materials in Fig. 6. The starting material for Patiño Douce & Beard's (1995) experiments was a mechanical mixture of hornblende, plagioclase (An_{38}), quartz and ilmenite. The hornblende in the starting material was separated from a calc-alkaline quartz diorite, and contains 0.7 wt% K_2O and 1.3 wt% Na_2O. Results from melting experiments on SQA are a model for melting of metaigneous rocks more evolved than basaltic amphibolites, such as meta-andesites and meta-dacites. Melts derived from SQA tend to resemble calc-alkaline granites more closely than melts of basaltic amphibolites, particularly in their K_2O/Na_2O ratios (Fig. 6c). However, compared to calc-alkaline granites SQA melts are still deficient in total alkalis, both in absolute terms and relative to lime (Figs 6b and c). SQA melts are also consistently peraluminous (Patiño Douce & Beard 1995), and hence do not reproduce the metaluminous compositions of many calc-alkaline granites.

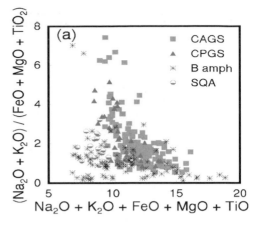

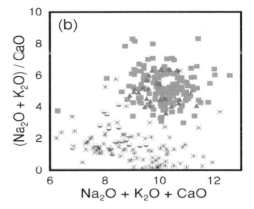

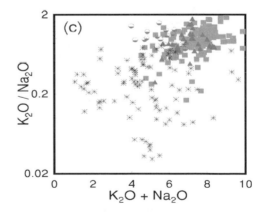

Fig. 6. Compositions of melts produced by experimental dehydration melting of basaltic amphibolites (asterisks) and of the quartz amphibolite studied by Patiño Douce & Beard (1995, half-filled circles), compared to those of calc-alkaline granites (squares) and Cordilleran Interior peraluminous granites (triangles)—sources of data are given in the Appendix.

Roberts & Clemens (1993) proposed that 'high-K calc-alkaline I-type granites' derive from melting of calc-alkaline andesites and basaltic andesites. Melts produced from calc-alkaline meta-andesites are unlikely to be high-K calc-alkaline granites, as shown by the melts produced from SQA. Melting of more evolved calk-alkaline rocks, such as tonalites, may produce calc-alkaline granitic liquids, but only over a rather restricted pressure range. At low pressure (5 kbar or less), the peritectic assemblage is dominated by Ca-rich plagioclase, and the melts are depleted in Ca and Al relative to calc-alkaline granites (Patiño Douce 1997). At pressures greater than 8 kbar, the peritectic assemblage is dominated by clinopyroxene $\pm$ garnet, and the melts become too peraluminous compared to calc-alkaline rocks (see also below). The experiments of Patiño Douce (1997) also showed that fertility of calc-alkaline rocks is limited, owing to their low H_2O contents. These are all important arguments against calc-alkaline andesites as protoliths for calc-alkaline granites, but the main flaw of the proposal of Roberts & Clemens (1993) is another one, namely, that it fails to explain the origin of the evolved calc-alkaline protoliths themselves. The model merely moves the problem into the background, rather than addressing it. In other words, an explanation must be sought for the origin of evolved calc-alkaline rocks, whether they are the final product that one is considering or a source that re-melts to yield even more silica-rich calc-alkaline granites. At some point in the history of evolved calc-alkaline igneous suites (i.e. those that comprise andesites and more siliceous rocks) interaction of mafic mantle-derived magmas with crustal rocks takes place (e.g. Grove & Baker 1984; Hildreth & Moorbath 1988). Assimilation of immature quartzofeldspathic metasediments by basaltic melts is a likely explanation for how this interaction takes place, given the abundance of such rocks in orogenic metasedimentary sequences. At least in some cases, the involvement of metasediments in the origin of calc-alkaline rocks has also been detected isotopically (e.g. Pickett & Saleeby 1994).

Cordilleran granites are plotted in Figs 7 and 8. These diagrams shows that neither CAGS nor CPGS are pure melts of metasedimentary sources. Both groups, however, include rocks that are similar to melts of metagreywacke sources, but different from melts of pelitic sources. Therefore, the curves shown in Figs 7 and 8 were constructed to model reaction of metagreywackes with basalts. In most cases, the natural data form well-defined trends. These trends generally follow the model reaction curves, albeit with some scatter and, in some cases, with somewhat different slopes. Given the wide assortment of data used in constructing the figures, and the fact that other magma-chamber processes in addition to hybridization almost certainly contribute to the final rock compositions, the agreement between natural data and calculated curves is remarkable, and is unlikely to be coincidental. This is underscored by the fact that another granite group (that of MAGS, see below) does not define any trends in some of these diagrams (Fig. 7c and d), implying that there is no reason why granitic rocks in general should define tight trends.

There are differences between CAGS and CPGS, some of them subtle, that are nonetheless fully consistent with one another. Compared to CAGS, CPGS are on average somewhat depleted in alkalis relative to ferromagnesian components (Fig. 7a), enriched in Al_2O_3 and CaO, both in absolute contents (Fig. 7d) and relative to ferromagnesian components (Fig. 7b, c), and depleted in CaO relative to Al_2O_3 (Fig. 7d). CPGS are also almost always peraluminous, whereas CAGS tend to straddle the metaluminous-peraluminous boundary and include many decidedly metaluminous rocks (Fig. 8).

The calculated reaction curves show that every single one of these differences can be explained if one assumes that CAGS and CPGS are hybrid melts produced from similar sources but at different pressures, with CPGS representing higher pressure hybrid melts than CAGS. The differences in melt composition can then be related to the progressive transformation, with increasing pressure, of orthopyroxene + plagioclase to clinopyroxene + garnet, and to the increasing Na contents of both plagioclase and clinopyroxene along this continuous reaction (see Patiño Douce 1995). The deep origin of CPGS is in some cases attested by their deep emplacement levels (up to 10 kbar, e.g. Zen & Hammarstrom 1984; Grover et al. 1992). More generally, CPGS have high Sr contents and steep rare-earth element patterns (LREE-enriched and HREE-depleted) with negligible Eu anomalies (e.g. Schuster & Bickford 1985; Foster & Hyndman 1990; Miller & Barton 1990). These trace element characteristics are consistent with a high-pressure, plagioclase-poor and garnet-rich residual assemblage. CPGS are also notable for having high initial $^{87}Sr/^{86}Sr$ ratios but low Rb/Sr ratios (e.g. Schuster & Bickford 1985; Miller & Barton 1990). This trait can be explained if the crustal components of CPGS are old plagioclase-rich metagreywackes that contribute large amounts of radiogenic Sr to the hybrid magmas as a result of near-complete breakdown of plagioclase at

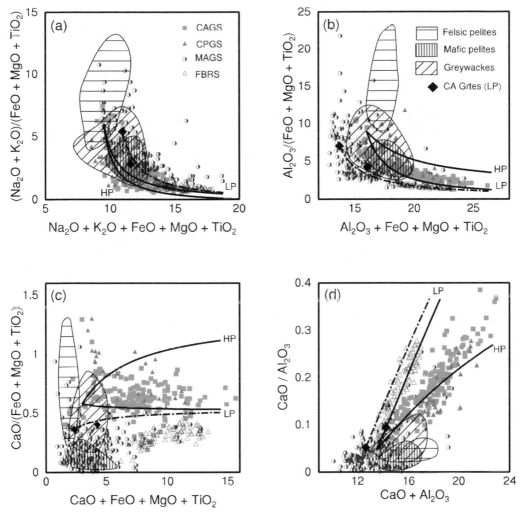

Fig. 7. Compositions of Cordilleran granites (symbols as in Fig. 6), chiefly metaluminous A-type granites (half-filled circles), and basalt plateau rhyolites (open triangles), compared to compositional ranges of experimental metasediment-derived melts (fields, taken from Fig. 1), and low-pressure melts of calc-alkaline granitoids (solid diamonds, also from Fig. 1). The thick solid lines are reaction curves that model the melt compositions that would be produced by hybridization of high-Al olivine tholeiite with metagreywacke (see also Patiño Douce 1995). At low pressure ('LP', $P \leqslant 5$ kbar) the reaction produces an assemblage dominated by plagioclase and orthopyroxene, whereas the dominant product of the reaction at high pressure ('HP', $P = 12$–15 kbar) is clinopyroxene, accompanied by lesser amounts of garnet and plagioclase, and minor orthopyroxene. Note that although Cordilleran Interior granites not always define tight trends, they are always displaced towards the high-pressure reaction curves compared to calc-alkaline granites. The dash-dot lines are reaction curves for low-pressure hybridization of calc-alkaline granites with high-Al olivine tholeiites, with production of plagioclase + orthopyroxene. Note that most A-type granites have lower CaO contents than low pressure melts of calc-alkaline granites, and that basalt plateau rhyolites generally follow trends that are consistent with low pressure reaction curves, although displaced to somewhat lower contents of CaO relative to mafic components (c). These discrepancies suggest that separation of plagioclase in nature may be more important than in the experiments of Patiño Douce (1997).

high pressure. In contrast, the isotopic and geochemical characteristics of CPGS are incompatible with plagioclase-poor metapelitic crustal precursors (see also Miller 1985).

CAGS and CPGS may not constitute two different groups, but may rather span a continuum defined by increasing depth of hybridization. Calc-alkaline granites formed at

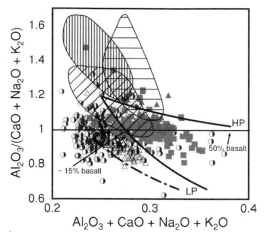

Fig. 8. Alumina saturation relationships (molar alumina over alkalis plus lime, abscissa shows moles per 100 g of rock) of Cordilleran granites, A-type granites, and basalt plateau rhyolites (symbols, fields and curves as in Fig. 7). All of the comments on Fig. 7 apply to this figure too. Hybrid melts produced at high pressure with up to 50% of basaltic material in them are peraluminous (see also Patiño Douce 1995). Note that the effect of pressure is opposite to that of pelite-hybridized melts (compare Fig. 4), because of the different high-pressure assemblages (dominated by garnet in one case and by clinopyroxene in the other).

convergent continental margins (e.g. the Sierra Nevada Batholith and the various Andean Batholiths) correspond to magmas formed by hybridization in continental crust of 'normal' thickness, at depths of 30 km or less. In contrast, the more aluminous granitic batholiths characteristic of Cordilleran Interior settings (e.g. Miller & Barton 1990) are formed by mantle–crust interactions at greater depths, within thickened continental crust. Examples of these rocks are the Cordilleran Interior granites of Western North America, of which the Idaho Batholith is the largest and best known example, and perhaps also the Lower Palaeozoic granitoid plutons of the eastern Blue Ridge of Georgia and North Carolina (e.g. Miller *et al.* 1997).

Hybridization of basaltic melts with relatively Al-poor immature quartzofeldspathic metasediments, i.e. metagreywackes, probably takes place at some point in the lineage of most Cordilleran granites. This may be a single stage process, or it may involve several stages separated in time. Thus, hybrid rocks produced by an initial episode of assimilation of metasediments could re-melt to produce younger and more silicic magmas without further mantle-crust interactions. Whether or not one chooses to call

these younger magmas 'I-type' (e.g. Roberts & Clemens 1993) is not important. The important point is that input of K from an exogenous weathering cycle is probably present in most Cordilleran granites, even if it is so removed from the final product that we observe that it is difficult to detect its presence isotopically or geochemically (see also Castro *et al.* 1999).

Silicic igneous rocks of shallow origin

The last two groups of silicic igneous rocks discussed here, MAGS and FBRS, are formed in areas where the continental crust is of normal, or less than normal, thickness. They typically occur as relatively shallow, often sub-volcanic intrusions, and as high-temperature lavas and ash-flows. Their compositions are also shown in Figs 7 and 8. In contrast to virtually every other group of rocks that have been discussed here, MAGS compositions tend to be notoriously scattered. They, and FBRS, are nevertheless distinctly different from every other group of granitic rocks.

Figure 7 shows that both MAGS and FBRS are depleted in Al_2O_3 and CaO (Fig. 7b, c), compared to other granite types. They are also more strongly metaluminous than other granites (Fig. 8). MAGS, but not FBRS, are enriched in alkalis relative to other granites (Fig. 7a). Although not shown in the figures, MAGS are also characteristically enriched in FeO/MgO, TiO_2/MgO (see Patiño Douce 1997), and Ga/Al, and depleted in Sr and Eu, compared to other granites (e.g. Collins *et al.* 1982).

Dehydration-melting of calc-alkaline granitoids at low pressure ($P \leqslant 4$ kbar) generates silica-rich liquids with all of these major and trace element characteristics (Patiño Douce 1997). The key aspect of this genetic model is the presence of hornblende + quartz in the protolith. Incongruent dehydration-melting of this assemblage at low pressure ($P \leqslant 4$ kbar) produces Ca-rich plagioclase + orthopyroxene. Crystallization of abundant peritectic plagioclase depletes low-pressure melts in Al_2O_3, CaO, Sr and Eu, and, because plagioclase excludes Ga relative to Al (Malvin & Drake 1987), also gives rise to the distinctively high Ga/Al ratios of MAGS. Depletion of MgO relative to both FeO and TiO_2 arises partly from crystallization of abundant peritectic orthopyroxene, and partly from pressure-dependent changes in Fe/Mg partitioning between orthopyroxene and melt, which reflect the stabilization of enstatite relative to ferrosilite with decreasing pressure (see Patiño Douce 1997). The abundance of peritectic plagioclase decreases strongly with increasing pressure of melting, until this phase is totally suppressed at pressures greater than approximately 8 kbar. Clinopyroxene, together with garnet at higher pressures (Patiño Douce & McCarthy 1998), then become the dominant peritectic phases. As a consequence of this change in peritectic assemblage, none of the distinctive major and trace element characteristics of MAGS are formed during dehydration-melting of calc-alkaline rocks at high pressure (see also Patiño Douce 1997).

Whether or not hybridization of basaltic melts with calc-alkaline rocks is important in the origin of MAGS is not clear from Fig. 7. Although MAGS are predominantly high-silica rocks (e.g. Collins et al. 1982), they do include rocks with SiO_2 contents of 65 wt% or less (see Fig. 2), that are not easily explained as pure crustal melts. The isotopic compositions of many MAGS are also inconsistent with them being pure crustal melts. Some MAGS may be the products of fractional crystallization of basalts with little or no crustal input. This may be especially true in the case of Na-rich MAGS (e.g. Shannon et al. 1997). In many other cases, however, isotopic compositions suggest that MAGS are the products of open-system mantle–crust interactions (e.g. Kerr & Fryer 1993; Poitrasson et al. 1995). Restite entrainment is not a viable explanation for the low SiO_2 contents of some MAGS, because it would erase the distinctive major and trace element signatures that result from low-pressure melting. Given also the fact that MAGS are high-temperature magmas (e.g. Clemens et al. 1986; Creaser et al. 1991), it is almost certain that basaltic magmas are involved in their origin, both physically and chemically. It is possible that magma-chamber processes are widespread and powerful enough in these high-temperature, low-viscosity magmas to obscure the effects of hybridization that are better preserved in other groups of granites.

In contrast to MAGS, FBRS do define tight trends in these compositional diagrams, that in every case emanate from the field of MAGS (Figs 7 and 8). These trends are reproduced by curves that model reaction of basaltic magma with calc-alkaline granitoids at low pressure, with crystallization of a solid assemblage composed of calcic plagioclase (An_{75}) and ortho-pyroxene, in proportions of 2 to 1. In some cases, most notably CaO relative to ferromagnesian components (Fig. 7c) and alumina saturation indices (Fig. 8), there are some disparities between natural FBRS compositions and model curves. These disparities suggest that plagioclase fractionation is an important process in nature (see also Skjerlie & Johnston 1993), that causes

additional Ca depletion relative to that observed in the experiments of Patiño Douce (1997), on which the model curves are based. Note that an excellent agreement between reaction curves and FBRS would be obtained in these diagrams (Figs 7c and 8) by starting from less calcic and less peraluminous crustal end-members, well within the field of MAGS. Thus, the data and models shown in Figs 7 and 8 strongly suggest that the rhyolites that are abundant in some basaltic plateaus (Paraná and Karoo) are the products of shallow hybridization of basalts with calc-alkaline rocks.

Shallow melting of calc-alkaline rocks that contain the assemblage hornblende + quartz is essential in the generation of those MAGS and FBRS that display crustal precursors in their isotopic compositions. This conclusion is supported by experimental results. It is also in good agreement with the facts that both of these groups of rocks are formed in regions of cratonized crust, where older calc-alkaline rocks are abundant (e.g. Collins et al. 1982; Anderson 1983; Creaser et al. 1991), and that they are never associated with thick continental crust. The difference between MAGS and FBRS may be merely the rate at which basaltic magma is supplied to the system. In continental flood basalt provinces, a large and fairly continuous flux of basaltic magma into the crust may generate hybrid rhyolitic magmas at such a fast rate that they are forced to erupt before there is much chance for magma chamber processes to modify their initial compositions. This would lead to the tight compositional trends seen in Figs 7 and 8. If basaltic magmas invade the crust sporadically, or at a much slower rate, then the hybrid silicic magmas may remain in the sub-surface long enough to undergo strong modifications by processes such as fractional crystallization. Fractional crystallization at low pressure, where plagioclase is likely to be an important fractionating phase, will exacerbate the extreme compositions of 'A-type' magmas, and would also lead to scattered compositions (Figs 7 and 8). A significant volume of such magmas may never erupt, but rather give rise to the relatively shallow 'anorogenic' intrusions in which MAGS are characteristically found (e.g. Anderson 1983).

Many A-type intrusive complexes, including rapakivi granites, are associated in space and time with anorthosite–norite–mangerite–charnockite complexes (e.g. Anderson 1983; Emslie 1991; Haapala and Rämö 1992). These rocks are characterized by the assemblage ortho-pyroxene + plagioclase, and often have positive Eu anomalies that are complementary to the Eu depletion in the granites (Emslie 1991). It is possible that many of these anorthositic-mangeritic rocks are cumulates of peritectic phases, formed during shallow melting of calc-alkaline rocks or shallow hybridization of basaltic magmas with calc-alkaline rocks.

Conclusions

The origin of any single granitic or rhyolitic rock is undoubtedly much more complex than the generalized models presented here. However, it is possible at this stage of our experimental understanding of the behavior of silica-rich systems to derive some conclusions that may have universal validity, and that may thus help us achieve some synthesis in our understanding of granitic rocks.

(1) Peraluminous leucogranites are the only granitic rocks that are undoubtedly pure crustal melts. They form by dehydration-melting of muscovite-rich metapelitic sources, most likely during fast adiabatic uplift and decompression of thickened crustal orogens.

(2) Other groups of strongly peraluminous granites and rhyolites, including 'S-type' granites and Cordilleran Interior granites, are not pure crustal melts. All of them are hybrid magmas that contain substantial basaltic material in their lineage. Both pressure and composition of the crustal assimilate have strong influence on magma composition. Thus, low pressure and a metapelitic crustal component give rise to 'S-type' magmas, whereas high pressure and a less aluminous, metagreywacke crustal component give rise to Cordilleran Interior peraluminous granites.

(3) Calc-alkaline Cordilleran granites and rhyolites, that are characteristic of convergent continental margins, include supracrustal (meta-sedimentary) precursors in their ancestry. The alkali contents, in particular K_2O, of these rocks are not easily explained without a metasedimentary component. Calc-alkaline granites may be the direct products of a single hybridization event, or they may derive from re-melting of older calc-alkaline rocks which were themselves the products of assimilation of metasediments by basaltic melts. The metasedimentary input may be masked by several generations of remelting and petrologic reprocessing. However, concentration of K by exogenous weathering processes is likely to be an essential step in the origin of calc-alkaline granites, as well as of the other K-rich granites discussed here. It is important to note in this respect that K-rich granites are almost entirely absent from the tonalite-trondhjemite-granodiorite suites that are characteristic of the earliest continental crust.

(4) 'Anorogenic', predominantly meta-luminous 'A-type' granites, rapakivi granites, and rhyolites associated with continental flood basalts may be the products of very shallow (15 km or less) melting of calc-alkaline rocks. Their origin probably entails widely variable degrees of hybridization with basalts.

(5) Generation of granitic magmas must always give rise to comparable volumes of refractory solid assemblages, composed of either peritectic phases or of solid assemblages produced by reaction of basalts with crustal rocks. For example, hybrid grabbronoritic and pyroxenitic cumulates form simultaneously with calc-alkaline magmas, and hybrid garnet-clinopyroxene residues form together with deeper Cordilleran Interior granites. These cumulate rocks are not abundant at the Earth's surface. However, lower crustal xenolith suites show that pyroxene- and garnet-rich rocks with igneous textures and with isotopic compositions indicative of substantial crustal contributions are abundant in the deep continental crust (e.g. Kempton *et al.* 1990; Rudnick 1990; Kempton & Harmon 1992; Hanchar *et al.* 1994). Patiño Douce (1995) pointed out that these rocks do not represent melt compositions, which would be impossibly mafic given their isotopic compositions (Kempton *et al.* 1990; Downes 1993), but are rather cumulates of solid phases formed by reaction of basalts with crustal rocks.

(6) The complementary cumulates formed together with 'A-type' magmas have the compositions of norites and anorthosites. Because 'A-type' magmas form at shallow depths, these noritic-anorthositic cumulates are found at the surface of the Earth more commonly than the clinopyroxene- and garnet-rich cumulates that form in conjunction with other types of granites.

(7) Except for peraluminous leucogranites, generation of granitic magmas of every other type discussed here is associated with growth of the continental crust. This is so because basaltic components are clearly present in S-type granites, Cordilleran granites, basalt plateau rhyolites and, very possibly, A-type granites as well. This underscores the fact that, throughout geologic time, the continental crust has grown not only at convergent continental margins, but also by dominantly vertical magmatic accretion in continental interiors. In all cases, however, the crust most likely grows with an already differentiated chemical structure, with hybrid granitic magmas rising to the top and hybrid mafic cumulates collecting near the bottom.

I thank A. Castro for organizing the conference that gave rise to this book, and for inviting me to be a part of it. Also, I thank him for many stimulating discussions—he pointed out to me the importance of surface weathering in concentrating K and in explaining the origin of K-rich granites. I also thank C. Miller and C. Barnes for very helpful and constructive reviews. They should not be held accountable, however, for any ideas included in this paper that they do not agree with. This work was supported by NSF grants EAR-9118418, EAR-9316304, EAR-9725190.

Appendix: sources of data on experimental melts and natural rocks

Experimental melt compositions

Felsic pelites: Patiño Douce & Harris (1998); Patiño Douce & McCarthy (1998).

Mafic pelites: Vielzeuf & Holloway (1988); Patiño Douce & Johnston (1991).

Metagreywackes: Skjerlie & Johnston (1993); Patiño Douce & Beard (1995, 1996); Montel & Vielzeuf (1997); Patiño Douce & McCarthy (1998).

Amphibolites: Beard & Lofgren (1991); Rapp et al. (1991); Rushmer (1991); Sen & Dunn (1994); Wolf & Wyllie (1994); Patiño Douce & Beard (1995).

Calc-alkaline granites: Patiño Douce (1997).

Natural rock compositions

Peraluminous leucogranites (PLGS): Le Fort (1981); Wickham (1987); France-Lanord & Le Fort (1988); Nabelek et al. (1992); Inger & Harris (1993).

S-type granites (PSGS): McKenzie & Clarke (1975); Clarke & Halliday (1980); White & Chappell (1988); Chappell & White (1992); Giménez de Patiño Douce (unpublished data from the Achala Batholith, Argentina).

Cordilleran Peraluminous Granites (CPGS): Hyndman (1983); Foster & Hyndman (1990); Norman et al. (1992).

Calc-alkaline granites (CAGS): Bateman & Chappell (1979); Ewart (1979); Pitcher et al. (1985); Davidson et al. (1988); Francis et al. (1989).

Metaluminous A-type granites (MAGS): Collins et al. (1982); Conrad (1984); Novak (1984); Rytuba & McKee (1984); Harris (1985); Kinnaird et al. (1985); Long et al. (1986); Whalen et al. (1987); Eby (1990); Rämö & Haapala (1990); Whalen & Currie (1990); Landenberger & Collins (1996); Tollo & Aleinikoff (1996).

Basalt plateau rhyolites (FBRS): Cleverly et al. (1984); Duncan et al. (1984); Piccirillo et al. (1988); Garland et al. (1995).

References

ANDERSON, J. L. 1983. Proterozoic anorogenic granite plutonism of North America. In: MEDARIS, L. G. ET AL. (eds) Proterozoic geology: selected papers from an international proterozoic symposium. Geological Society of America Memoirs, 161, 133–154.

BATEMAN, P. C. & CHAPPELL, B. W. 1979. Crystallization, fractionation, and solidification of the Tuolumne intrusive series, Yosemite National Park, California. Geological Society of America Bulletin, 90, 465–482.

BEARD, J. S. & LOFGREN, G. E. 1991. Dehydration melting and water-saturated melting of basaltic and andesitic greenstones and amphibolites at 1, 3 and 6.9 kb. Journal of Petrology, 32, 365–401.

BERGANTZ, G. W. 1989. Underplating and partial melting: Implications for melt generation and extraction. Science, 245, 1093–1095.

BOWEN, N. L. 1922. The behavior of inclusions in igneous magmas. Journal of Geology, 30, 513–570.

CASTRO, A., PATIÑO DOUCE, A. E., CORRETGÉ, L. G., DE LA ROSA, J. D., EL BIAD, M. & EL-HMIDI, H. 1999. Origin of peraluminous granites and granodiorites, Iberian massif, Spain. An experimental test of granite petrogenesis. Contributions to Mineralogy & Petrology, 135, 255–276.

CHAPPELL, B. W. & WHITE, A. J. R. 1974. Two contrasting granite types. Pacific Geology, 8, 173–174.

—— & —— 1992. I- and S-type granites in the Lachlan Fold Belt. Royal Society of Edinburgh Transactions, Earth Sciences, 83, 1–26.

CLARKE, D. B. & HALLIDAY, A. N. 1980. Strontium isotope geology of the South Mountain Batholith, Nova Scotia. Geochimica et Cosmochimica Acta, 44, 1045–1058.

——, —— & HAMILTON, P. J. 1988. Neodymium and strontium isotopic constraints on the origin of the peraluminous granitoids of the South Mountain Batholith, Nova Scotia, Canada. Chemical Geology, 73, 15–24.

CLEMENS, J. D., HOLLOWAY, J. R. & WHITE, A. J. R. 1986. Origin of an A-type granite: Experimental constraints. American Mineralogist, 71, 317–314.

CLEVERLY, R. W., BETTON, P. J. & BRISTOW, J. W. 1984. Geochemistry and petrogenesis of the Lebombo Rhyolites. In: ERLANK, A. J. (ed.) Petrogenesis of the volcanic rocks of the Karoo Province. Geological Society of South Africa Special Publications, 13, 171–194.

COLLINS, W. J. 1996. Lachlan Fold Belt granitoids: products of three-component mixing. Transactions of the Royal Society of Edinburgh: Earth Sciences, 87, 171–181.

—— & VERNON, R. H. 1994. A rift drift delamination model of continental evolution—Paleozoic tectonic development of Eastern Australia. Tectonophysics, 235, 249–275.

——, BEAMS, S. D., WHITE, A. J. R. & CHAPPELL, B. W. 1982. Nature and origin of A-type granites with particular reference to Southeastern Australia.

Contributions to Mineralogy and Petrology, **80**, 189–200.

CONRAD, W. K. 1984. The mineralogy and petrology of compositionally zoned ash flow tuffs, and related silicic volcanic rocks, from the McDermitt caldera complex, Nevada-Oregon. *Journal of Geophysical Research*, **89**, 8639–8664.

——, NICHOLLS, I. A. & WALL, V. J. 1988. Water-saturated and -undersaturated melting of metaluminous and peraluminous crustal compositions at 10 kb: evidence for the origin of silicic magmas in the Taupo Volcanic Zone, New Zealand, and other occurrences. *Journal of Petrology*, **29**, 765–803.

CREASER, R. A., PRICE, R. C. & WORMALD, R. J. 1991. A-type granites: Assessment of a residual-source model. *Geology*, **19**, 163–166.

DAVIDSON, J. P., FERGUSON, K. M., COLUCCI, M. T. & DUNGAN, M. A. 1988. The origin and evolution of magmas from the San Pedro-Pellado volcanic complex, S. Chile: Multicomponent sources and open system evolution. *Contributions to Mineralogy and Petrology*, **100**, 429–445.

DOWNES, H. 1993. The nature of the lower continental crust of Europe: petrological and geochemical evidence from xenoliths. *Physics of the Earth and Planetary Interiors*, **79**, 195–218.

DUNCAN, A. R., ERLANK, A. J. & MARSH, J. S. 1984. Regional geochemistry of the Karoo igneous province. *In*: ERLANK, A. J. (ed.) *Petrogenesis of the volcanic rocks of the Karoo Province*. Geological Society of South Africa Special Publications, **13**, 355–388.

EBY, G. N. 1990. The A-type granitoids: A review of their occurrence and chemical characteristics and speculations on their petrogenesis. *Lithos*, **26**, 115–134.

ELBURG, M. A. 1996a. Evidence of isotopic equilibration between microgranitoid enclaves and host granodiorite, Warburton granodiorite, Lachlan Fold Belt, Australia. *Lithos*, **38**, 1–22.

—— 1996b. Genetic significance of multiple enclave types in a peraluminous ignimbrite suite, Lachlan Fold Belt, Australia. *Journal of Petrology*, **37**, 1385–1408.

EMSLIE, R. F. 1991. Granitoids of rapakivi granite-anorthosite and related associations. *Precambrian Research*, **51**, 173–192.

EWART, A. 1979. A review of the mineralogy and chemistry of Tertiary-Recent dacitic, latitic, rhyolitic and related salic volcanic rocks. *In*: BARKER, F. (ed.) *Trondhjemites, dacites and related rocks*. Elsevier, Amsterdam, 13–122.

FOSTER, D. A. & HYNDMAN, D. W. 1990. Magma mixing and mingling between synplutonic mafic dikes and granite in the Idaho-Bitterroot batholith. *In*: ANDERSON, J. L. (ed.) *The Nature and Origin of Cordilleran Magmatism*. Geological Society of America Memoirs, **174**, 347–358.

—— & GRAY, D. R. 1998. Chronology, kinematics and tectonics in an accretionary wedge-type thrust system, Western Lachlan Orogen, Australia. *Geological Society of America Abstracts with Programs*, **30**, A-235.

FRANCE-LANORD, C. & LE FORT, P. 1988. Crustal melting and granite genesis during the Himalayan collision orogenesis. *Transactions of the Royal Society of Edinburgh: Earth Sciences*, **79**, 183–195.

FRANCIS, P. W., SPARKS, R. S. J., HAWKESWORTH, C. J., THORPE, R. S., PYLE, D. M., TAIT, S. R., MANTOVANI, M. S. & MCDERMOTT, F. 1989. Petrology and geochemistry of volcanic rocks of the Cerro Galán caldera, Northwest Argentina. *Geological Magazine*, **126**, 515–547.

GARLAND, F., HAWKESWORTH, C. J. & MANTOVANI, M. S. M. 1995. Description and petrogenesis of the Paraná Rzhyolites, Southern Brazil. *Journal of Petrology*, **36**, 1.193–1.227.

GRAY, C. M. 1984. An isotopic mixing model for the origin of granitic rocks in southeastern Australia. *Earth and Planetary Science Letters*, **70**, 47–60.

—— 1990. A strontium isotopic traverse across the granitic rocks of southeastern Australia: Petrogenetic and tectonic implications. *Australian Journal of Earth Sciences*, **37**, 331–349.

GROVE, T. L. & BAKER, M. B. 1984. Phase equilibrium controls on the tholeiitic versus calc-alkaline differentiation trends. *Journal of Geophysical Research*, **89**, 3253–3274.

——, GERLACH, D. C. & SANDO, T. W. 1982. Origin of calc-alkaline series lavas at Medicine Lake volcano by fractionation, assimilation and mixing. *Contributions to Mineralogy and Petrology*, **80**, 160–182.

GROVER, T. W., RICE, J. M. & CAREY, J. W. 1992. Petrology of aluminous schist in the Boehls Butte region of northern Idaho: Phase equilibria and P-T evolution. *American Journal of Science*, **292**, 474–507.

GRUNDER, A. L. 1995. Material and thermal roles of basalt in crustal magmatism: Case study from eastern Nevada. *Geology*, **23**, 952–956.

HAAPALA, I. & RÄMÖ, O. T. 1992. Tectonic setting and origin of the Proterozoic rapakivi granites of southeastern Fennoscandia. *Royal Society of Edinburgh Transactions, Earth Sciences*, **83**, 165–171.

HANCHAR, J. M., MILLER, C. F., WOODEN, J. L., BENNETT, V. C. & STAUDE, J. 1994. Evidence from xenoliths for a dynamic lower crust, eastern Mojave desert, California. *Journal of Petrology*, **35**, 1377–1415.

HARRIS, N. B. W. 1985. Alkaline granites from the Arabian shield. *Journal of African Earth Sciences*, **3**, 83–88.

HARRIS, N. & MASSEY, J. 1994. Decompression and anatexis of the Himalayan metapelites. *Tectonics*, **13**, 1537–1546.

HARRISON, T. M., GROVE, M., LOVERA, O. M. & CATLOS, E. J. 1998. A model for the origin of Himalayan anatexis and inverted metamorphism. *Journal of Geophysical Research*, **103**, 27017–27032.

HILDRETH, W. & MOORBATH, S. 1988. Crustal contribution to arc magmatism in the Andes of central Chile. *Contributions to Mineralogy and Petrology*, **98**, 455–489.

HUPPERT, H. E. & SPARKS, R. S. J. 1988. The generation of granitic magmas by intrusion of basalt into continental crust. *Journal of Petrology*, **29**, 599–632.

HYNDMAN, D. W. 1983. The Idaho Batholith and associated plutons, Idaho and western Montana. *In*: RODDICK, J. A. (ed.) *Circum-Pacific Plutonic Terranes*. Geological Society of America Memoirs, **159**, 213–240.

INGER, S. & HARRIS, N. 1993. Geochemical constraints on leucogranite magmatism in the Langtang Valley, Nepal Himalaya. *Journal of Petrology*, **34**, 345–368.

KEMPTON, P. D. & HARMON, R. S. 1992. Oxygen isotope evidence for large-scale hybridization of the lower crust during magmatic underplating. *Geochimica et Cosmochimica Acta*, **56**, 971–986.

——, ——, HAWKESWORTH, C. J. & MOORBATH, S. 1990. Petrology and geochemistry of lower crustal granulites from the Geronimo Volcanic Field, southeastern Arizona, *Geochimica et Cosmochimica Acta*, **54**, 3401–3426.

KERR, A. & FRYER, B. J. 1993. Nd isotopic evidence for crust-mantle interaction in the generation of A-type granitoid suites in Labrador, Canada. *Chemical Geology*, **104**, 39–60.

KILPATRICK, J. A. & ELLIS, D. J. 1992. C-type magmas: Igneous charnockites and their extrusive equivalents. *Transactions of the Royal Society of Edinburgh: Earth Sciences*, **83**, 155–164.

KINNAIRD, J. A., BOWDEN, P., IXER, R. A. & ODLING, N. W. A. 1985. Mineralogy, geochemistry, and mineralization of the Ririwai complex, northern Nigeria. *Journal of African Earth Sciences*, **3**, 185–222.

LANDENBERGER, B. & COLLINS, W. J. 1996. Derivation of A-type granites from a dehydrated charnockitic lower crust: Evidence from the Chaelundi Complex, Eastern Australia. *Journal of Petrology*, **37**, 145–170.

LE FORT, P. 1981. Manaslu leucogranite: a collision signature of the Himalaya, a model for its genesis and emplacement. *Journal of Geophysical Research*, **86**, 10545–10568.

——, CUNEY, M., DENIEL, C., FRANCE-LANORD, C., SHEPPARD, S. M. F., UPRETI, B. & VIDAL, P. 1987. Crustal generation of the Himalayan leucogranites. *Tectonophysics*, **134**, 39–57.

LONG, L. E., SIAL, A. N., NEKVASIL, H. & BORBA, G. S. 1986. Origin of granite at Cabo de Santo Agostinho, Northeast Brazil. *Contributions to Mineralogy and Petrology*, **92**, 341–350.

MALVIN, D. J. & DRAKE, M. J. 1987. Experimental determination of crystal/melt partitioning of Ga and Ge in the system forsterite-anorthite-diopside. *Geochimica et Cosmochimica Acta*, **51**, 2117–2128.

McCARTHY, T. C. & PATIÑO DOUCE, A. E. 1997. Experimental evidence for high-temperature felsic melts formed during basaltic intrusion of the deep crust. *Geology*, **25**, 463–466.

McKENZIE, C. B. & CLARKE, D. B. 1975. Petrology of the South Mountain batholith, Nova Scotia.

Canadian Journal of Earth Sciences, **12**, 1209–1218.

MILLER, C. F. 1985. Are strongly peraluminous granites derived from pelitic sedimentary sources? *Journal of Geology*, **93**, 673–689.

—— & BARTON, M. D. 1990. Phanerozoic plutonism in the Cordilleran Interior, U.S.A. *In*: KAY, S. M. & RAPELA, C. W. (eds) *Plutonism From Antarctica to Alaska*. Geological Society of America Special Papers, **241**, 213–232.

——, FULLAGAR, P. D., SANDO, T. W., KISH, S. A., SOLOMON, G. C., RUSSELL, G. S. & WOOD STELTENPOHL, L. F. 1997. Low-potassium, trondhjemitic to granodioritic plutonism in the Eastern Blue Ridge, southwestern North Carolina—northeastern Georgia. *In*: SINHA, S. K., WHALEN, J. B. & HOGAN, J. P. (eds) *The Nature of Magmatism in the Appalachian Orogen*. Geological Society of America Memoirs, **191**, 235–254.

MONTEL, J.-M. & VIELZEUF, D. 1997. Partial melting of metagreywackes, Part II. Compositions of minerals and melts. *Contributions to Mineralogy and Petrology*, **128**, 176–196.

NABELEK, P. I., RUSS-NABELEK, C. & DENISON, J. R. 1992. The generation and crystallization conditions of the Proterozoic Harney Peak Leucogranite, Black Hills, South Dakota, USA: Petrologic and geochemical constraints. *Contributions to Mineralogy and Petrology*, **110**, 173–191.

NORMAN, M. D., LEEMAN, W. P. & MERTZMAN, S. A. 1992. Granites and rhyolites from the northwestern U.S.A.: temporal variation in magmatic processes and relations to tectonic setting. *Transactions of the Royal Society of Edinburgh: Earth Sciences*, **83**, 71–81.

NOVAK, S. W. 1984. Eruptive history of the Rhyolitic Kane Springs Wash Volcanic Center, Nevada. *Journal of Geophysical Research*, **89**, 8603–8615.

PATIÑO DOUCE, A. E. 1995. Experimental generation of hybrid silicic melts by reaction of high-Al basalt with metamorphic rocks. *Journal of Geophysical Research*, **100**, 15623–15639.

—— 1996. Effects of pressure and H_2O content on the compositions of primary crustal melts. *Transactions of the Royal Society of Edinburgh: Earth Sciences*, **87**, 11–21.

—— 1997. Generation of metaluminous A-type granites by low-pressure melting of calc-alkaline granitoids. *Geology*, **25**, 743–746.

—— & BEARD, J. S. 1995. Dehydration-melting of biotite gneiss and quartz amphibolite from 3 to 15 kbar. *Journal of Petrology*, **36**, 707–738.

—— & —— 1996. Effects of P, $f(O_2)$ and Mg/Fe ratio on dehydration-melting of model metagreywackes. *Journal of Petrology*, **37**, 999–1024.

—— & HARRIS, N. 1998. Experimental constraints on Himalayan Anatexis. *Journal of Petrology*, **39**, 689–710.

—— & JOHNSTON, A. D. 1991. Phase equilibria and melt productivity in the pelitic system: implications for the origin of peraluminous granitoids and aluminous granulites. *Contributions to Mineralogy and Petrology*, **107**, 202–218.

—— & McCarthy, T. C. 1998. Melting of crustal rocks during continental collision and subduction. *In*: Hacker, B. R. & Liou, J. G. (eds) *When continents collide: Geodynamics and Geochemistry of Ultra-high Pressure Rocks*. Kluwer Academic Publishers, 27–55.

Picirillo, E. M., Melfi, A. J., Comin-Chiaramonti, P., Belleni, G., Ernesto, M., Marques, L. S., Nardy, A. J. R., Pacca, I. G., Roisenberg, A. & Stolfa, D. 1988. Continental Flood Volcanism from the Paraná Basin (Brazil). *In*: McDougall, J. D. (ed.) *Continental Flood Basalts*. Kluwer Academic Press, Dordrecht, 195–238.

Pickett, D. A. & Saleeby, J. A. 1994. Nd, Sr, and Pb isotopic characteristics of Cretaceous intrusive rocks from deep levels of the Sierra Nevada batholith, Tehachapi Mountains, California. *Contributions to Mineralogy and Petrology*, **118**, 198–215.

Pitcher, W. S., Atherton, M. P., Cobbing, E. J. & Beckinsale, R. D. 1985. *Magmatism at a plate edge. The Peruvian Andes*. Blackie, Glasgow.

Poitrasson, F., Duthou, J.-L. & Pin, C. 1995. The relationship between petrology and Nd isotopes as evidence for contrasting anorogenic granite genesis: Example of the Corsican Province (SE France). *Journal of Petrology*, **36**, 1251–1274.

Rämö, O. T. & Haapala, I. 1990. The rapakivi granites of eastern Fennoscandia: A review with insights into their origin in the light of new Sm-Nd isotopic data. *In*: Gower, C. F. *et al*. (eds) *Mid-Proterozoic Laurentia–Baltica: Geological Association of Canada Special Paper*, **38**, 401–416.

Rapp, R. P., Watson, E. B. & Miller, C. F. 1991. Partial melting of amphibolite/eclogite and the origin of Archean trondhjemites and tonalites. *Precambrian Research*, **51**, 1–25.

Roberts, M. P. & Clemens, J. D. 1993. Origin of high-potassium, calc-alkaline I-type granitoids. *Geology*, **21**, 825–828.

Rudnick, R. L. 1990. Nd and Sr isotopic compositions of lower-crustal xenoliths from north Queensland, Australia: Implications for Nd model ages and crustal growth processes. *Chemical Geology*, **83**, 195–208.

—— 1992. Xenoliths — Samples of the lower continental crust. *In*: Fountain, D. M., Arculus, R. J. & Kay, R. W. (eds) *The Lower Continental Crust*. Elsevier, New York, 269–316.

—— & Fountain, D. M. 1995. Nature and composition of the continental crust: A lower crustal perspective. *Reviews of Geophysics*, **33**, 267–309.

Rushmer, T. 1991. Partial melting of two amphibolites: contrasting experimental results under fluid-absent conditions. *Contributions to Mineralogy and Petrology*, **107**, 41–59.

Rytuba, J. J. & McKee, E. H. 1984. Peralkaline ash flow tuffs and calderas of the McDermitt Volcanic Field, Southeast Oregon and North Central Nevada. *Journal of Geophysical Research*, **89**, 8616–8628.

Schuster, R. D. & Bickford, M. E. 1985. Chemical and isotopic evidence for the petrogenesis of the Northeastern Idaho Batholith. *Journal of Geology*, **93**, 727–742.

Sen, C. & Dunn, T. 1994. Dehydration melting of a basaltic composition amphibolite at 1.5 and 2.0 Gpa: implications for the origin of adakites. *Contributions to Mineralogy and Petrology*, **117**: 394–409.

Shannon, W. M., Barnes, C. G. & Bickford, M. E. 1997. Grenville magmatism in west Texas: Petrology and geochemistry of the Red Bluff granitic suite. *Journal of Petrology*, **38**, 1279–1305.

Skjerlie, K. P. & Johnston, A. D. 1993. Fluid-absent melting behavior of an F-rich tonalitic gneiss at mid-crustal pressures: implications for the generation of anorogenic granites. *Journal of Petrology*, **34**, 785–815.

—— & Patiño Douce, A. E. 1998. Fluid-absent melting of a zoisite-bearing eclogite from the Western Gneiss Region of Norway. *EOS*, **79**, F994.

Thompson, A. B. 1982. Dehydration melting of pelitic rocks and the generation of H_2O-undersaturated granitic liquids. *American Journal of Science*, **282**, 1567–1595.

Tollo, R. P. & Aleinikoff, J. N. 1996. Petrology and U-Pb geochronology of the Robertson River igneous suite, Blue Ridge province, Virginia-Evidence for multistage magmatism associated with an early episode of Laurentian rifting. *American Journal of Science*, **296**, 1045–1090.

Vielzeuf, D. & Clemens, J. D. 1992. The fluid-absent melting of phlogopite + quartz: experiments and models. *American Mineralogist*, **77**, 1206–1222.

—— & Holloway, J. R. 1988. Experimental determination of the fluid-absent melting relations in the pelitic system. Consequences for crustal differentiation. *Contributions to Mineralogy and Petrology*, **98**, 257–276.

—— & Montel, J. M. 1994. Partial melting of metagreywackes. 1. Fluid-absent experiments and phase relationships. *Contributions to Mineralogy and Petrology*, **117**, 375–393.

Whalen, J. B. & Currie, K. L. 1990. The Topsails igneous suite, western Newfoundland; fractionation and magma mixing in an 'orogenic' A-type granite suite. *In*: Stein, H. J. & Hannah, J. L. (eds) *Ore bearing granite systems; petrogenesis and mineralizing processes*. Geological Society of America Special Papers, **246**, 287–300.

——, —— & Chappell, B. W. 1987. A-type granites: Geochemical characteristics, discrimination and petrogenesis. *Contributions to Mineralogy and Petrology*, **95**, 407–419.

White, A. J. R. & Chappell, B. W. 1977. Ultrametamorphism and granitoid genesis. *Tectonophysics*, **43**, 7–22.

—— & —— 1988. Some supracrustal (S-type) granites from the Lachlan Fold Belt. *Transactions of the Royal Society of Edinburgh: Earth Sciences*, **79**, 169–181.

Wickham, S. M. 1987. Crustal anatexis and granite petrogenesis during low-pressure regional metamorphism: The Trois Seigneurs Massif, Pyrenees, France. *Journal of Petrology*, **28**, 127–169.

WOLF, M. B. & WYLLIE, P. J. 1991. Dehydration-melting of solid amphibolite at 10 kbar: Textural development, liquid interconnectivity and applications to the segregation of magmas. *Mineralogy & Petrology*, **44**, 151–179.

—— & —— 1994. Dehydration-melting of amphibolite at 10 kbar: the effects of temperature and time. *Contributions to Mineralogy and Petrology*, **115**, 369–383.

WYLLIE, P. J., WOLF, M. B. & VAN DER LAAN, S. R. 1996. Conditions for formation of tonalites and trondh-jemites: magmatic sources and products. *In*: DE WIT, M. J. & ASHWAL, L. D. (eds) *Tectonic evolution of greenstone belts*. Oxford University Press, 258–267.

YARDLEY, B. W. D. & VALLEY, J. W. 1997. The petrologic case for a dry lower crust. *Journal of Geophysical Research*, **102**, 12173–12185.

ZEN, E. & HAMMARSTROM, J. M. 1984. Magmatic epidote and its petrologic significance. *Geology*, **12**, 515–518.

Geometry of granite emplacement in the upper crust: contributions of analogue modelling

TERESA ROMÁN-BERDIEL

Departamento de Geodinámica, Facultad de Ciencias, Universidad del Pais Vasco, apartado 644, 48080 Bilbao, Spain (e-mail: acasas@posta.unizar.es)

Abstract: Granite emplacement in the brittle crust can be modelled by means of the injection of a Newtonian fluid (low-viscosity silicone putty) into sandpacks. This paper describes dynamically scaled analogue models of granite intrusions in the upper crust under different tectonic regimes. Experiments analyse three boundary conditions: (1) static conditions, with different rheological profiles (single sand-layer system, two-layer silicone–sand system, three-layer sand–silicone–sand system and five-layer sand–silicone–sand–silicone–sand system), (2) extensional regime, including gravitational sliding and divergent basal plate, with both mobile and fixed velocity discontinuities and (3) strike-slip regime, induced by two mobile basal plates. The results obtained indicate that: (1) a soft level between two competent units is necessary for laccolith formation—the critical thickness of the soft layer necessary for laccolith formation decreases with increasing depth, (2) when there are two soft layers in the brittle crust, laccolith emplacement occurs in the deeper soft level, even when this is thin and the overburden does reach its critical thickness, (3) in extensional regimes the geometry of intrusions is mainly controlled by normal faults and at the same time intrusions determine the location of faults within the cover and (4) in strike-slip zones intrusions are elongate and their long axis tends to track the principal stretching direction associated with the strike-slip regime. Some natural examples of granitic bodies were considered to test the applicability of experimental results.

Granitic massifs intruded in the upper crust are very common in orogenic zones. When a granitic massif intrudes, its internal structures and the country-rock structures record the interference between the regional strain (tectonic) and the local strain field associated with the intrusion effect (Brun & Pons 1981; Lagarde *et al.* 1990*a*). The nature of tectonic processes (e.g. extension, thickening, strike-slip) contemporary with the intrusion of a granitic massif can be inferred from structural analysis (Bouchez *et al.* 1981; Hutton 1988; Vigneresse 1988; Vernon 1989; Aranguren *et al.* 1997; Castro & Fernandez 1998). However, despite the large number of studies, the nature of intrusion emplacement mechanisms at upper structural levels remains controversial (Paterson *et al.* 1991; Pitcher 1992; Brown & Solar 1998; Cruden 1998; Paterson & Miller 1998). Different proposals have been put forward: diapirism (Brun 1981; Ramberg 1981), magma ascent via fractures and lateral spreading (Dehls *et al.* 1988; Brun *et al.* 1990; Lagarde *et al.* 1990*b*; Roig *et al.* 1998), fill in of extensional jogs of pull-apart type formed along strike-slip zones (Castro 1986; Hutton 1988; Schmidt *et al.* 1990;

Lacroix *et al.* 1998), magma accommodation in the local tectonic context (Brun & Pons 1981). Several recent works focus on the problem of why magma rises along fractures and dykes (Castro 1987; Brun *et al.* 1990; Lagarde *et al.* 1990*b*; Clemens & Mawer 1992; Clemens 1998), why it stops, and how horizontal crustal discontinuities influence the interruption of magma ascent and spreading (Hogan & Gilbert 1995; Hogan *et al.* 1998).

A question extensively debated about granitic intrusions is the space needed for their emplacement (Hutton *et al.* 1990; Tikoff & Teyssier 1992; Paterson & Fowler 1993*a*). Recently, authors have proposed models of pluton emplacement that rely on local or regional extension (e.g. Hamilton 1981; Tobisch *et al.* 1986; Hutton 1988; Glazner 1991). Such emplacement models are widely accepted because it is easy to imagine wall rock being pulled apart and a gap opening that can be filled by magma, and also because of the belief that intrusions are harder to emplace during contraction. Recent brittle–ductile analogue models have shown that pluton emplacement does not require local extensional

From: CASTRO, A., FERNÁNDEZ, C. & VIGNERESSE, J. L. (eds) *Understanding Granites: Integrating New and Classical Techniques.* Geological Society, London, Special Publications, **168**, 77–94. 1-86239-058-4/99/$15.00 © The Geological Society of London 1999.

regimes, even under pure strike-slip boundary conditions (Román-Berdiel *et al.* 1997). In the last years a great variety of field studies have provided structural and geophysical data that have been interpreted as indicating intrusive bodies emplaced in contractional stress/strain fields (Brown & Solar 1998; Paterson & Miller 1998; Roig *et al.* 1998; Benn *et al.* 1999). On the other hand, in an anisotropic and fractured crust, Paterson and Fowler (1993*b*) see little mechanical difference between pluton emplacement during contraction (during which vertical extension occurs), and extension (during which vertical contraction occurs).

Many granitoid plutons that are emplaced at rather shallow crustal levels are commonly laccolithic or tabular in shape with horizontal dimension far in excess of vertical thickness (Clemens & Mawer 1992). Several experimental studies of laccolithic intrusions have used rather stiff materials to model the country-rocks (Howe 1901; McCarthy 1925; Hurlbut & Griggs 1939; Pollard & Johnson 1973; Dixon & Simpson 1987). Only a few recent experiments have allowed situations where faulting of country-rocks could accompany intrusion (Merle & Vendeville 1992, 1995; Román-Berdiel *et al.* 1995, 1997; Benn *et al.* 1998). Merle & Vendeville (1992, 1995) placed a soft level under the sand pack to obtain laccolithic bodies, but their study focuses on deformation induced in the cover by the intrusion, rather than in analysing the geometry of intrusions. Several experiments showed that the occurrence of weak levels, such as shales, pelites or salt, within the brittle crust was conveniently modelled using silicone layers (Vendeville 1987; Richard *et al.* 1989, 1991; Basile 1990; Nalpas & Brun 1993). Román-Berdiel *et al.* (1995) have shown that properly scaled sand–silicone putty models were convenient to study intrusion processes in the upper crust. Benn *et al.* (1998) made experiments on intrusions in transpressional conditions, but only in a homogeneous brittle crust, and they obtained funnel-shaped intrusions. Studies by Román-Berdiel *et al.* (1995, 1997) focused on laccolithic-type intrusions in the upper crust under different tectonic regimes (static conditions, extensional conditions and strike-slip regime). These experiments have shown that intrusion geometries and their relations with faults developed in the brittle layers were diagnostic of both crustal rheological profile and boundary conditions.

In this paper, we present some new models of laccolithic-type intrusions based on analogue modelling. Particularly, the new models outline the role of several soft levels within the brittle crust in the conditions of laccolith formation. Moreover new experiments in extensional conditions, but with different boundary conditions, constitute an important contribution in defining the common features on the geometry of intrusions in extensional tectonic regimes. The new results obtained corroborate that the mechanical behaviours of the country rocks, and hence the rheological layering of the crust, are first-order factors controlling the emplacement mechanism. In order to test the applicability of these experimental results to natural conditions we will refer to some granitic bodies emplaced in different geological settings.

Experimental procedures

Experiments were made at the Analogue Modelling Laboratory of Rennes University (France). Models are 50×50 cm squares built to different heights on a fixed basal plate, and centred over an injection point (Fig. 1a). The injection tube in all models is 10 mm in diameter. The fluid is pushed up in the tube by a piston driven at constant velocity by a stepper motor (Fig. 1a). Various tectonic regimes can be simulated by a variety of experimental apparatus.

(1) Static regime testing different rheological profiles.

(2) The extensional regime has been simulated by three different procedures. (i) Models were tilted a few degrees in order to examine the effects of syn-injection deformation by gravitational sliding (Fig. 1a). (ii) A thin divergent basal plate lying on the fixed plate where the injection tube is located (Fig. 1b). The mobile plate was pulled away at constant rate. (iii) A plastic sheet attached to a vertical end wall sliding along a rigid horizontal plate, and diverging from a narrow space (velocity discontinuity, V.D.) at constant rate (Fig. 1c). To ensure distributed deformation in a wide zone of the model surface, a layer of silicone putty ($50 \times 30 \times 1$ cm) was placed at the base of the models (Tron & Brun 1991).

(3) Strike-slip regime. The strike-slip motion was imposed at the base of the model by two thin mobile rigid plates lying on the fixed plate, and separated from each other by 3 cm (Fig. 1d). Mobile plates were moved in opposite directions at a constant rate. This configuration with two basal discontinuities induces a relatively wide strike-slip shear zone in the models (>5 cm). It thus allowed intrusions within the zone, above the fixed injection tube located between the two mobile plates (Fig. 1d). To ensure distributed shearing between the mobile plates at the base of the models, a band of silicone putty ($80 \times 6 \times 1$ cm) was placed on the basal velocity discontinuities (Fig. 1d) (Basile 1990). The models are 80 cm in length, 30 cm in width, and have a variable height.

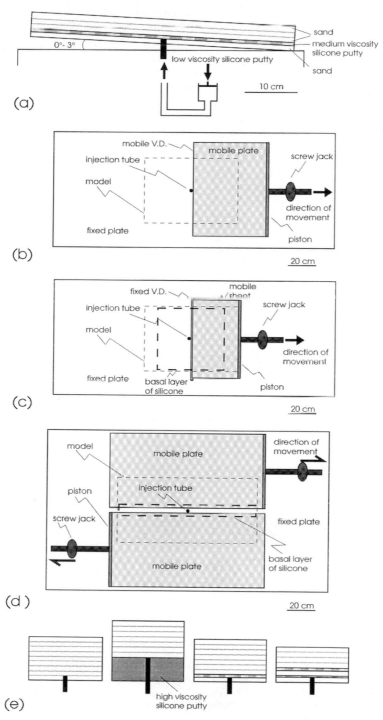

Fig. 1. Sketch of experimental settings. (**a**) Vertical view of three-layer experiments with or without syn-injection gravitational sliding. (**b**) Horizontal view of syn-injection extensional regime produced by basal mobile velocity discontinuity. (**c**) Horizontal view of syn-injection extensional regime produced by basal fixed velocity discontinuity. (**d**) Horizontal view of syn-injection strike-slip experimental setting. (**e**) Types of models layering: single-layer models, two-layer models, three-layer models and five-layer models.

In our experiments the rheological profile of the crust was modelled by sand and silicone (Table 1). For the brittle crust, we used dry sand, which is almost pure quartz, with a maximum grain size of 500 μm. For the layer thicknesses used in our experiments, the mean density of the sand is approximately 1500 kg m^{-3}. Dry sand has an internal friction angle of about 30°, and a negligible cohesion (Mandl *et al.* 1977). For ductile materials, we used silicone putty (GS1R gum of Rhône-Poulenc), which is almost perfectly Newtonian (Vendeville *et al.* 1987). Three different types of silicone have been used. A standard, medium-viscosity silicone ($\mu = 7.5 \times 10^3$ Pa s, $\rho = 1270$ kg m^{-3}) was used to introduce soft ductile layers within the sandpack. A low-viscosity silicone ($\mu = 2.5 \times 10^3$ Pa s, $\rho = 1330$ kg m^{-3}), obtained by adding violet basonyl powder to the standard silicone, was used for the intrusion. For the ductile crust, we used a higher viscosity silicone ($\mu = 1.5 \times 10^4$ Pa s, $\rho = 1400$ kg m^{-3}), obtained by adding galena powder to the standard silicone. In models intrusion is pressurized by moving a piston. Because rigid plates supported the overburden in these experiments, the density difference did not affect the pressure through differential loading and is irrelevant. Density difference between injected silicone and sand is 170 kg m^{-3}, sufficient to ensure that injected silicone does not let down. Sand and silicone putty are suitable materials for dynamically scaled modelling in the normal gravity field if the experiments last a few hours (Faugère & Brun 1984; Vendeville *et al.* 1987; Davy & Cobbold 1991; Nalpas & Brun 1993). The room temperature was 30 °C in all experiments.

In all models, the sandpack contained horizontal passive markers made of coloured sand, and the silicone layer contained vertical passive markers. These were obtained by building the layer with vertical bands containing a slight amount of blue methylene powder mixed with the silicone, alternating with bands of methylene-free silicone. At the end of each experiment, these markers allowed us to observe the deformation on cross-sections. A grid of square passive sand markers drawn on the upper surface of the models allowed examination of surface deformation during experiments.

Scaling

Models have been scaled following principles discussed by Hubbert (1937) and Ramberg (1981). The scaling used here is described in detail by Román-Berdiel *et al.* (1995). We used a length ratio of 10^5, so that 1 cm in models is equivalent to 1 km in nature. Given the ratio of density of natural rocks relative to the analogue materials (of the order of magnitude of 1), the length ratio used yields a stress ratio of the order of 10^5 of nature relative to model. In others words, the models are about 10^5 weaker than the natural crust. The time ratio is 10^9 (1 h represents 115 000 years). According to

the experimental scaling of Román-Berdiel *et al.* (1995), magma driving pressures of the order of 50 MPa, comparable to those inferred for natural situations (Johnson & Pollard 1973), are properly scaled using linear injection rates of the order of 31 cm/h (Table 1), which scale to 10^{-8} m s^{-1} (27 cm/year) in nature. This magma ascent rate is contained inside the gap done for granitic magmas by the different authors (see Clemens & Mawer 1992; Paterson & Tobisch 1992).

Modelling of syntectonic intrusion requires a convenient balance between injection dynamics and large-scale boundary conditions. Rather rapid cooling rates are expected for a magma emplaced at high crustal levels, i.e. in a low-temperature environment (e.g. Spera 1980). Once substantially cooled, the intrusion, and its hornfelsed mantle, can tend to behave like a resistant object during subsequent deformation. This cannot be modelled in our experiments. Consequently, the bulk extensional regime and the strike-slip motion were only applied during injection, in order to avoid post-emplacement deformation. Our experimental study of interactions between pluton emplacement and tectonics requires that (1) the overall syn-intrusion deformation is sufficiently large, and (2) injection rates are sufficiently slow to ensure proper scaling. For these reasons ratios between injection rate and bulk strike-slip motion are relatively small (Table 1) compared with what might be expected in many natural situations (Paterson & Tobisch 1992).

Experimental results

Emplacement in static conditions

Injection of silicone in single layer sand models can be related with magma intrusion in a homogeneous brittle crust. In these experiments, the injection was made 1 cm above the base of the sandpack (Fig. 1e). For single layer experiments, injection within the sandpack results in more or less cylindrical piercing intrusions surrounded by reverse faults that propagate upwards within the cover (Fig. 2a, model 1) (Román-Berdiel *et al.* 1995).

Two-layer models consist of a basal layer of high viscosity silicone putty overlain by a sandpack. The low-viscosity silicone putty was injected at the sand–silicone interface in order to examine intrusion kinematics at the crustal brittle–ductile transition (Fig. 1e). Mushroom-type intrusions within the basal silicone layer were obtained (Fig. 2b, model 2). Intrusions localized at the upper boundary of the silicone. Locally they cut across the bottom of the sandpack but do not expand within the brittle layer.

Injection of silicone in a three-layer system (Fig. 1e), sand–silicone–sand, simulating a

Table 1. *Model characteristics and experimental conditions for the different experiments performed*

Series	Model	Tectonic regime	Thickness of high viscosity silicone layer (cm)	Thickness of sand layer below injection point (cm)	Thickness of décollement level (cm)	Thickness of sand overburden (cm)	Volume of injected silicone (cm³)	Brittle/ductile ratio	Dip angle (°)	Total displacement rate (cm/h)	Injection velocity (cm/h)	Total displacement at model surface (cm)
Single layer	1	static	—	1	—	8.0	98	—	0	—	38.0	—
Two layer	2	static	5.5	—	—	7.7	98	1.40	0	—	38.0	—
Three layer-I	3	static	—	1	0.2	9.5	98	52.50	0	—	34.5	—
	4	static	—	1	0.2	8.0	98	45.00	0	—	37.5	—
	5	static	—	1	0.2	6.0	98	35.00	0	—	41.5	—
	6	static	—	1	0.2	5.0	98	30.00	0	—	39.2	—
	7	static	—	1	0.2	4.0	98	25.00	0	—	37.5	—
Three layer-II	8	static	—	1	0.6	5.6	98	11.00	0	—	34.0	—
	9	static	—	1	0.6	4.3	98	8.80	0	—	30.0	—
	10	static	—	1	0.6	3.0	98	6.60	0	—	32.5	—
	11	static	—	1	0.6	2.0	98	5.00	0	—	33.5	—
	12	static	—	1	0.6	1.0	98	3.30	0	—	33.2	—
Five layer	13	static	—	1	0.2 & 0.6	1.0 & 4.0	49	7.50	0	—	38.0	—
Three layer-III	14	extensional	—	1	0.6	4.0	98	8.30	3	—	31.2	4.8
	15	extensional	—	1	0.6	3.0	98	6.60	3	—	31.2	5.0
	16	extensional	—	1	0.6	2.0	98	5.00	3	—	31.2	4.8
	17	extensional	—	1	0.6	1.0	98	3.30	3	—	31.2	5.0
Three layer-IV	18	extensional	—	1	0.6	2.0	—	5.00	0	0.5	—	2.2
	19	extensional	—	1	0.6	2.0	98	5.00	0	0.5	31.2	2.0
	20	extensional	—	1	0.6	2.0	98	5.00	0	0.5	31.2	2.1
Three layer-V	21	extensional	—	1	0.6	2.0	49	1.87	0	2.5	31.5	5.0
	22	extensional	—	1	0.6	3.0	49	2.50	0	2.5	31.5	5.0
	23	extensional	—	1.5	0.6	3.0	32	2.81	0	5.0	41.5	5.0
Three layer-VI	24	strike-slip	—	1	0.6	1.0	—	1.25	0	10.0	—	20.0
	25	strike-slip	—	1	0.6	4.0	49	3.12	0	10.0	31.2	19.4
	26	strike-slip	—	1	0.6	3.0	49	2.50	0	10.0	31.2	19.2
	27	strike-slip	—	1	0.6	2.0	49	1.87	0	10.0	31.2	20.0
	28	strike-slip	—	1	0.6	1.0	49	1.25	0	10.0	31.2	20.0
Three layer-VII	29	strike-slip	—	1	0.6	2.0	98	1.87	0	5.0	31.2	19.8
	30	strike-slip	—	1	0.6	2.0	12	1.87	0	10.0	31.2	9.6

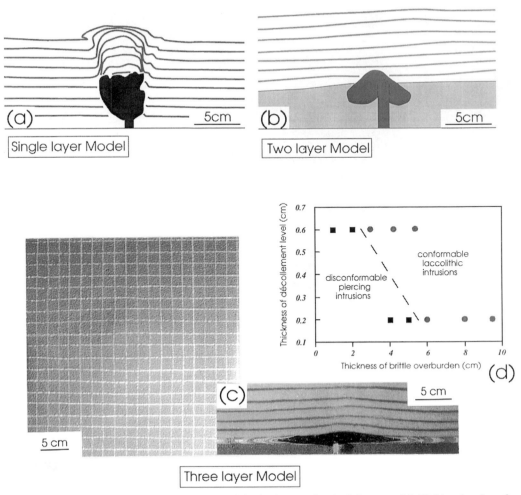

Fig. 2. (a) Line drawing of central cross-section of the final stage of a single layer model. **(b)** Line drawing of central cross-section of the final stage of a two layer model. **(c)** Photographs of surface view and central cross-section of the final stage of three layer model 8 with a 5.6 cm sand overburden. **(d)** Fields of conformable laccoliths and of piercing intrusions according to thicknesses of soft layer and sand overburden (experiment series three layer I and II).

low-strength level within the upper crust has been largely analysed by Román-Berdiel *et al.* (1995). In that work (see experimental conditions summarized in Table 1) it is concluded that for a given thickness of the soft layer (2 or 6 mm), there is a critical threshold of overburden thickness above which laccoliths spread laterally and lift the overburden (Fig. 2c, model 8), and below which intrusions pierce their roof. With a thin soft layer (2 mm), the threshold of overburden thickness for laccolith development was found to be around 6 cm. This value dropped to 2 cm for a thicker soft layer (6 mm). Thus, the deeper the injection, the smaller the thickness of the soft layer necessary to form a laccolith.

With two soft layers of different thicknesses (2 and 6 mm) (five-layer system, Fig. 1e), if the injection point is located at the base of the deeper soft layer, injection results in a lens-shaped intrusion located in the thin soft layer (Fig. 3b, model 13). If we compare this result with a three-layer model (one décollement 2 mm thick), we find that in the latter, intrusions push aside and fracture the sand overburden producing a ring-dyke geometry (Fig. 3a, model 5). In the experiment with two soft layers a laccolith forms because the shallower thick soft layer absorbs strain plastically and the brittle sand overburden does not fracture (Fig. 3b, model 13). Thus, the threshold overburden thickness for laccolith

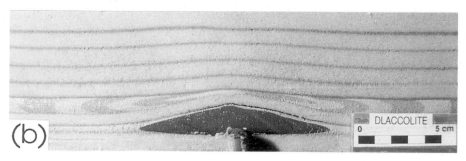

Fig. 3. (a) Photograph of central cross-section of the final stage of three layer 5. (b) Photograph of central cross-section of the final stage of five layer model.

development with a thin soft layer (2 mm) decreases when there is a second shallow soft layer interbedded within the overburden. Another consequence is that laccolithic intrusion occurs in the first soft layer encountered by a rising magma during rising, even if the overburden has not the critical thickness necessary to form laccoliths in the case of only one soft layer.

Emplacement in extensional conditions

The extensional regime has been simulated by three different procedures: (1) gravitational sliding (see Fig. 1b), (2) a divergent basal plate with a mobile velocity discontinuity (see Fig. 1c) and (3) a divergent plastic sheet with a fixed velocity discontinuity (see Fig. 1d). Three-layer models were used, because this is the simplest rheological profile that led to laccoliths.

In gravitational sliding (Román-Berdiel et al. 1995), the soft layer acts as an active décollement. The main extensional structure that develops in the overlying brittle cover is a central graben, located above the deep injection pipe, which elongates perpendicular to the dip-slip direction (Fig. 4a, model 15). Intrusions produced in a gravitational sliding regime show rather flat roofs

(Fig. 4d, model 15). In addition, a systematic geometrical feature observed in these experiments is a strong asymmetry of the intrusion consistent with the shear sense along the décollement. This asymmetry is marked by (Fig. 4d, model 15) (1) an offset of the intrusion downslope of the feeding conduit, as a result of bulk dip-slip motion of the sandpack, (2) a tendency for the frontal boundary of the intrusion to dip more steeply than the rear boundary and (3) an asymmetric spreading of the intrusion along the décollement (also Fig. 4c). This asymmetric spreading is expressed by the development of a sheared basal tongue trailing behind the intrusion, linking it to its feeder.

In extensional regimes imposed by a divergent basal plate with mobile V.D. (Fig. 5) the main extensional structures which develop in the overlying brittle cover are (Fig. 5a, model 18): (1) a central gentle synform (7 cm wide), located above the deep basal velocity discontinuity and parallel to it and (2) two symmetric grabens, one on each side of the synform, parallel to the basal discontinuity. In models with injection these structures are modified. Injection controls the structures that develop in the overburden (Fig. 5b and c, models 19 and 20). When injection is

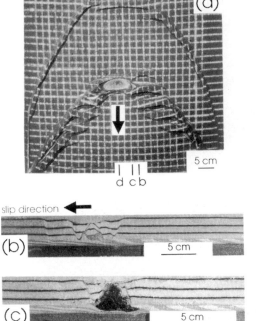

Fig. 4. Photographs of the final stage of three layer model 15 with syn-injection gravitational sliding. (**a**) Surface-view with location of different cross-sections (**b, c** and **d**) within the model; arrow points towards the slip direction.

located next to the basal discontinuity (Fig. 5b, model 19) the strike of normal faults is deflected in the vicinity of intrusions, and the synform disappears. When injection is located 6 cm from the basal discontinuity (Fig. 5c, model 20) intrusion occurs in the graben located above the fixed plate. The synform above the discontinuity is maintained, but the graben above the mobile plate disappears. Intrusions formed in this type of extensional regime (Fig. 5b and c, models 19 and 20) show rather flat roofs, as for intrusions produced in gravitational sliding (see Fig. 4d, model 15). Intrusions are also asymmetric. This asymmetry is marked by (Fig. 5b and c, models 19 and 20): (1) an offset of the intrusion towards the mobile basal plate and (2) a more important spreading of the intrusion along the soft layer located above the fixed plate.

In extensional regimes imposed by a divergent basal sheet with fixed V.D. (Fig. 6a and b, model 23) experiments are characterized by a graben on each side on the injection point separated by a gentle syncline, as for extensional regime with mobile V.D., but the separation between the grabens is greater (10 cm) and there is an offset of the syncline with respect to the V.D. towards the mobile basal sheet. Intrusions occur close to the graben developed over the fixed basal sheet, and deflect the orientation of normal faults. Intrusions show convex roofs in central cross section (Fig. 6d). Intrusions are also asymmetric, with offset of the intrusion towards the mobile basal sheet and a more important spreading of the intrusion along the soft layer located above the fixed plate. In lateral cross section intrusions are cross-cut by normal faults (Fig. 6c).

Emplacement in strike-slip regime

Injection in the strike-slip regime has been analysed by Román-Berdiel *et al.* (1997). In this section I will only summarize some aspects of the shape of intrusions and their relationships with the geometries obtained in other tectonic regimes.

In all strike-slip experiments, the fault pattern developed in the sand overburden consists of synthetic faults showing two main sets of orientation. Numerous faults at low angles to the bulk shearing direction (R and Y faults) (Fig. 7a and b, models 24 and 28) occur within the main deformation zone. This pattern results from early development of Riedel faults, which are subsequently overprinted by faults at low angle to the bulk shearing direction during progressive strike-slip and strain localization. The fault pattern developed in the sand overburden of experiments without injection is homogeneous along the shear band (Fig. 7a, model 24). When there is injection, the intrusion modifies the fault pattern, deformation is less homogeneous, the Y faults are more developed, and they limit the intrusion boundaries (Fig. 7b, model 28).

Horizontal cross-sections show that intrusions are always elongate in the extensional axis defined by the strike-slip deformation (Fig. 7b and c, models 28 and 29). The angle between the long axis of intrusions and the shear direction decreases consistently with increasing shear strain. From this, we infer that the long axis of model intrusions tends to track the long axis of the bulk strain ellipsoid (Fig. 7c and d). The irregular geometry of the intrusions (Fig. 7e, model 26) is probably due to interactions between emplacement and bulk strike-slip motion. In particular, the strike-slip environment favoured the formation of dykes and stocks

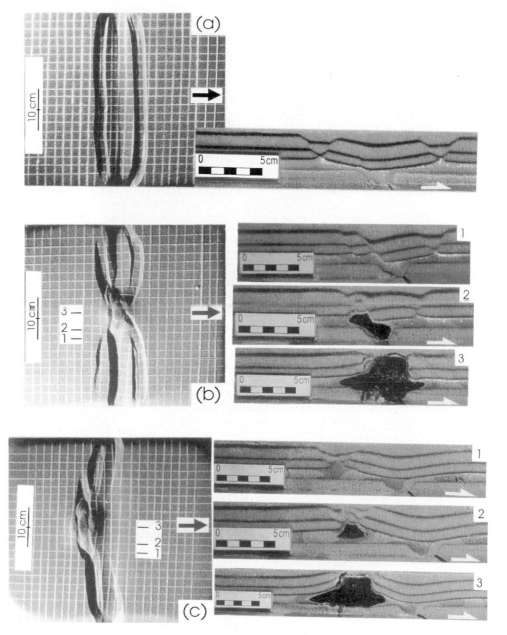

Fig. 5. Photographs of the final stage of three layer-IV series with extensional regime produced by a basal mobile velocity discontinuity. (**a**) Surface-view and central cross-section of final stage of model 18. (**b**) Surface-view with location of different sections (1, 2 and 3) within model 19. (**c**) Surface-view with location of different sections (1, 2 and 3) within model 20. Arrows point towards the direction of movement of the basal plate.

along faults throughout the sand overburden (Fig. 7e, model 26). Faults control the local shape of intrusion boundaries (Fig. 7b and c, models 28 and 29). Thus, the orientations of many segments of intrusion boundaries, as seen on horizontal cross-sections (angle ϕ, Fig. 7c), are subparallel to synthetic faults observed on the surface of the models (Fig. 7b). Irregular shapes of intrusion walls result from local variations in the effects of faults, from a minimum in the case of convex towards the weak layer (right-hand side of intrusion in Fig. 7e) to a maximum

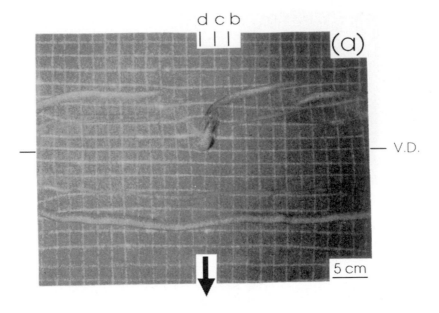

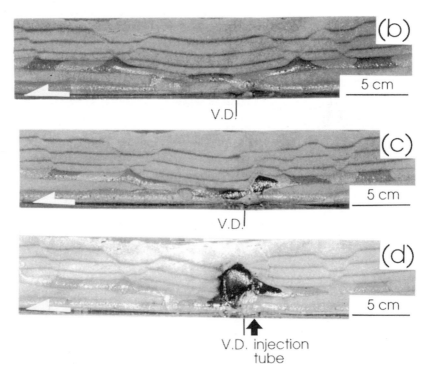

Fig. 6. Photographs of the final stage of three layer model 23 with extensional regime produced by a basal fixed velocity discontinuity. (**a**) Surface-view with location of different cross-sections (**b**, **c** and **d**) within the model. Arrows point towards the direction of movement of the basal sheet. V.D. indicates the position of the velocity discontinuity.

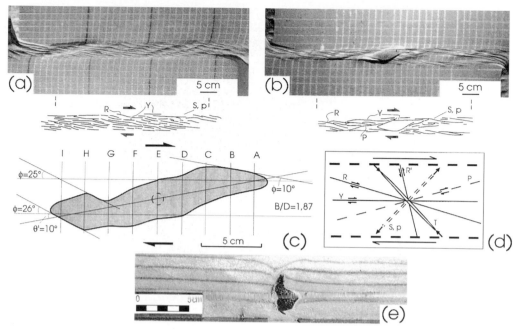

Fig. 7. Surface photographs and line drawings of three layer models 24 and 28 with strike-slip regime (**a**) without injection, and (**b**) with injection. (**c**) Horizontal section across model 29. Letters A to I refer to serial cross-sections made at final stage. B/D ratio is the thickness ratio between sand and silicone layers. θ', angle between long axis of intrusion and bulk shearing direction. Φ, local angle between intrusion boundary and bulk shearing direction. The two horizontal lines are 3 cm apart and correspond to the traces of the two basal velocity discontinuities. Dotted circle is the trace of the underlying feeding pipe. (**d**) Different structures that can appear in a strike-slip shear zone. (**e**) Photograph showing a section of model 26 containing the intrusion and perpendicular to a strike-slip zone.

in the case of subvertical contacts (left-hand side of intrusion in Fig. 7e).

Discussion

Although analogue modelling produces realistic geometric and kinematic models of granitic bodies, and although the modelling materials simulate the deformation mechanics of brittle sedimentary rocks in the upper crust, of weak décollements in the brittle crust, of ductile crust and of granitic melt, important limitations must be borne in mind when interpreting the results.

In our experiments, space for intrusion is obtained by pushing aside the weak layer (see Figs 2c, 3b and 7e) of country-rocks and lifting the overburden (see Figs 2a, 3a and 7e). Our models are thus consistent with inferences of Paterson & Fowler (1993*a*) who argued that a way of making space for magmas in the crust is to displace the Earth's surface. Dyking also occurs locally along faults in our models in the strike-slip regime (Fig. 7e). Horizontal cross-sections (Fig. 7c) further suggest that dyking occurred preferentially along Riedel faults. This could reflect preferential dyking along zones of

maximum dilatancy, thus suggesting some control of the local extensional stress field on model intrusions, as often invoked in natural situations (e.g. Hutton 1988; Paterson & Fowler 1993*b*).

The effects of parameters such as the diameter of the feeding pipe, or the relative rates of injection and tectonic motion, have not been examined. Nor have I modelled changes of viscosity and strength which accompany synemplacement crystallisation and cooling of natural intrusions, as well as possible effects of non-Newtonian magma rheologies (Johnson & Pollard 1973).

However, bearing these provisos in mind, analogue models are powerful tools for understanding the mechanics of granite emplacement in the upper crust.

Emplacement in static conditions

From the experimental results listed above we can infer that: (1) cylindrical piercing intrusions occur when there is a brittle country rock, (2) mushroom-type intrusions occur when there is a ductile country rock and an ascending magma meets a barrier, brittle crust or more

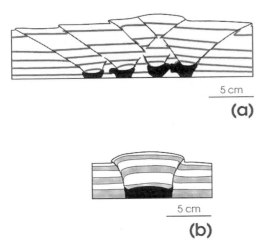

5 cm

(a)

5 cm

(b)

Fig. 8. (**a**) Line-drawing of cross-section of the final stage of a single layer model of intrusions in transpression (from a photograph, after Benn *et al.* 1998). (**b**) Line-drawing of cross-section of a single layer model of forced salt piercement (from a photograph, after Schultz-Ela *et al.* 1993).

competent lithology, (3) laccoliths were only obtained when there are soft layers interbedded in a competent cover.

More or less cylindrical piercing intrusions (Fig. 2a) and ring-dyke geometries (Fig. 3a) with steeply dipping walls resulted in static conditions when intrusions occur in a brittle crust and the injected material not spread in any soft level or discontinuity. Intrusions pierce and push upwards the overburden leading to development of cylindrical reverse faults linked to the extremities of intrusions. Local rising of the injected fluid is produced along reverse faults (Fig. 3a). In static conditions deformation of overburden is entirely produced by kinematic intrusion.

Similar intrusion geometries in cross-section, defined as funnel shapes, have been obtained in models of intrusions in a homogeneous brittle crust in transpressional regime (Fig. 8a) (Benn *et al.* 1998). The differences with the static regime lose in the amplification of deformation in the overburden and the shallower fault dips. These differences must be due to the contribution of the transpressional regime, and to the interference of reverse faults formed by two adjacent intrusions. Nevertheless the style of deformation is similar to the static conditions. The 3D geometry of intrusions is determined in these experiments by the boundary conditions (length of the feeding pipe equal to the model width). Benn *et al.* state that faults have a critical role in determining the emplacement mechanisms and the final shapes of plutons in contractional belts. From the

experiments cited above we infer that similar features can be obtained in static conditions and so they are not necessarily unique to transpressional conditions. These results suggest that the rheology of the crust is a main factor in controlling the shape of intrusions, independent of the tectonic regime. Field studies by Corriveau *et al.* (1998) and Benn *et al.* (1999) conclude that the nature of the host and the crustal structure/ rheology, and not the tectonic regime, control the mechanics of emplacement, what agrees with the experimental result.

Similar features are also obtained in analogue models of forced salt diapiric piercement performed with sand and silicone (Fig. 8b) (Schultz-Ela *et al.* 1993). Similarities exist mainly in the earliest stages, where lifting and angular arching of the diapir roof occur. Reverse faults extended steeply upward from the top corners of the diapir, then curved outward to shallower dips near the surface. The top layers in the centre of the roof block thinned diffusely to make a relatively flat surface (similar features can be observed in our single layer model and three layer model-5, Figs 2a and 3a). Comparisons between salt and granite diapirs have been made since Sorgenfrei (1971), who showed that there is a close correspondence in size and shape, which is possibly due to the similarity in the density contrasts involved. Schultz-Ela *et al.* (1993) stated that the most characteristic and fundamental process during active diapirism is the elevation of the diapiric roof above its regional datum. Where the effects of active diapirism and regional shortening constructively interfere, it is difficult to separate these effects of purely geometric grounds. The study of intrusions in a brittle crust in contractional belts is a very broad and complicated task needing a great quantity of analogue models.

In nature laccoliths are located in low-strength zones along which they are emplaced by lateral spreading. Low strength zones can be of two kinds. (1) Lithological discontinuities. They can correspond to easily deformed rocks (e.g. pelite), or to a low-strength bed underlying a level that constitutes a barrier limiting the upward migration of the magma (Mudge 1968; Johnson & Pollard 1973). (2) Tectonic discontinuities. They correspond either to ductile shear zones that become zones of weakness or to initially soft lithologies. Examples are abundant: contact between the Higher Himalayan crystalline rocks and the Tethyan sediments (Herren 1987; Pêcher 1991; Gapais *et al.* 1993), the Variscan front in southern England (Shackleton *et al.* 1982) and the bottom of the Montana nappe (Schmidt *et al.* 1990).

These lithological and tectonic discontinuities can be compared with the soft level (silicone) interbedded in the sand (more competent) in our analogue models.

Román-Berdiel *et al.* (1995) underlined that their laccolith models have shapes similar to natural laccoliths (Gilbert 1877; Johnson & Pollard 1973; Talbot 1993) and that the diameter/thickness ratios of model laccoliths show values of the same order of magnitude (0.1 to 0.3) as the mean aspect ratio of 0.14 reported by Gilbert (1877) for the Henry Mountains laccoliths. Moreover, from experimental results, Román-Berdiel *et al.* (1995) deduced that strength discontinuities alone can be expected to be as efficient in laccolith localization where the weight of the overburden is high, i.e. that they are deep seated (Fig. 2d). In the light of the new results presented in this paper, we can state that thinner soft levels or weak discontinuities can be efficient in laccolith localization, even when overburden is not significantly thick, if there is another thick soft level interbedded in the overlying cover (Fig. 3b). These results suggest that laccoliths may provide information not only about the structure of the underlying crust (Román-Berdiel *et al.* 1995), but also about the eroded overlying crust. This can have a direct implication for structural geologists when looking for deformation produced by a granitic body in its surrounding wall-rock during intrusion. Deformation could concentrate in an overlying soft level eroded at present, and evidence of deformation during intrusion may then be absent. This can be the case of many granitic bodies of the Variscan belt of Spain considered classically as post-kinematic plutons emplaced in permissive conditions.

Emplacement in extensional conditions

In extensional regime (gravitational sliding or basal V.D.) laccoliths elongate in the direction of grabens when they intrude inside them (Figs 4a and 5c, models 15 and 20) and, in this case, show rather flat roofs in cross section (Figs 4d and 5c, section 3). When intrusions occur without but close to the graben, then they are cross-cut by normal faults (Figs 6c and 5b, section 2). Normal faults cross-cutting the injected material are directly related to the existence of a basal V.D. close to the injection point (Figs 6 and 5b). A common feature of all intrusions in extensional regime is the asymmetry of intrusions in the central vertical section. Thus, the asymmetry of laccolithic intrusions could be diagnostic of syn-emplacement bulk kinematics. In all these models faults are located over the injection

point, this indicates that the process of magma intrusion may be an important factor in nucleating normal faults, as it has been reported in natural examples (Aranguren & Tubía 1992).

In the preceding sections we discussed that some natural laccoliths are located along tectonic discontinuities. These tectonic discontinuities can correspond to extensional ductile shear zones. Examples include leucogranites of southern Brittany (western France), which have been interpreted as intruded in a crustal-scale extensional shear zone that separates the lower and upper crust (Bouchez *et al.* 1981; Audren 1987; Gapais *et al.* 1993; Richer 1994). In the Himalayas, the Manaslu granite was emplaced during regional shearing (Le Fort *et al.* 1987; France-Lanord & Le Fort 1988; Pêcher 1991; Guillot *et al.* 1993). Sections across the Manaslu granite (Le Fort *et al.* 1987; Guillot *et al.* 1993) show a strongly asymmetric laccolith that can be compared with our syn-extensional intrusions (Fig. 4) (Román-Berdiel *et al.* 1995).

Emplacement in strike-slip regime

Relationships of plutons to strike-slip crustal-scale shear zones are particularly well documented. Some of these intrusions are lozenge-shaped like our models built in a strike-slip regime. Examples include the Saraya batholith in Eastern Senegal (Pons *et al.* 1992), which can be interpreted as intruded in a large shear zone, or the Mortagne granite in southern Brittany (leucogranite outcropping along the South Armorican Shear Zone, Guineberteau *et al.* 1987). In the light of the experimental results, a broad discussion has been made by Román-Berdiel *et al.* (1997) about the emplacement model of the Mortagne granite. These authors emphasized that the room problem is not a critical factor in explaining the shape of many plutons emplaced along strike-slip zones. In particular, intrusion between two faults can yield typical lozenge-shaped plutons for which emplacement mechanisms like those involving pull-aparts are irrelevant.

The Central Extremadura Batholith, located in the southern part of the Central Iberian zone in the Variscan belt of Spain (Fig. 9), provides another natural example showing structural resemblances with some of our analogue models of intrusions under strike-slip conditions. The Central Extremadura Batholith constitutes an alignment of several plutons and numerous minor intrusions of quartz-diorite to alkali-feldspar granites emplaced in low-grade metamorphic rocks (epizonal domain), schists and greywackes of Precambrian age. The plutons are

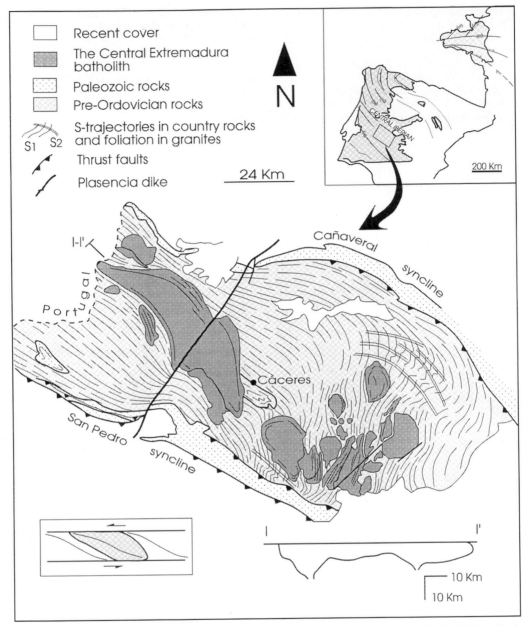

Fig. 9. Geological and structural sketch map of the area around the Central Extremadura Batholith (modified after Castro 1986) and cross-section of the Cabeza de Araya pluton inferred from 3D inversion of gravity data (after Vigneresse 1995). Inset shows general sketch of model kinematic context which can compare to the natural example.

elongated bodies parallel to the regional tectonic foliation. Their internal structure, analysed in detail by Castro (1986) and Castro & Fernández (1998), is mainly determined by preferable orientation of K-feldspar megacrysts and C–S structures. The Cabeza de Araya pluton, the largest of the batholith, shows a lozenge shape and a concentric, steeply-dipping foliation parallel to the pluton elongation (Fig. 9). Its eastern border is affected by thrust shear zones. Traces of foliation planes indicate a bulk NE–SW shortening of the massif. The occurrence of thrust shear zones along the eastern border of the pluton (Castro 1986) is also compatible with NE–SW shortening. Most magnetic lineations strike consistently NW–SE and are generally gently plunging (Vigneresse 1995). Thus, geological and geometric evidence suggests that the Cabeza de Araya pluton was emplaced during left-lateral strike-slip motion. This interpretation agrees with the large-scale, sinistral shear-zones which constitute one of the most representative structural features of the Iberian Massif (Fig. 9). They are usually related to the late stages of the compressional deformation within the Iberian Massif.

A detailed gravimetric study (Vigneresse 1995) has shown that the floor of the pluton is generally flat, except at the southeastern and northwestern terminations of the massif where floor depths exceed 10 and 12 km (Fig. 9). These zones are interpreted as feeding pipes (Vigneresse 1995).

Experiments by Román-Berdiel et al. (1997) showed that model intrusions which spread preferentially in the extension direction associated with the strike-slip motion yield overall patterns similar to those observed for the Cabeza de Araya pluton. This massif compares particularly well to lozenge-shaped model intrusions located between two major faults (Fig. 7c, and Román-Berdiel et al. 1997).

Conclusions

In this paper I show the results of an experimental study on the intrusion of granitic magmas in a brittle crust with different rheologic profiles and under different tectonic regimes. The applicability of these experimental results to natural conditions is illustrated by comparing with some granitic bodies emplaced in different geological settings. The more important aspects of the experimental study are as follows.

(1) Laccolith formation requires the occurrence of a soft layer or a potential décollement level, between two competent units. The critical thickness of the soft layer necessary for laccolith formation decreases with increasing depth (overburden thickness) (Román-Berdiel et al. 1995).

(2) When there are two soft layers in the brittle crust, laccolith emplacement occurs in the deeper soft level, even when this is thin and the overburden does reach its critical thickness. Strain produced by magma intrusion in country rock is absorbed by a nearly overlying soft level.

(3) Extensional regimes produce asymmetric laccoliths with offset of the intrusion with respect to the feeding pipe, and with asymmetric spreading of the intrusion in the soft level. Moreover, intrusions control the position of normal faults developed in the overburden. Different boundary conditions result in different features on the geometry of intrusions, as elongation of intrusions in the direction of grabens, flat or convex roofs, or intrusions affected by normal faults in vertical section.

(4) Intrusions into strike-slip zones are elongated as a result of preferential lateral expansion in the principal extension direction determined by the bulk strain field. Faults formed in the overburden locally limit the lateral expansion of intrusions, leading to local parallelism between faults and intrusion boundaries. The strike-slip environment allowed local rising of the injected fluid along faults formed in the overburden (Román-Berdiel et al. 1997).

I thank J. J. Kermarrec (Géosciences Rennes, France) who made and maintained the experimental apparatus. This paper has benefited greatly from discussion with A. Casas, to whom I am grateful. A. Castro, C. Talbot and K. Benn are thanked for their constructive reviews. This work forms part of the research project PB96-1452-C03-03 (DGICYT).

References

ARANGUREN, A. & TUBÍA, J. M. 1992. Structural evidence for the relationship between thrusts, extensional faults and granite intrusions in the Variscan belt of Galicia (Spain). *Journal of Structural Geology*, **14**, 1229–1237.

——, LARREA, F. J., CARRACEDO, M., CUEVAS, J. & TUBÍA, J. M. 1997. The Los Pedroches Batholith (Southern Spain): Polyphase interplay between shear zones in transtension and setting of granites. *In*: BOUCHEZ, J. L., HUTTON, D. H. W. & STEPHENS, W. E. (eds) *Granite: From Segregation of Melt to Emplacement Fabrics*. Kluwer Academic Publishers, Dordrecht, 215–229.

AUDREN, C. 1987. Evolution structurale de la Bretagne méridionale au Paléozoïque. *Mémoires de la Societé Géologique et Minéralogique de Bretagne*, **31**, 365.

BASILE, C. 1990. Analyse structurale et modélisation analogique d'une marge transformante. Exemple de la marge de Cote d'Ivoire-Ghana. *Memoires et Documents du Centre Armoricain d'Etude Structurale des Socles*, **39**, 220.

BENN, K., ODONNE, F. & SAINT BLANQUAT, M. 1998. Pluton emplacement during transpression in brittle crust: New views from analogue experiments. *Geology*, **26**, 12, 1079–1082.

——, ROEST, W. R., ROCHETTE, P., EVANS, N. G. & PIGNOTTA, G. S. 1999. Geophysical and structural signatures of syntectonic batholith construction: the South Mountain Batholith, Meguma Terrane, Nova Scotia. *Geophysical Journal International*, **136**, 144–158.

BOUCHEZ, J. L., GUILLET, P. & CHEVALIER, F. 1981. Structures d'écoulement liées à la mise en place du granite de Guérande (Loire-Atlantique, France).

Bulletin de la Société géologique de France, **7**, 387–399.

BROWN, M. & SOLAR, G. S. 1998. Granite ascent and emplacement during contractional deformation in convergent orogens. *Journal of Structural Geology*, **20**, 1365–1394.

BRUN, J. P. 1981. *Instabilités gravitaires et déformation de la croûte continentale. Aplication au développement des dômes et des plutons.* Thèse, Université de Rennes I.

—— & PONS, J. 1981. Strain patterns of pluton emplacement in a crust undergoing non-coaxial deformation, Sierra Morena, Southern Spain. *Journal of Structural Geology*, **3**, 219–229.

——, GAPAIS, D., COGNÉ, J. P., LEDRU, P. & VIGNERESSE, J. L. 1990. The Flamanville granite (Northwest France): an unequivocal example of a syntectonically expanding pluton. *Geological Journal*, **25**, 271–286.

CASTRO, A. 1986. Structural pattern and ascent model in the Central Extremadura batholith, Hercynian belt, Spain. *Journal of Structural Geology*, **8**, 633–645.

—— 1987. On granitoid emplacement and related structures. A review. *Geologische Rundschau*, **76**, 101–124.

—— & FERNÁNDEZ, C. 1998. Granite intrusion by externally induced growth and deformation of the magma reservoir, the example of the Plasenzuela pluton, Spain. *Journal of Structural Geology*, **20**, 1219–1228.

CLEMENS, J. D. & MAWER, C. K. 1992. Granitic magma transport by fracture propagation. *Tectonophysics*, **204**, 339–360.

—— 1998. Observations on the origins and ascent mechanisms of granitic magmas. *Journal of the Geological Society, London*, **155**, 843–851.

CORRIVEAU, L., RIVARD, B. & BREEMEN, O. V. 1998. Rheological controls on Grenvillian intrusive suites: implications for tectonic analysis. *Journal of Structural Geology*, **20**, 1191–1204.

CRUDEN, A. R. 1998. On the emplacement of tabular granites. *Journal of the Geological Society, London*, **155**, 853–862.

DAVY, P. & COBBOLD, P. R. 1991. Experiments on shortening of 4-layer model of continental lithosphere. *Tectonophysics*, **188**, 1–25.

DEHLS, J. F., CRUDEN, A. R. & VIGNERESSE, J. L. 1998. Fracture control of late Archean pluton emplacement in the northern Slave Province, Canada. *Journal of Structural Geology*, **20**, 1145–1154.

DIXON, J. M. & SIMPSON, D. G. 1987. Centrifuge modelling of laccolith intrusion. *Journal of Structural Geology*, **9**, 87–103.

FAUGÈRE, E. & BRUN, J. P. 1984. Modélisation expérimentale de la distension continentale. *Comptes Rendus de l'Academie des Sciences, Paris*, **229**, 365–370.

FRANCE-LANORD, C. & LE-FORT, P. 1988. Crustal melting and granite genesis during the Himalayan collision orogenesis. *Transactions of the Royal Society of Edinburgh: Earth Sciences*, **79**, 183–195.

GAPAIS, D., LAGARDE, J. L., LECORRE, C., AUDREN, C., JEGOUZO, P., CASAS-SAINZ, A. & VAN-DEN-DRIESSCHE, J. 1993. La zone de cisaillement de Quiberon: témoin d'extension de la chaîne varisque en Bretagne méridionale au Carbonifère. *Comptes Rendus de l'Academie des Sciences, Paris*, **316**, 1123–1129.

GILBERT, G. K. 1877. *Report on the Geology of the Henry Mountains, Utah.* US Geographical and Geological Surveys, Rocky Mountain Region, **170**.

GLAZNER, A. F. 1991. Plutonism, oblique subduction, and continental growth: An example from the Mesozoic of California. *Geology*, **19**, 784–786.

GUILLOT, S., PECHER, A., ROCHETTE, P. & FORT, P. L. 1993. The emplacement of the Manaslu granite of Central Nepal: field and magnetic susceptibility constraints. *In*: TREOLAR, P. J. & SEARLE, M. P. (eds) *Himalayan Tectonics.* Geological Society, London, Special Publications, **74**, 413–428.

GUINEBERTEAU, B., BOUCHEZ, J. L. & VIGNERESSE, J. L. 1987. The Mortagne granite pluton (France) emplaced by pull-apart along a shear zone: structural and gravimetric arguments and regional implication. *Geological Society of America Bulletin*, **99**, 763–770.

HAMILTON, W. 1981. Crustal evolution by arc magmatism. *Philosophical Transactions of the Royal Society, London*, **302**, 279–291.

HERREN, E. 1987. Zanskar shear zone: Northeast-southwest extension within the Higher Himalayas (Ladakh, India). *Geology*, **15**, 409–413.

HOGAN, J. P. & GILBERT, M. C. 1995. The A-type Mount Scott Granite sheet: Importance of crustal magma traps. *Journal of Geophysical Research*, **100**, B8, 15779–15792.

HOGAN, J. P., PRICE, J. D. & GILBERT, M. C. 1998. Magma traps and driving pressure: consequences for pluton shape and emplacement in an extensional regime. *Journal of Structural Geology*, **20**, 155–1168.

HOWE, E. 1901. Experiments illustrating intrusion and erosion. *United States Geological Survey 21st Annual Report*, 291–303.

HUBBERT, M. K. 1937. Theory of scale models as applied to the study of geologic structures. *Geological Society of America Bulletin*, **48**, 1459.

HURLBUT, C. S. & GRIGGS, D. T. 1939. Igneous rocks of the Highwood Mountains, Montana-I. The laccoliths. *Geological Society of America Bulletin*, **50**, 1043–1112.

HUTTON, D. H. W. 1988. Granite emplacement mechanisms and tectonic controls: inferences from deformation studies. *Transactions of the Royal Society of Edinburgh: Earth Sciences*, **79**, 245–255.

——, DEMPSTER, T. J., BROWN, P. E. & BECKER, S. D. 1990. A new mechanism of granite emplacement: intrusion in active extensional shear zones. *Nature*, **343**, 452–455.

JOHNSON, A. M. & POLLARD, D. D. 1973. Mechanics of growth of some laccolithic intrusions in the Henry Mountains, Utah, I. Field observations, Gilbert's model, physical properties and flow of the magma. *Tectonophysics*, **18**, 261–309.

LACROIX, S., SAWYER, E. W. & CHOWN, E. H. 1998. Pluton emplacement within an extensional transfer zone during dextral strike-slip faulting: an example from the late Archaean Abitibi Greenstone Belt. *Journal of Structural Geology*, **20**, 43–59.

LAGARDE, J. L., AIT-OMAR, S. & RODDAZ, B. 1990a. Structural characteristics of granitic plutons emplaced during weak regional deformation: examples from late Carboniferous plutons, Morocco. *Journal of Structural Geology*, **12**, 805–821.

——, BRUN, J. P. & GAPAIS, D. 1990b. Formation des plutons granitiques par injection et expansion latérale dans leur site de mise en place: une alternative au diapirisme en domaine épizonal. *Comptes Rendus de l'Academie des Sciences de Paris*, **310**, 1109–1114.

LE FORT, P., CUNEY, M., DENIEL, C., FRANCE-LANORD, C., SHEPPARD, S. M. F., UPRETI, B. N. & VIDAL, P. 1987. Crustal generation of the Himalayan leucogranites. *Tectonophysics*, **134**, 39–57.

MANDL, G., DE-JONG, L. N. J. & MALTHA, A. 1977. Shear zones in granular material. *Rock Mechanics, Wien*, **9**, 95–144.

McCARTHY, G. R. 1925. Some facts and theories concerning laccoliths. *Journal of Geology*, **33**, 1–18.

MERLE, O. & VENDEVILLE, B. 1992. Modélisation analogique de chevauchements induits par des intrusions magmatiques. *Comptes Rendus de l'Academie des Sciences de Paris*, **315**, 1541–1547.

—— & —— 1995. Experimental modelling of thin-skinned shortening around magmatic intrusions. *Bulletin of Volcanology*, **57**, 33–43.

MUDGE, M. R. 1968. Depth control of some concordant intrusions. *Geological Society of America Bulletin*, **79**, 315–332.

NALPAS, T. & BRUN, J.-P. 1993. Salt flow and diapirism related to extension at crustal scale. *Tectonophysics*, **228**, 349–362.

PATERSON, S. R. & FOWLER, T. K. 1993a. Re-examining pluton emplacement processes. *Journal of Structural Geology*, **15**, 191–206.

—— & —— 1993b. Extensional pluton-emplacement models: Do they work for large plutonic complexes? *Geology*, **21**, 781–784.

—— & MILLER, R. B. 1998. Mid-crustal magmatic sheets in the Cascades Mountains, Washington: implications for magma ascent. *Journal of Structural Geology*, **20**, 1345–1364.

—— & TOBISCH, O. T. 1992. Rates of processes in magmatic arcs: implications for the timing and nature of pluton emplacement and wall rock deformation. *Journal of Structural Geology*, **14**, 291–300.

——, BRUDOS, T., FOWLER, K., CARLSON, C., BISHOP, K. & VERNON, R. H. 1991. Papoose Flat pluton: forceful expansion or postemplacement deformation? *Geology*, **19**, 324–327.

PÊCHER, A. 1991. The contact between the Higher Himalayan crystallines and the Tibetan Sediment-

ary series: Miocene large-scale dextral shearing. *Tectonics*, **10**, 587–589.

PITCHER, W. S. 1992. *The nature and origin of granite*. Blackie Academic & Professional, London.

POLLARD, D. D. & JOHNSON, A. M. 1973. Mechanics of growth of some laccolithic intrusions in the Henry Mountains, UTAH, II. Bending and failure of overburden layers and sill formation. *Tectonophysics*, **18**, 311–354.

PONS, J., OUDIN, C. & VALERO, J. 1992. Kinematics of large syn-orogenic intrusions: example of the Lower Proterozoic Saraya batholith (Eastern Senegal). *Geologische Rundschau*, **81**, 473–486.

RAMBERG, H. 1981. *Gravity, Deformation, and the Earth's Crust in Theory, Experiments and Geological Applications*. Academic Press, New York.

RICHARD, P., LOYO, B. & COBBOLD, P. 1989. Formation simultanée de failles et de plis au-dessus d'un décrochement de socle: modélisation expérimentale. *Comptes Rendus de l' Academie des Sciences de Paris*, **309**, 1061–1066.

RICHARD, P., MOCQUET, B. & COBBOLD, P. R. 1991. Experiments on simultaneous faulting and folding above a basement wrench fault. *Tectonophysics*, **188**, 133–141.

RICHER, C. 1994. *Tectonique extensive dans la Chaîne Varisque de Bretagne Méridionale (Pays de Retz, St-Brévin-les-Pins)*. D.E.A., Univ. Rennes 1.

ROIG, J.-Y., FAURE, M. & TRUFFERT, C. 1998. Folding and granite emplacement inferred from structural, strain, TEM and gravimetric analyses: the case study of the Tulle antiform, SW French Massif Central. *Journal of Structural Geology*, **20**, 1169–1190.

ROMÁN-BERDIEL, T., BRUN, J. P. & GAPAIS, D. 1995. Analogue models of laccolith formation. *Journal of Structural Geology*, **17**, 1337–1346.

——, —— & —— 1997. Granite intrusion along strike-slip zones in experiment and nature. *American Journal of Science*, **297**, 651–678.

SCHMIDT, C. J., SMEDES, H. W. and O'NEILL, J. M. 1990. Syncompressional emplacement of the Boulder and Tobacco Root batholiths along old fault zones (Montana-USA) by pull-apart. *Geological Journal*, **25**, 305–318.

SCHULTZ-ELA, D. D., JACKSON, M. P. A. & VENDEVILLE, B. C. 1993. Mechanics of active salt diapirism. *Tectonophysics*, **228**, 275–312.

SHACKLETON, R. M., RIES, A. C. & COWARD, M. P. 1982. An interpretation of the Variscan structures in SW England. *Journal of the Geological Society, London*, **139**, 533–541.

SORGENFREI, T. 1971. On the granite problem and the similarity of salt and granite structures. *Forhandlingar Geoliscka Foreningens, Stockholm*, **93**, 371–435.

SPERA, F. J. 1980. Aspects of magma transport, *In*: HARGRAVES, R. B. (ed.) *Physics of Magmatic Processes*. Princeton University Press, NY, 265–314.

TALBOT, C. J. 1993. Spreading of salt structures in the Gulf of Mexico. *Tectonophysics*, **228**, 151–166.

TIKOFF, B. & TEYSSIER, C. 1992. Crustal-scale, en echelon 'P-shear' tensional bridges: A possible solution to the batholithic room problem. *Geology*, **20**, 927–930.

TOBISCH, O. T., SALEEBY, J. B. & FISKE, R. S. 1986. Structural history of continental volcanic arc rocks, eastern Sierra Nevada, California: A case for extensional tectonics. *Tectonics*, **5**, 65–94.

TRON, V. & BRUN, J.-P. 1991. Experiments on oblique rifting in brittle–ductile systems. *Tectonophysics*, **188**, 71–84.

VENDEVILLE, B. 1987. Champs de failles et tectonique en extension: Modélisation expérimentale. Chapitre I: Méthodologie. Thèse de Doctorat de l'Université de Rennes I. *Memoires et Documents du Centre Armoricain d'Etude Structurale des Socles*, **15**, 5–17.

——, COBBOLD, P. R., DAVY, P., BRUN, J. P. & CHOUKROUNE, P. 1987. Physical models of extensional tectonics at various scales. *In*: COWARD, M. P., DEWEY, J. F. & HANCOCK, P. L. (eds) *Continental Extensional Tectonics*. Geological Society, London, Special Publications, **28**, 95–107.

VERNON, R. H. 1989. Evidence of syndeformational contact metamorphism from porphyroblast-matrix microstructural relationships. *Tectonophysics*, **158**, 113–126.

VIGNERESSE, J. L. 1988. Géophysique appliquée aux granitoïdes. *In*: CREGU, (ed.) *Uranium et granitoïdes*. Géologie et Géochimie de l'Uranium, Memoires Nancy.

—— 1995. Far- and near-field deformation and granite emplacement. *Geodinamica Acta*, **8**, 211–227.

A multidisciplinary approach combining geochemical, gravity and structural data: implications for pluton emplacement and zonation

L. HECHT[1,2] & J. L. VIGNERESSE[2]

[1]*Lehrstuhl für Angewandte Mineralogie und Geochemie, Technische Universität München, Lichtenbergstr. 4, D-85747 Garching, Germany (e-mail: lutz.hecht@geo.tum.de)*
[2]*CREGU, UMR 7566 G2, BP 23, F-54501 Vandoeuvre, France*

Abstract: Granitic pluton emplacement and zonation are controlled, among others, by regional deformation and the rate of magma supply. The latter has consequences for the disposition of successively emplaced, more chemically evolved, batches of magma. Our general interpretation is based on a multidisciplinary approach combining field observations, gravity data, internal structures and geochemical variations. Magma feeders are identified in the plutons as the deepest zones, inferred from gravity measurements, when they also correspond to vertical lineations. Correlation of the root zone location with compositional zoning indicate how the magma evolved during emplacement. Two case studies of Hercynian granite plutons illustrate the interpretations: the normally zoned Cabeza de Araya pluton (Spain), and the multiphase Fichtelgebirge pluton (Germany) which displays both normal and reverse zoning. It is proposed that reverse zoning reflects discontinuous magma injection due to a tectonic rate slower than the rate of magma supply. Conversely, normal zoning can occur when magma injection is continuous in time, with successive magma batches entering within not yet crystallized magma. The two case studies illustrate how the understanding of compositional zoning and emplacement of granitic plutons can be improved by multidisciplinary approaches combining classical and modern techniques.

Granitic intrusions represent a large part of the continental crust and occur in numerous tectonic environments (Pitcher 1993). These environments are sometimes addressed by geochemical diagrams, which attempt to elucidate the respective role of temperature, degree of melting, chemical composition of the source in terms of tectonic settings (Pearce *et al.* 1984). Other studies have focused on the relation between large scale structures and granite generation (Pitcher 1993). Such approaches are commonly carried out at large scale over several intrusions. Bulk trends provide information reflecting the relationships between granite formation and crustal deformation. To infer small-scale variations within a single massif, more detailed surveys are required. Geological observations commonly describe the structure, mineralogy and composition of outcrops of rock. Geophysical measurements help to determine buried structures by addressing a physical parameter that directly depends on the shape or on the volume of the geological object. Both disciplines are complementary and should therefore be combined in simultaneous studies.

Two case studies of Hercynian plutons, in which geophysical data are correlated with structural measurements and with geochemical data, are presented. The two examples show how data may be interpreted at different scales and how broader interpretations may be developed when different techniques are correlated. This paper aims to demonstrate how multidisciplinary approaches can improve our understanding of the mechanism of both compositional zoning and emplacement of granitic plutons.

The Cabeza de Araya granitic pluton

Geological setting

The geology of the granitic pluton of Cabeza de Araya, Spain, has recently been compiled and reviewed by Vigneresse & Bouchez (1997), based on original petrological and geochemical (Corretgé *et al.* 1985; Perez del Villar 1988), structural (Castro 1985; 1986; Amice & Bouchez 1989) and gravity studies (Audrain *et al.* 1989).

This late Hercynian (*c.* 300 Ma) pluton, with its sigmoidal shape is about 70 by 18 km, and intrudes late Precambrian mica schists.

From: CASTRO, A., FERNÁNDEZ, C. & VIGNERESSE, J. L. (eds) *Understanding Granites: Integrating New and Classical Techniques.* Geological Society, London, Special Publications, **168**, 95–110. 1-86239-058-4/99/$15.00 © The Geological Society of London 1999.

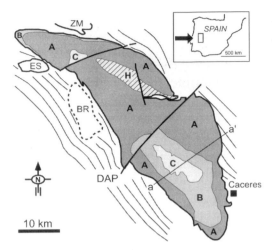

Fig 1. Geological sketch map of the granitic massif of Cabeza de Araya, Spain. Satellite granite bodies are Zarza la Major (ZM), Estorninos (ES), and Brozas (BR). The Cretaceous dike Plasencia-Alentejo that crosscuts throughout Spain is indicated (DAP). Cleavage trajectories in the country rocks are also represented (redrawn from Vigneresse & Bouchez 1997).

The country rocks, principally sandstones and schists, are folded with vertical, N130° striking axial planes, parallel to the regional schistosity. A second schistosity, with the same general trend, but locally wrapping around the granitic intrusions, has been dated as Late Carboniferous and is synchronous with the emplacement of the plutonic massif (Castro 1985, 1986). The massif displays a normal petrographic zoning from periphery to centre, i.e. with decreasing iron content. Three more-or-less concentric facies can be mapped (Fig. 1): type A is a porphyritic granite, mostly found at the periphery of the pluton; type B is a coarse-grained, two-mica granite, with biotite being dominant; and type C is a leucocratic, fine-grained two-mica granite, located mainly in the core of the pluton. A heterogeneous type (H), formed by mixing of type C and B, occurs in the northern part of the pluton. All facies are peraluminous.

Gravity data

To achieve the required resolution, 760 gravity stations were recorded in the massif and its immediate surroundings (Audrain et al. 1989). The close isovalue contour lines of the Bouguer anomaly map define a steep inward gradient, indicating that the granite body is bounded by inward dipping surfaces. Two regions of pronounced minima with relative amplitudes of −11 mgal are present northwest and southeast of the pluton. They are not connected to each other. The densities of 72 rock samples were measured from all granite types and the surrounding schists. A 3D gravity inverse modelling technique (Vigneresse 1990; Améglio et al. 1997) was used, adapted to cases in which surface data can constrain the geometry and distribution of the sources to the gravity field. Based on the variations in depth, obtained by using different density contrasts, it is estimated that the total uncertainty on the calculated depths does not exceed 20% (Améglio et al. 1997). This constrains the density contrast used, and consequently the depth values presented. Because of the good coverage of gravity measurements, the overall shape of the pluton is certainly better evaluated and constrained than the depth values themselves.

The depth map of the pluton's floor (Fig. 2a) indicate that the pluton is relatively thin, since 75% of its volume lies within 4.5 km below the present surface. Two deeper regions roughly represent the gravity lows on the gravity map. Their presence, however, does not directly depend on the choice of the density contrasts used in the model (Vigneresse & Bouchez 1997).

Structural data

Field observations were collected and incorporated with measurements on cored samples, resulting in 230 stations of measurement, regularly spaced throughout the pluton. At each cored site, samples were measured for the anisotropy of magnetic susceptibility. This yields the bulk magnetic susceptibility, the values and orientations of magnetic anisotropy tensor. The long axis of the magnetic ellipsoid (K_{max}) indicates the magnetic lineation, and the short axis (K_{min}) is normal to the magnetic foliation plane. Both usually mimic the lineation and foliation recorded by the average orientation of biotite (Bouchez 1997), the main iron-bearing silicate in this granite, but also of the K-feldspars megacrysts which characterize facies types A and B. From this data set, a map of the foliation planes and lineations was compiled (Vigneresse & Bouchez 1997). In the Cabeza de Araya pluton, foliation planes present directions generally oriented around N139°. Most foliation planes have a steep dip, particularly along the periphery of the pluton. The steepest dips are recorded near the border of subtype C, in the south. Carried along the foliation planes, the lineations present two main groups of

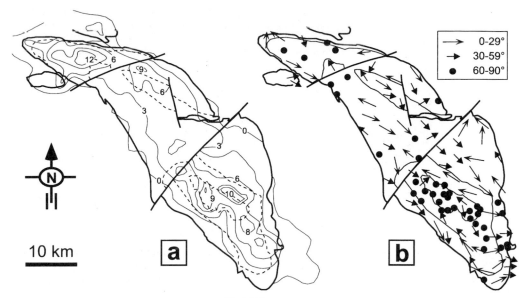

Fig. 2. Internal structures of the granitic massif of Cabeza de Araya, Spain. (**a**) Depth contours map obtained from gravity data inversion. Dashed lines show granite facies boundaries. (**b**) Magnetic lineation map showing NW–SE-trending lineations with low plunges in facies A (and partly B) and mainly subvertical plunges within facies C. Modified and redrawn from Vigneresse & Bouchez (1997). The heavy lines (late faults and Cretaceous dyke) are included, though they have no importance for emplacement.

orientation (Fig. 2b). One with steep plunges (>60°), whereas the other group is sub-horizontal (<30°). Intermediate values are restricted to regions of interference between the groups. Subvertical lineations are mostly restricted to subtype C (Fig. 2b) in the southern part of the pluton. They also form clusters in the north, within subtypes C and B. The subset with low plunges occurs within subtypes A and B. The mean azimuth of the subhorizontal lineations, at about N137°, is similar to the trend displayed by foliation planes, and is parallel to the elongation of the pluton in plan view. Some lineations, oblique to the A–B contact, however, suggest a discontinuity in magma flow between these facies.

Geochemistry

The massif has a similar bulk composition as other Variscan granites of the region and is formed by crustal anatexis (Corretgé *et al.* 1985). Whole-rock analysis with all major and 19 trace elements were obtained with an average spacing of about 2 km between samples (Perez del Villar 1988). Most analyses were performed on the southern part of the pluton where facies A, B and C are well separated in the field. The southern part was sampled in a regular grid whereas the northern part was sampled only locally. Thus,

only data from the southern part of the pluton are presented in this paper.

According to Corretgé *et al.* (1985) and Perez del Villar (1988), the facies A to C show a successive magmatic differentiation pattern indicated by decreasing contents of $Fe_2O_3^T$, MgO, TiO_2, Zr, Th, Ba and Sr, and increasing contents of Rb, Li and Sn. Since facies A, B, and C show considerable chemical overlap (Fig. 3), however, they probably do not represent simple *in situ* fractionation of a common batch of magma. This is particularly evident from a Th versus $Fe_2O_3^T$ diagram (Fig. 3), which indicates a separate fractionation path for facies A and B + C. Nevertheless it is possible that all three facies were derived from a common source at depth.

Shape at depth and emplacement

The major structural features of the pluton can be summarized as follows. The average azimuth of the magmatic lineation (N130°) is parallel to the bulk elongation of the pluton. Two zones of steeply plunging lineations are observed, mainly in the southern part of the massif, but also less distinctively in the northern part. In the majority of its surface area, the massif is rather thin (75% in volume <4.5 km), but two restricted zones are deeper than 6 km from the present surface, and there, the pluton walls dip inward at 30° to 70°

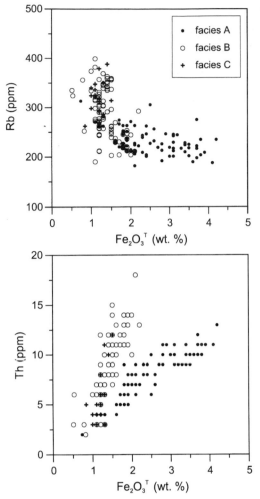

Fig. 3. Geochemical data from the massif of Cabeza de Araya showing the evolution of the magma. In the first diagram (Rb vesus $Fe_2O_3^T$) data overlap, which obliterates specific evolution trends. However, on a Th vesus $Fe_2O_3^T$ diagram, a separate fractionation path appears for facies A compared to facies B + C. Nevertheless it cannot be ruled out that all three facies are derived from a common source at depth. Because of analytical procedures and transcription (Perez del Villar 1988), Th values are given without decimals, but this does not alter interpretations.

The major zone of subvertical lineations, in the south, correlates with the deepest part of the pluton floor (compare Fig. 2a and b) and with the more differentiated subtype C (compare Figs 1, 2a and b). In the north a correlation between lineation plunge and depth of the floor also exists, although less clearly expressed. No specific correlation with petrography is observed. Regions of deep floor and vertical lineations are

interpreted as root zones for the magma. They are represented at the surface as outcrops of late facies.

In the northern half of the pluton, the northeastern concave contact has a dip of 70°, whereas the southwestern wall has a much shallower dip of 30° (Fig. 2). The same pattern is valid for the southern root zone where the concave southeastern wall is very steep, in contrast with the gentle dip of the facing wall. This change in wall dip is interpreted as the level of brittle–ductile rheological transition of the crust (Améglio *et al.* 1997). The root zones themselves, located below the latter transition, correspond to an abrupt increase in dips of their walls, up to 70° and more, at depths larger than 6 km. They become a small conduit with a diameter of 4–5 km at a depth of more than 8 km. The asymmetry of the root reflects a double gash opening system (Vigneresse & Bouchez 1997), similar as the Donegal granite proposed by Hutton (1982).

Compositional zoning in relation to pluton shape

The compositional zoning in relation to the pluton shape will be discussed with reference to profiles. In both case studies, the lines of the section were selected to cut the root zone and all major facies of the granite, where possible. Data from two sectors of about 2 km on either side of the section were projected on the profile to delineate the general trend of chemical data with respect to the pluton shape at depth. The chemical variation of the granites along the profiles is represented by $Fe_2O_3^T$ and Rb (Fig. 4).

The profile across the southern part of the Cabeza de Araya pluton shows typically normal zoning with increasing differentiation towards the root zone which is located near the centre of the pluton. The regular increase in Rb contents suggests a more-or-less continuous increase in fractionation from the margin towards the centre of the pluton. Variations in $Fe_2O_3^T$, however, indicate discontinuous zonation between facies A and B. This could reflect either differences in temperature and/or fluid conditions during melting in the source region, or separate fractionation trends. Combining the Rb, Th and total iron data, as well as the distribution of other trace elements (not discussed here), it is concluded that facies A and B fractionated as at least two separate magma batches (compare Figs 3 and 4).

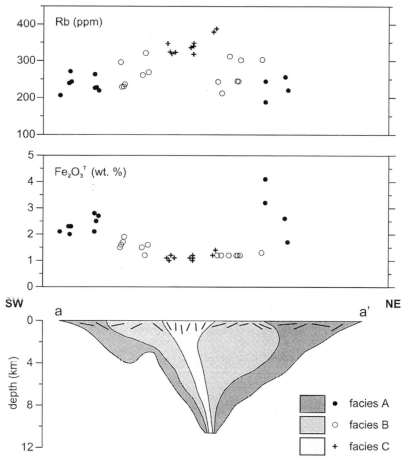

Fig. 4. Cross section (A–A′, see Fig. 1) of the southern part of the Cabeza de Araya massif, obtained from gravity data inversion, and correlated to chemical data (Rb and $Fe_2O_3^T$). The plunging of the magnetic lineation is schematically indicated by several black lines in the upper part of the cross section. While Rb content increases slowly toward the root zone, large variations in content and in trend of $Fe_2O_3^T$ clearly reflect the existence of distinct magmas.

Origin of compositional zoning and mode of emplacement

Corretgé *et al.* (1985) concluded that the normal zoning of the massif of Cabeza de Araya is due to a 'normal orthomagmatic' evolution and all facies are genetically related. Based on structural field observation, a model of emplacement was proposed by Castro (1985, 1986), invoking the filling of a regional extensional gash. This will be retained as a first order approximation. Instead of a single and very long gash, two smaller and unconnected tension gashes are proposed. They extend to depths of between 6 and 13 km and form arcuate shapes. As proposed for the Donegal granite (Hutton 1982), a double gash opening fits the data better. Both gashes opened

in symmetrical positions with respect to the centre of the pluton, as a consequence of regional shear, parallel to the regional trend of foliation in the country rocks. The roots have slightly arcuate shapes, with large radii of curvature of 35 to 50 km compared to their length of 20 km. They have opposite concavities and asymmetrical walls on cross-sections.

As the two gashes opened at the transition from ductile to brittle crust, facies A intrudes the upper crust. Space was progressively provided by the regional tectonics, and guided by the incremental opening along N130°. The equi-granular coarse grained facies (B) followed probably with a short delay compared to the intrusion of facies A. However, facies A was not yet crystallized at the time of emplacement of facies B, as indicated by the gradational contacts, but in general, both facies are structured concordantly. Locally, some obliquity between fabrics across domains also reflects a time lag. With successive opening, a more evolved facies (C) intrudes the former facies in the south. Not enough crystals have formed to reach the stage of a rigid framework (rigid particle threshold, or RPT as defined in Vigneresse *et al.* 1996). Consequently, the new magma does not encounter resistance and intrudes the former facies. Sharp contacts result, reflecting the change in fabrics between facies.

The three facies display the surface pattern of a normal zoned pluton. When correlating surface data with structures at depth, the late facies C is located right above the feeder. It displays vertical lineations indicating that the magma has not been displaced laterally. Its concordant structures at the contact with the older facies also reveal intrusion into an unconsolidated magma. Consequently, the emplacement of a normally zoned pluton can be associated with a succession of chemically evolving magma, coming out from one single root zone, and pushing aside the former batches of magma. This model of magma emplacement differs from those of a more static viewpoint, like the development of normal

zoning by *in situ* assimilation of the surroundings. In our case, magma evolves by fractionation, mostly in the middle part of the crust, followed by a time delayed emplacement reflected by the slight difference in the fractionation trend between each facies. Unfortunately, sufficient precise dating of the three facies is not presently available to confirm our model.

The Fichtelgebirge granite pluton

Geological setting and petrography

The late Hercynian granites of the Fichtelgebirge are located at the northwestern border of the Bohemian Massif (Fig. 5). The country rocks of the granites are mainly composed of early Palaeozoic metasediments (phyllites, micaschists, quartzites, etc.). They underwent low- to medium-temperature and low-pressure metamorphism during the Hercynian orogeny (Mielke *et al.* 1979). In the southwest, the basement rocks of the Fichtelgebirge are bounded against a Mesozoic sequence by the Franconian line, a major southeast dipping fault.

The granites crop out over 300 km^2 in the German part of the Fichtelgebirge and about 50 km^2 in the northeastern part, within the Czech

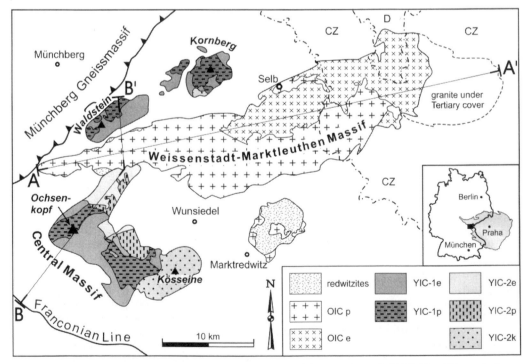

Fig. 5. Geological sketch map of Fichtelgebirge pluton, modified from Hecht *et al.* (1997). The different facies are indicated, as well as the position of the two profiles used in Figs 8 and 9.

Republic (Fig. 5). Based on the petrography, chemistry, and radiometrical (Rb/Sr and K/Ar) age determinations the granites are generally subdivided into an older intrusive complex (OIC, 326 ± 2 Ma) and a younger intrusive complex (YIC, 286–306 Ma). According to Carl & Wendt (1993) the YIC is further subdivided into two intrusive events, hereafter named YIC-1 (305 ± 5 Ma) and YIC-2 (286–289 Ma). The OIC and YIC are both composed of several granite facies defined by Richter & Stettner (1979). In this paper a simplified notation of the granite facies, which reflects the most recent age determinations and the major petrographic characteristics (see also Hecht et al. 1997), is used.

The OIC forms a large elongated east–west trending body called the Weissenstadt–Marktleuthen Massif (Fig. 5). Two major granite types are distinguished, the OICp and OICe. The OICp is a biotite-dominated, medium- to coarse-grained porphyritic granite with large K-feldspar phenocrysts. The western part of the OIC is entirely composed of the OICp which also extends much further to the east (Fig. 5). The OICe is equigranular, fine- to coarse-grained granite types with varying biotite and muscovite contents. There is a transition from fine-grained and more biotite-rich types in the western part (southwest of Sclb) towards medium- to coarse-grained, more muscovite-rich types in the eastern part of the OICe. In the most eastern part of the granite massif (within the Czech Republic) the OICe consists of a nearly pure muscovite granite (Stemprok 1992). Intrusive contacts between the OICp and OICe indicate that the emplacement of the OICe followed the emplacement of the OICp. This is confirmed by inclusions of OICp within OICe.

The YIC crops out in several intrusive bodies, the so-called Waldstein Massif and Kornberg Massif north of the OIC and the Central Massif (including the Kösseine sub-massif), south of the OIC (Fig. 5). The YIC formed during at least two major intrusive events, YIC-1 and YIC-2. Although they differ in age, the granites formed during these two intrusive events are similar, regarding their petrography, geochemistry and spatial distribution, and are thus grouped together as the YIC. The major part of the YIC was formed during the first intrusive event (YIC-1), dated at about 306 Ma (Carl & Wendt 1993). The YIC-1 consists of two-mica granites with two textural varieties. An equigranular type YIC-1e and a porphyritic type YIC-1p, the latter in general located at the margins of the granite massifs. The YIC-1p has a rather fine–grained matrix with feldspar, mica and quartz pheno-

crysts (up to 4 cm). The YIC-1e is medium- to coarse-grained and occasionally contains inclusions of YIC-1p close to the margin or roof of the granite massifs. The granites developed during the second intrusive event, YIC-2, are similar to the granite types of the first intrusive event: an equigranular coarse- to medium-grained type YIC-2e, and a more-or-less porphyritic fine-grained type YIC-2p (marginal facies). Although the petrographic characteristics are very much the same, the YIC-2p and YIC-2e are more fractionated than their equivalents of the YIC-1.

In the Kösseine area at the southeast of the Central Massif, another inhomogeneous group of granite types occurs (YIC-2k). The YIC-2k belongs to the second intrusive event (YIC-2) and is characterized by peraluminous minerals such as garnet, cordierite and silliminate, and abundant enclaves (mainly metasediments) from different sources (Schödlbauer et al. 1997).

Gravity survey and the shape of the pluton

The depth contours of the granites of the German part of the Fichtelgebirge were modelled, based on a detailed gravity survey (Fig. 6a) with more than one measurement per km^2 (Hecht et al. 1997). Density measurements were taken on samples representing all outcroping facies and surrounding rocks (Hecht et al. 1997). The western part of the OIC (Weissenstadt-Marktleuthen Massif) is rather thin with an average depth of less than 2 km. The OIC deepens towards the east to at least 6 km within the German territory (Fig. 6b). The gravity map of the adjacent Czech Republic (Polansky & Skvor 1975; Plaumann 1986) was joined to our gravity map of the German territory. This joined gravity map suggests that the granite floor below the OICe type deepens even more to the east, up to the point of the minimum gravity value (Fig. 6a). In the western part of the OIC, a further deepening of 3 to 4 km occurs (Fig. 6b) which is attributed to the presence of rocks of the underlying YIC (see below).

The main body of the YIC (Central Massif) in the south has a great average thickness of 6 to 7 km (Fig. 6b). The two smaller granite bodies of the YIC, located north of the OIC (Waldstein and Kornberg) have a restricted thickness of less than 2 km. No clear separation between the granite facies is observed in terms of thickness variations. The contours of the Waldstein Massif and Kornberg Massif are separated from the southern unit (Central Massif). It is, however, suggested that the YIC of the Waldstein in the north and the Central Massif in the south are

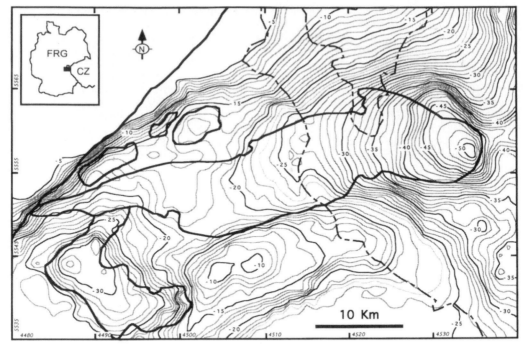

Fig. 6. (a)

connected to each other below the flat OIC. Evidence for this concealed north–south connection of the YIC is given by an anomalous increase of the depth contours of the OIC between the Waldstein and Central Massif (Fig. 6b).

Geochemistry

The granites of the Fichtelgebirge have a large variation in chemical composition: $SiO_2 = 65.5$–76.5 wt%, $Fe_2O_3^T = 0.4$–4.5 wt%, $CaO = 0.06$–2.5 wt%, $Rb = 180$–900 ppm, $Sr = 4$–290 ppm, $Th = 2.3$–48 ppm and $Zr = 10$–500 ppm (see also Fig. 7). Both the OIC and YIC are peraluminous with molar $Al_2O_3/(K_2O + Na_2O + CaO)$ increasing from 1.05 towards 1.4 with increasing fractionation. The most fractionated samples of the YIC are more strongly enriched in some granitophile trace elements (e.g. Rb and F) compared to the OIC (Fig. 7a). The OICp and the YIC differ significantly in their Sr- and Nd-isotope composition. The OICp with $\varepsilon_{Nd(T)} = -3.1$ to -3.8 and $Sr_{ini} = 0.7082$ suggest lower crustal source rocks, or the involvement of some mantle material during magma generation (Siebel *et al.* 1997). The YIC have isotope signatures of $\varepsilon_{Nd(T)} = -5.4$ to -8.4 and $Sr_{ini} = 0.7106$ to 0.7202), typical of late Hercynian peraluminous

granites, which were most probably derived from the partial melting of Proterozoic crust (Siebel *et al.* 1997).

The OICe is more fractionated than the OICp. The variation of many major and trace elements suggest continuous magmatic fractionation from the OICp towards the OICe (Fig. 7a). However the OICe differs from the OICp in the fractionation pattern of some trace elements like Th (Fig. 7a) and the REE (Hecht *et al.* 1997). Furthermore the $\varepsilon_{Nd(T)}$ values of the OICe ($\varepsilon_{Nd(T)}$ between -5 and -6) are significantly lower than those of the OICp ($\varepsilon_{Nd(T)}$ ranging from -3 to -4, data from Holl *et al.* 1989; Siebel *et al.* 1997; and unpublished data). The OICe and OICp most likely represent separate intrusions that differ in their protoliths. Consequently, they are not related by differentiation of a common parental magma.

The YIC-1 displays the same fractionation pattern for the porphyritic and equigranular facies except for the Ochsenkopf area (Fig. 5). At the Ochsenkopf the marginal porphyritic facies YIC-1p is clearly less fractionated than the equigranular facies YIC-1e, located in the central part of the pluton. This holds true for some other single outcrops in the YIC, but to a much weaker extent. It is suggested that the two granite facies of the YIC, the porphyritic and equigranular type, were produced by sidewall crystallization of

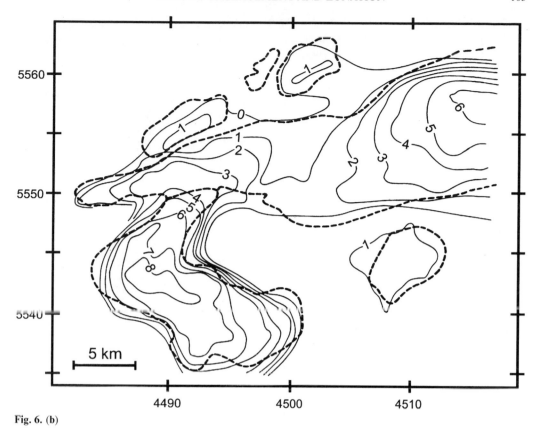

Fig. 6. (b)

Fig. 6(a) and (b). (a) Bouguer anomaly map (with contours in mgal) and (b) depth (in km) contours of the floor of the Fichtelgebirge granitic complex, modified from Hecht *et al.* (1997). A clear dichotomy in the shape at depth of the two units (OIC and YIC) appears in the map of the floor.

a common magma batch (Hecht *et al.* 1997). The YIC-2p and YIC-2e are more fractionated than their equivalent of the first intrusive event (YIC-1), which is evident from a stronger enrichment of granitophile elements, like Rb (Fig. 7b), Li and F. The granites of the Kösseine area (YIC-2K) can be clearly distinguished from the other granites of the YIC (Fig. 7b). The YIC-2K most probably represents a separate intrusion with a distinct genesis characterized by a combination of restite unmixing, contamination and fractional crystallization (Hecht *et al.* 1997; Schödlbauer *et al.* 1997).

Compositional zoning in relation to pluton shape

Two profiles show the compositional zoning of both the OIC and YIC in relation to their shape at depth (Figs 8 and 9). They cross the major

facies and the location of the inferred root zone. The variation in chemical composition of the granite along the profile is shown in the same form as the previous example, by projecting values measured on samples 2 km apart onto the line of profile.

The OIC presents a rather simple compositional zoning with increasing degree of fractionation eastwards in the direction of the root zone (Profile A–A', Fig. 8). Although this simple zoning is disturbed by the eastern- and western-most samples of the OICp, the general increase in the degree of fractionation towards the root zone also holds true for the OICp. This is reflected in mineralogical differences. From east to west the muscovite content of the OICp increases from about 0.5 to approximately 3 vol%. The overall increasing degree of fractionation of the OIC towards the root zone reflects a normal zoning.

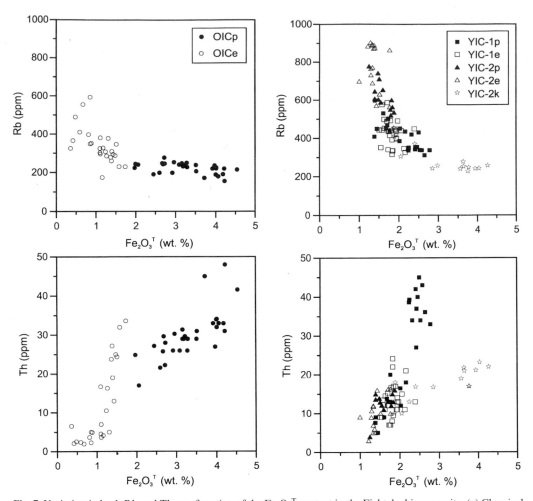

Fig. 7. Variation in both Rb and Th as a function of the $Fe_2O_3^T$ content in the Fichtelgebirge granite. (**a**) Chemical data for the OIC with the unconclusive answer about the existence of internal variations within OIC. (**b**) Conversely, YIC shows a opposite trend in fractionation (Rb versus $Fe_2O_3^T$) and a different trend on the diagram of Th versus total iron.

Only one root zone can be identified for the YIC, but its compositional zoning is much more complex compared to the OIC (Profile B–B', Fig. 9). The YIC-1p and YIC-1e make up the main part of the YIC and occur in all its individual granite bodies (Fig. 5). In contrast to the OIC, the two facies of the YIC-1, the porphyritic YIC-1p and the equigranular YIC-1e are not related by a simple zonation pattern. Two compositional trends are evident, (1) a general trend with respect to the whole granite complex, and (2) a local trend:

(1) The general trend is characterized by decreasing differentiation of both the YIC-1p and YIC-1e towards the Ochsenkopf area in the

Central Massif (Figs 5 and 9) close to the root zone. The compositional variation is relatively small (especially in YIC-1e) compared to the OIC (Fig. 7). Although this zonation pattern is in the reverse sense compared to the OIC, it is also asymmetric with respect to the surface outline of the granite complex.

(2) The second trend indicates more local compositional zoning between the YIC-1p and YIC-1e. Because of the marginal or roof position of the YIC-1p with respect to the core position of the YIC-1e, a zonation pattern of increasing differentiation inwards is indicated for single granite bodies or parts of them. With increasing distance from the inferred root zone, e.g. from

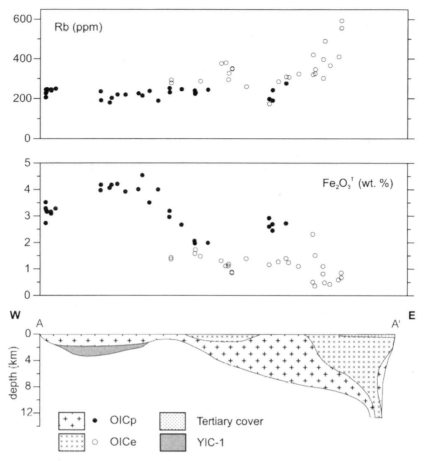

Fig. 8. Cross section (A–A', see Fig. 5) of the OIC of the Fichtelgebirge pluton, obtained from gravity data inversion. A bulk increase in fractionation towards the root zone is evident from the Rb and Th variations, reflecting a normal zoning.

the Ochsenkopf towards the Waldstein (Figs 5 and 9) the compositional difference between the YIC-1p and YIC-1e decreases or is largely blurred. The YIC-2e and YIC-2p, which comprise the most differentiated granites of the YIC, occur only in the Central Massif, at the north and northeast of the Ochsenkopf area (Fig. 5). Within the YIC this is a more-or-less central position close to the inferred root zone. The spatial occurrence of the YIC-2e and YIC-2p does not fit the continuous zonation pattern of the YIC-1 (Fig. 9). The latter was emplaced discordantly by YIC-2e + YIC-2p, relatively close to the inferred root zone. The YIC-2K are of the least differentiated granites of the YIC-2 and intruded at a greater distance from the root zone, compared to the YIC-2p and YIC-2e (Fig. 5). The bulk zonation pattern of the YIC-2 shows an increasing differentiation towards the

root zone which is the reverse to the YIC-1 but the same as for the OIC (Hecht *et al.* 1997).

Origin of compositional zoning and mode of emplacement

Although a time gap of about 30 Ma exists between the intrusion of an OIC and the YIC, both units are distinct from each other in terms of magma origin, differentiation pattern, shape at depth and petrochemical zonation pattern (Hecht *et al.* 1997). With respect to the root zones, as detected by gravity data inversion and field mapping, the OIC shows normal zoning. In contrast, the YIC displays initially reverse zoning during the first intrusive event (YIC-1) and thereafter normal zoning during the second intrusive event (YIC-2).

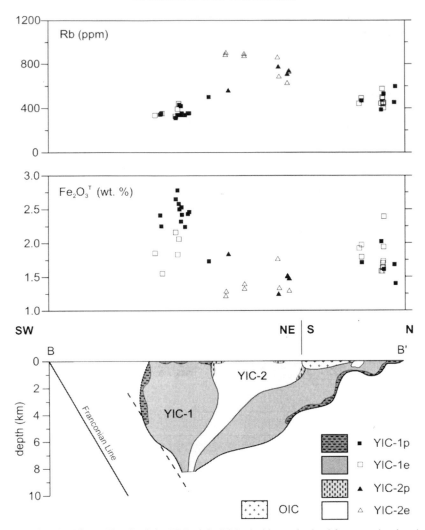

Fig. 9. Cross section (B–B′, see Fig. 5) of the YIC of the Fichtelgebirge, obtained from gravity data inversion on which the compositional zoning appears more complex. Differentiation decreases for YIC-1p and YIC-1e towards the root zone. Conversely, the trend for YIC-2e and YIC-2p is not continuous and shows increasing differentiation towards the root zone.

Normal zoning in both the OIC and YIC-2 was caused by intrusion of successively more fractionated magma batches that remained close to the root zone. *In situ* fractionation by sidewall crystallization during emplacement may be proposed as a possible explanation for the reverse zoning of the YIC-1. Furthermore, this mechanism can also explain the local zonation pattern of the YIC-1, characterized by a less differentiated porphyritic marginal facies (YIC-1p) and a more differentiated equigranular core facies (YIC-1p).

The rheological properties of granite magma and the deformation regime during emplacement are considered to play an important role in the type of compositional zoning that develops during pluton formation (Stephens 1992; Vigneresse 1995; Hecht *et al.* 1997). The complex zonation pattern of the Fichtelgebirge massif suggests that both the rheological properties of magmas and the deformation regime has changed considerably during the approximately 30 Ma that brackets the late Hercynian granitic magmatism. The rather flat shape of the OIC in the west with a strong deepening towards one root in the east suggest that it has been emplaced in a plastic deforming crust (Vigneresse 1995). The schistosity in the surrounding is well developed, which also corresponds to granites emplaced in a plastically deformed crust in the same region (Behrmann & Tanner 1997). In contrast, the YIC displays a deep floor with a less sharply defined root and steep walls that trend at a high angle. On its western side the deepest part of the YIC (>6 km) is controlled by a structure plunging 60–70° to the northeast and trending N160°. This structure parallels the Franconian Line, a major dextral fault which plunges at about 70° northeastward, observed at the surface and determined at depth by both seismic profiles and deep drilling at the KTB site (Durbaum et al. 1993; Harjes et al. 1997). The strong control of the western wall by the Franconian Line suggests that the emplacement of the YIC is probably related to a major stress field approximately oriented north–south, which locally induced a pull-apart system between anastomosing shear zones (Hecht et al. 1997).

Discussion

Pluton zonation is generally interpreted as the result of complex geochemical processes that involve wall rock assimilation, preferential accretion of minerals, fractional crystallization or mechanical interactions between magma batches (Ayrton 1988; Fridrich & Mahood 1984; Allen 1992; Stephens 1992; Brandon & Lambert 1994). Granitic plutons have normal zoning in most common cases, and reverse zoning in some specific cases, without any evidence of control, either by the tectonic, or geochemical setting. In the Fichtelgebirge, for instance, two distinct units are observed, one (OIC) with normal zoning, followed by reverse zoning in the YIC-1 (about 30 Ma later), followed again by a return to normal zoning (YIC-2). On the other hand, differences in the shape of both the units, were also observed. It is therefore suggested that the differences in the shape, zonation and magma evolution may serve as indicators of the interaction between granitic emplacement, magma evolution and tectonic setting.

In the Fichtelgebirge, the OIC unit displays a discrete normal zoning. The succession of magma pulses in time was fast enough so that the magma had no time to crystallize, and thus did not respond to the ambient stress field. In this case, the more evolved and viscous magma was hindered to flow far away from the root zone during emplacement. As a result, the successive batches of magma evolved towards the root zone of the massif. In Cabeza de Araya, no major break exists between the different facies as reflected by their concordant structures, excepted locally between facies A and B. Normal zoning

of the pluton resulted. In the case of a stronger influence by tectonic deformation, as observed during transtension, it is suggested that the zoning of the pluton is related to the rate of magma supply.

Conversely, when the volume of magma supply is small, with breaks in between magma pulses, reverse zonation may develop. In the case of YIC-1, the small volume (120 km² at the outcrop) resulted in a fast crystallization of magma. It thus had time to develop its own strength. Therefore, the later delivered, more evolved magma was emplaced at successively increasing distance from the root zone. This resulted in reverse zoning of YIC-1. The YIC is characterized by two major stages of emplacement. During the second stage (YIC-2) the highly evolved magma has discordantly emplaced the YIC-1, and led to a final, overall normal, zonation of the YIC.

The observed character of zonation in granitic plutons may therefore be related to the magma supply rate and regional tectonics. To test this idea, some data from the Sierra Nevada, where many normal and reverse zoned granitic massifs intruded at variable rates, were examined. When magma is delivered at a fast rate, as in the Tuolumne series and Mount Givens, the zoning is normal (Bateman & Chappell 1979). In both cases the volume of emplaced magma is large (800 and 1400 km² at the surface outcrop), and the time of emplacement (5–6 Ma in both cases) is short (Kistler et al. 1986; Hill et al. 1988). Conversely, in the nearby Lamark granodiorite, Sierra Nevada, the volume is smaller (400 km² at outcrop) and the time of emplacement longer (9 Ma), so we would expect a reverse zonation, although it has been attributed to magma mixing and mingling (Frost & Mahood 1987; Coleman et al. 1992). Reverse zoned massifs are less abundant and often lack detailed ages. However, reversely zoned plutons appear at the Smartville complex, northern California (Beard & Day 1988) which is of very small size (about 25 km² at outcrop).

It is suggested, that in addition to chemical processes that could modify the nature of a magma (differentiation, modification of the parameters of the source), the interaction between magma supply and deformation of the crust can govern bulk zoning in granitic plutons. Depending on the rate at which a certain amount of magma is delivered, and thus cools down, the rate at which the next pulse of magma is emplaced can determine the chemical zonation in granitic plutons. In the case of rapidly delivered magma batches, the former magma has insufficient time to solidify, and thus later pulses of magma are emplaced into the core, and are able to push aside the former batches. When magma batches consist of magma fractionated at depth, the resulting zonation of the pluton is normal, the later intrusive being more felsic towards the core. But, when magma is delivered at a sufficiently reduced rate, to allow it to cool and crystallize before the arrival of the next batch, the latter cannot develop enough force to overcome the strength of the now solid earlier magma batch. New and more evolved magma is thus forced to intrude the border regions of the former intrusion. Thus, reverse zoning can be induced, with the first emplaced more mafic magma in the core, close to the root zone. An intermediate stage occurs when the rate of magma supply is just as fast as the opening rate, or when magma is reheated, thus inducing magma mixing or mingling.

The example of Cabeza de Araya shows a unique zonation pattern, with possibly different source conditions. The Fichtelgebirge example demonstrates, that both types of zonation patterns may develop within a relatively short time period. A combination of different approaches such as petrography, chemistry and geophysics may be necessary to provide the essential information to understand such zonation patterns in plutonic complexes.

Conclusions

In the above mentioned examples, an interactive process resulted from combining geophysical data, namely depth values inferred from gravity measurements, and classical geological or petrological studies, using either structural or geochemical data.

The simultaneous occurrence of greater depth and vertical lineations suggests the existence of a feeder zone at depth from which magma emerged. The orientation of the root zone, the bulk trend of the massif, and general magmatic flow fabrics emphasize the control of granite emplacement by deformation (Bouchez 1997). The correlation between magma evolution and the path of magma from its root zone is more complex. In most cases, granitic plutons show an inward chemical evolution, with the more fractionated magma less able to flow, remaining close to the root. This is confirmed by the large number of normal zoned plutons. In some other cases, however, the reverse petrographic zonation reflects a slow rate of magma supply playing a less significant role than the rate at which magma can be emplaced into the upper crust.

Several approaches for studying granite evolution during emplacement are therefore recommended. Firstly, field studies provide constraints for all further studies. They include mapping of the facies, mineralogical observations, and mapping of magmatic structures. The latter requires mineralogical determination as well as AMS-based measurements (Bouchez 1997). Such study provides a fast determination of the Fe-content through the amplitude of the magnetic susceptibility value. At the same time, it provides valuable information about the structures (foliations and lineations) (see Amice & Bouchez 1989 for the massif of Cabeza de Araya). Contacts between facies are of primary interest. Detailed mapping of the structures (concordant or discordant, flattening or stretching ellipsoid), as well as detailed geochemical investigations (e.g. trace elements, isotopes) are suggested. Both yield important information necessary to distinguish between *in situ* differentiation and multiple magma injections.

Correlation of structural with geophysical data (mainly gravity) assists the determination of the deep structures of a granitic intrusion. In particular, the number, position and orientation of the feeder zones are major parameters for understanding how the magma intruded and formed the massif.

Finally, correlation between the shape at depth and the chemical characteristics of the magma indicates how the magma evolved during its emplacement. This has implications for the way in which petrographical zoning develops in a granitic massif, as shown by the two preceding examples of Cabeza de Araya and Fichtelgebirge.

We are grateful to A. R. Cruden and W. E. Stephens for constructive criticism and improvements to the manuscript. Part of this work was funded by the European Community (MA1-0039-D/BA) and the 'Deutsche Forschungsgemeinschaft' (Mo 232/24).

References

ALLEN, C. M. 1992. A nested diapir model for the reversely zoned Turtle Pluton, southeastern California. *Transactions of the Royal Society of Edinburgh, Earth Science*, **83**, 179–190.

AMÉGLIO, L., VIGNERESSE, J. L. & BOUCHEZ, J. L. 1997. Granite pluton geometry and emplacement mode inferred from combined fabric and gravity data. In: BOUCHEZ, J. L., HUTTON, D. H. W. & STEPHENS, W. E. (eds) *Granite: from segregation of melt to emplacement fabrics*. Kluwer Academic Publishers, Dordrecht, 295–318.

AMICE, M. & BOUCHEZ, J. L. 1989. Susceptibilité magnétique et zonation du batholithe granitique de Cabeza de Araya (Extremadura, Espagne). *Comptes Rendus de l'Académie des Sciences Paris II*, **308**, 1171–1178.

AUDRAIN, J., AMICE, M., VIGNERESSE, J. L. & BOUCHEZ, J. L. 1989. Gravimétrie et géométrie tri-dimensionelle du pluton granitique de Cabeza de Araya (Extrémadure, Espagne). *Comptes Rendus de l'Académie des Sciences Paris II*, **309**, 1757–1764.

AYRTON, S. 1988. The zonation of granitic plutons: the 'failed ring-dyke' hypothesis. *Schweizerische Mineralogische und Petrographische Mitteilungen*, **68**, 1–19.

BATEMAN, P. C. & CHAPPELL, B. W. 1979. Crystal-lization, fractionation, and solidification of the Tuolumne Intrusive Series, Yosemite National Park, California. *Geological Society of America Bulletin*, **90**, 465–482

BEARD, J. S. & DAY, H. D. 1988. Petrology and emplacement of reversely zoned gabbro-diorite plutons in the Smartville Complex, Northern California. *Journal of Petrology*, **29**, 965–995.

BEHRMANN, J. H. & TANNER, D. C. 1997. Carboniferous tectonics of the Variscan basement collage in eastern Bavaria and western Bohemia. *Geologische Rundschau*, **86**, Suppl., S15–S27.

BOUCHEZ, J. L. 1997. Granite is never isotropic. *In*: BOUCHEZ, J, L,, HUTTON, D. H. W. & STEPHENS, W. E. (eds) *Granite: from segregation of melt to emplacement fabrics.* Kluwer Academic Publishers, Dordrecht, 95–112.

BRANDON, A. D. & LAMBERT, R. J. 1994. Crustal melting in the Cordillera interior: The mid-Cretaceous White Creek batholith in the southern Canadian cordillera. *Journal of Petrology*, **35**, 239–269.

CARL C. & WENDT, I. 1993. Radiometrische Datierung der Fichtelgebirgsgranite. *Zeitschrift der Geolo-gischen Wissenschaften*, **21**, 49–72.

CASTRO, A. 1985. The Central Extremadura batholith: geotectonic implications (European Hercynian belt). An outline. *Tectonophysics*, **120**, 57–68.

—— 1986. Structural pattern and ascent model in the central Extremadura batholith, Hercynian belt, Spain. *Journal of Structural Geology*, **8**, 633–645.

COLEMAN, D. S., FROST, T. P. & GLAZNER, A. F. 1992. Evidence from the Lamark granodiorite for rapid Late Cretaceous crust formation in California. *Science*, **258**, 1924–1926.

CORRETGÉ. L. G., BEA, F. & SUAREZ, O. 1985. Las características geoquímicas del batolito de Cabeza de Araya (Cáceres, España): implicaciones petro-genéticas. *Trabajos de Geología de la Universidad de Oviedo*, **15**, 219–238.

DURBAUM, H. J., HIRSCHMANN, G., REICHERT, C., SADOWIAK, P., STILLER, M., WIEDERHOLD, H. & THE DEKORP RESEARCH GROUP 1993. The nature of seismic reflections at the KTB location, Oberpfalz Germany. *Terra Abstract*, **5**, 129–130.

FRIDRICH, C.J. & MAHOOD, G. A. 1984. Reverse zoning in the resurgent intrusions of the Grizzly Peak cauldron, Sawatch Range, Colorado. *Geological Society of America Bulletin*, **95**, 779–787.

FROST, T. P. & MAHOOD, G. A. 1987. Field, chemical, and physical constraints on mafic–felsic magma interaction in the Lamarck granodiorite, Sierra Nevada, California. *Geological Society of America Bulletin*, **99**, 272–291.

HARJES, H. P., BRAM, K., DURBAUM, H. J., GEBRANDE, H., HIRSHCMANN, G., JANIK, M., KLÖCKNER, M., LÜSCHEN, E., RABBEL, W., SIMON, M., THOMAS, R., TORMANN, J. & WENZEL, F. 1997. Origin and nature of crustal reflections: results from inte-grated seismic measurements at the KTB super-deep drilling site. *Journal of Geophysical Research*, **102**, 18267–18288.

HECHT, L., VIGNERESSE, J. L. & MORTEANI, G. 1997. Constraints on the origin of zonation of the granite complexes in the Fichtelgebirge (Germany and Czech Republik): Evidence from a gravity and geochemical study. *Geologische Rundschau*, **86**, Suppl., S93–S109.

HILL, M., O'NEIL, J. R., NOYES, H., FREY, F. A. & WONES, D. R. 1988. Sr, Nd and O isotope variations in compositionally zoned and unzoned plutons in the Central Sierra Nevada batholith. *American Journal of Science*, **288A**, 213–241.

HOLL, P. K., DRACH, VON V., MÜLLER-SOHNIUS, D. & KÖHLER, H. 1989. Caledonian ages in Variscan rocks: Rb–Sr and Sm–Nd isotopic variations in dioritic intrusives from the northwestern Bohe-mian Massif, West Germany. *Tectonophysics*, **157**, 179–194.

HUTTON, D. H. W. 1982. A tectonic model for the emplacement of the main Donegal granite, NW Ireland. *Journal of the Geological Society, London*, **139**, 615–631.

KISTLER, R. W., CHAPPELL, B. W., PECK, D. L. & BATEMAN, P. C. 1986. Isotopic variation in the Tuolumne intrusive suite, central Sierra Nevada, California. *Contributions to Mineralogy and Petrology*, **94**, 205–220.

MIELKE, H. J., BLÜMEL, P. & LANGER, K. 1979. Regional low pressure metamorphism of low and medium grade in metapelites and psammites of the Fichtelgebirge area, NE-Bavaria. *Neues Jahrbuch Mineralogische Abhandlungen*, **137**, 83–112.

PEARCE, J. A., HARRIS, N. B. W. TINDLE, A. G. 1984. Trace element discrimination diagrams for tec-tonic interpretation of granitic rocks. *Journal of Petrology*, **25**, 956–983.

PEREZ DEL VILLAR, L. 1988. El uranio en el batolito de Cabeza de Araya y en el CEG del borde septentrional (Prov. de Cáceres). *Prospección, Geoquímica, Mineralogía y Metalogenia*. Thesis, Universidad de Salamanca.

PITCHER, W. S. 1993. *The nature and origin of granite*. Chapman and Hall, London.

PLAUMANN, S. 1986. Die Schwerekarte der Oberpfalz und ihre Bezüge zu Strukturen der oberen Erdkruste. *Geologisches Jahrbuch*, **E33**, 5–13.

POLANSKY, J. & SKVOR, V. 1975. Strukturnetektonicka problematika severozapadnich. *Cech. Sbornik Geolockyck*, **13**, 47–64.

RICHTER, P. & STETTNER, G. 1979. Geochemische und petrographische Untersuchungen der Fichtelge-birgsgranite. *Geologica Bavarica*, **78**, 1–127.

SCHÖDLBAUER, S., HECHT, L., HÖHNDORF, A. & MORTEANI, G. 1997. Enclaves in the S-type granites of the Kösseine massif (Fichtelgebirge, Germany): implications for the origin of granites. *Geologische Rundschau*, **86**, Suppl., S125–S140.

SIEBEL, W., TRZEBSKI, R., STETTNER, G., HECHT, L., CASTEN, U., HÖHNDORF, A. & MÜLLER, P. 1997. Granitoid magmatism of the NW Bohemian Massif revealed: gravity data, composition, age relations and phase concept. *Geologische Rundschau*, **86**, Suppl., S45–S63.

STEMPROK, M. 1992. The geochemistry of the Czechoslovak part of the Smrciny/Fichtelgebirge granite pluton. *Casopis Pro Mineralogii a Geologii*, **37**, 1–19.

STEPHENS, W. E. 1992. Spatial, compositional and rheological constraints on the origin of zoning in the Criffel pluton, Scotland. *Transactions of the Royal Society of Edinburgh, Earth Science*, **83**, 191–199.

VIGNERESSE, J. L. 1990. Use and misuse of geophysical data to determine the shape at depth of granitic intrusions. *Geological Journal*, **25**, 248–260.

—— 1995. Crustal regime of deformation and ascent of granitic magma. *Tectonophysics*, **249**, 187–202.

—— & BOUCHEZ, J. L. 1997. Successive granitic magma batches during pluton emplacement: the case of Cabeza de Araya (Spain). *Journal of Petrology*, **38**, 1767–1776.

—— BARBEY P. & CUNEY, M. 1996. Rheological transitions during partial melting and crystallization to felsic magma segregation and transfer. *Journal of Petrology*, **37**, 1579–1600.

The Coastal Batholith and other aspects of Andean magmatism in Peru

E. J. COBBING

British Geological Survey, Keyworth, Nottingham NG12 5GG, UK

Abstract: The opening of the South Atlantic ocean at 110 Ma triggered the inversion of the Casma basin and the switch from marine volcanicity to plutonism, which evolved through three distinct phases. The first was the intrusion of the Coastal Batholith which forms a well-defined linear structure over the whole coastal region and which, in the Lima segment, endured from 100 to 60 Ma, and was terminated by the formation of the ring complexes. The second was the post Incaic development of the andesitic terrestrial plateau volcanics, the Calipuy group with associated scattered plutons of tonalite and granodiorite, which extended from perhaps 50 Ma to 20 Ma, and the third was the emplacement of the high level stocks and associated ignimbrite sheets from 20 to 6 Ma. Of these the Cordillera Blanca batholith which is of trondhjemitic affinity is the most important, and it is the intrusives of this zone which are the most important economically. Many plutons within the Batholith were emplaced into the brittle crust by processes of magmatic stoping and some of the evidence for this process is presented.

The Coastal Batholith (Fig. 1) from Pisco as far north as Trujillo is hosted by the volcanogenic deposits of the Lower Cretaceous Huarmey–Canete basin, which are mainly of Albian age. Similar deposits of Jurassic to Cretaceous age provide a high proportion of the envelope from Lima southwards as far as Pisco and also occur at intervals along the entire Andean margin. From Trujillo northwards the batholithic line crosses the folded Jurassic and Cretaceous sedimentary formations of the Chicama basin and continues northwards through the Mesozoic and Tertiary shelf facies sedimentary rocks and Tertiary volcanic deposits which overlie the Palaeozoic Olmos arch. From Pisco southwards the batholithic intrusions were emplaced within sedimentary and volcanic formations, which are mainly of Mesozoic age and which overlie the metamorphic and igneous rocks of the Arequipa massif. The Coastal Batholith is remarkable because it was emplaced mainly into volcanogenic host rocks affected only by brittle fracture with little indication of any other structural complexity. For this reason it is of great interest for geologists as it provides an opportunity for studying granite emplacement in this tectonic setting. This matter provided one of the themes investigated during the study of the batholith conducted under the leadership of Wallace Pitcher. The themes of that investigation were necessarily wide ranging, including the basic mapping of the plutons and the identification of their geological and temporal characteristics,

their petrology, geochemistry and isotope geology. All of these themes were reported in Pitcher *et al.* (1985). The particular question of emplacement in this environment was conducted mainly by Wallace Pitcher and Andrew Bussell on particular plutons, some of which were either within, or were in close proximity to the ring complexes. The task of the regional mapping of the Batholith fell largely to myself in conjunction with other research students from Liverpool University.

Because of its size and internal complexity the Coastal batholith provides an exceptional range of situations in which the problem of pluton emplacement in this environment can be examined. The remarkable result is that although the component plutons were emplaced over a time span of as much as 40 million years, the mode of emplacement for all the plutons was mainly by magmatic stoping, accompanied in some cases by roof lifting and cauldron subsidence. Most of the structural investigations were conducted in the Lima segment of the Batholith, but the mechanisms identified there are considered to be applicable to the entire batholith and, in some degree to the later plutons further inland.

The Huarmey–Canete Marginal Basin: its formation and connection with batholith emplacement

The volcanic rocks of the Huarmey–Canete basin formed within a rift or pull apart in

From: CASTRO, A., FERNÁNDEZ, C. & VIGNERESSE, J. L. (eds) *Understanding Granites: Integrating New and Classical Techniques.* Geological Society, London, Special Publications, **168**, 111–122. 1-86239-058-4/99/$15.00 © The Geological Society of London 1999.

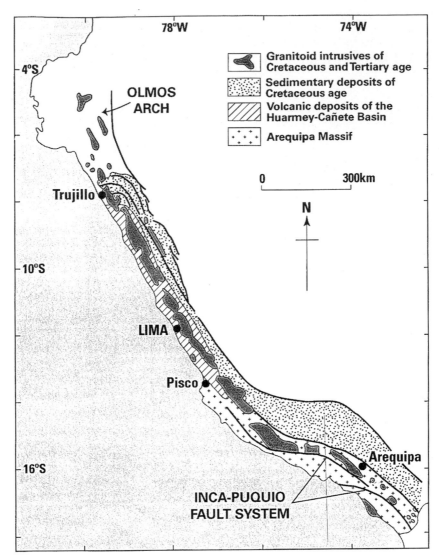

Fig. 1. Distribution of Mesozoic and Tertiary granitoids and of Cretaceous sedimentary and marine volcanic formations.

continental crust along the Andean margin. The basin that developed over this structure was rapidly filled with marine volcanic rocks consisting of pillow lavas, hyaloclastites, volcaniclastics and tuffs, closely associated with dykes and sills, and reached a thickness of greater than 9000 m (Bussell 1975; Myers 1980; Atherton 1990). They form the Casma Group (Cossio 1964). Compositionally, many are high alumina basalts that plot in the tholeiitic and calc-alkaline fields on geochemical discriminant diagrams (McCourt 1978). The basin is compositionally asymmetric with basalts predominating in the west and with andesites, dacites and rhyolites occurring

towards the east. Geophysical evidence (Couch *et al.* 1981) has shown that the basin is underlain by a basic arch of denser material (3.0 g cm^{-3}, $V_p = 6.66$ km s^{-1}), which is presumably a mantle wedge.

The basin developed by processes of incipient spreading-subsidence in which a structure analogous to that of a mid ocean ridge was active in the axial zone, and which contributed the bulk of the material to the basin fill (Atherton 1990). Although it is not known when the basin was initiated, Soler & Bonhomme (1990) have suggested that the initial rifting was synchronous with rift development along the proto Atlantic at

the Africa–South America join in the Neocomian. Rare fossils all provide Albian ages,
but structurally contiguous sequences around
Lima are known to be of Berriasian age.
Volcanic rocks of the Rio Grande Formation
south of Pisco are of Callovian age.

The volcanic deposits were affected by burial
metamorphism in which metamorphic isograds
developed progressively as the thickness of the
basin fill increased. These isograds were disposed
approximately subhorizontally and were of
decreasing grade upwards and towards the
margins forming, in cross section, a metamorphic dome focussed upon the central rift
which provided the heat. Aguirre & Offler (1985)
have pointed out that the Andean margin is
characterized by different degrees of marginal

basin development: Central Chile with an
aborted marginal basin, Peru with an extreme
form of an aborted basin, and Patagonia with a
fully developed marginal basin. Burial metamorphism similar to that in Peru is a feature of
these basins.

The eastern part of the basin, which is now
almost entirely separated from the western part
by the Batholith, differs from the western part in
that it consists predominantly of andesites,
dacites and subordinate rhyolites. These volcanic
rocks may have been partially erupted on to the
continental margin and they pass eastwards
through a very rapid transition into the Carhuaz
Formation of Neocomian age. Figure 2 shows
the chronology and distribution of geological
events in the Western Cordillera and High

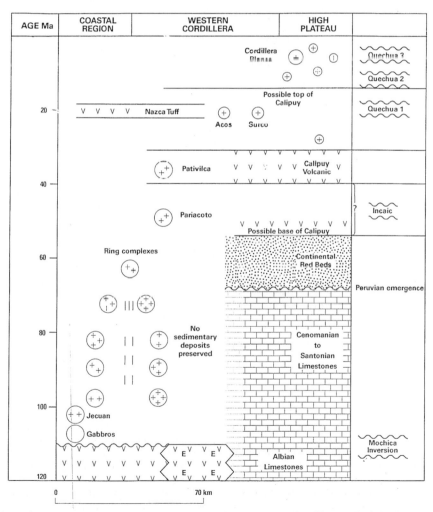

Fig. 2. Chronology and distribution of geological events in the Western Cordillera and High Plateau in Central
Peru, indicating the eastward migration of the magmatic arc.

Plateau in Central Peru, indicating the eastward migration of the magmatic arc. Spreading and subsidence in the basin ended towards the end of the Albian when extensional tectonics were supplanted by a compressional regime, which resulted in basin inversion producing a shallow elongate dome, conforming approximately to the shape of the former basin. This event, the Mochica phase, has been correlated with an increase in the spreading rate and the opening of the Atlantic Ocean (Soler & Bonhomme 1990).

It was at this stage that several large basic intrusions, the Patap gabbros, were emplaced along a linear zone that marks the western limit of the later batholith. These are of high alumina composition and their close temporal association with the western Casma marginal basin may suggest an ophiolitic relationship. The gabbros and some of the earliest granite intrusions are also of Albian age, showing that the reversal from a depositional volcano sedimentary mode in an extensional environment, to one of uplift and granitoid intrusion was extremely rapid. The timing of this event is constrained to the Upper Albian by a U/Pb age of 101 Ma on zircons from the Atocongo pluton (Mukasa & Tilton 1985) which cuts a gabbro that had previously intruded the basinal volcanic rocks. The period of batholith emplacement from 100 to 60 Ma and later (Fig. 2) was concentrated along the axial region of the former Huarmey–Canete basin and the extension of this lineament both northwards and southwards.

The predominantly basaltic volcanics of the Casma group were generated by the partial melting of a mantle source by a linear heat plume within the mantle (Soler & Bonhomme 1990). This resulted in the rapid building up of submarine volcanic rocks with a thickness of more than 9000 m and the development within the pile of a dome of burial metamorphism with the temperature isograds increasing with depth. The metamorphic gradient was high, and the amphibolite facies was established at depths of 4–6 km, with a zone of permitted melting at depths of 8–10 km. According to Atherton (1990), the switch from the generation of basaltic to tonalitic magmas was caused by the same linear plume initiating partial melting in the lower part of the basin fill, where the temperatures were high enough. Thus, the sequence— rift, pull apart, basin fill, burial metamorphism, basin inversion, batholith generation—are considered to be aspects of the same system generated by a mantle plume at the continental margin.

Atherton (1990) attributed the inversion of the Huarmey–Canete basin, with associated resurgent movement and fold development, to a major change in Pacific plate motion. Similarly, Soler & Bonhomme (1990) correlated the inversion of the basin and the generation of the batholithic magmas along the axial zone, to the opening of the Atlantic Ocean and the onset of rapid spreading, resulting in increased rates of ocean–continent convergence at the Pacific margin. It may be noted that batholithic granitoids of essentially similar character were emplaced both to the North and South of the Huarmey–Canete basin into a non-rifted continental margin with a sedimentary–volcanic cover. There does not, therefore, seem to be a necessary association of batholithic granitoids with marginal basin vulcanicity. This would seem to favour a more general subduction model for the generation of the granitoid magmas. In that scenario the remnant high heat flow along the axial zone of the marginal basin would have provided a highly favourable zone for the ascent and emplacement of granitoid magmas.

The emplacement of the Coastal Batholith

In all plutons, the contacts of the batholithic with the volcanic host rocks are extremely sharp and with no evidence of structural distortion. Stratification in both sedimentary and volcanic formations is regular and undisturbed right up to the contacts with nearly all the granitoid plutons. This suggests that the mechanisms of emplacement were passive rather than forceful.

The batholith immediately to the north of Lima is particularly interesting because of the remarkable structural symmetry displayed by the granitoid plutons. The structural pattern here is so well developed that it enables the formulation of a general model for pluton emplacement in this tectonic setting: (1) there is a bilateral symmetry with major tonalite plutons developed in equal proportions on either side of the central axis; (2) the presence of a chain of ring complexes spaced equidistantly at 35 km intervals in the axial zone; (3) the presence of a swarm of mafic dykes in the axial zone, which mostly predate the ring complexes; (4) the presence of a conjugate set of dextral and sinistral strike slip faults which intersect in the axial zone of the Batholith; (5) the well defined and linear box like shape of many of the plutons suggesting fracture control of their emplacement.

The two particular controlling factors which these phenomena reflect are the influence of an axial zone for magma generation approximately parallel to the plate edge, and the influence of fracture control at all scales of the emplacement of plutons.

The influence of the axial zone

Some kind of deep structure seems to have controlled the channelling of the granite magmas into the axial zone of the former basin over a lengthy period. The linear plume which provided the heat source for the generation of the basaltic magmas of the Huarmey basin and for their burial metamorphism, may have persisted, providing the energy for the partial melting and segregation of tonalitic melts within a relatively narrow zone, and focussing the ascent of magmas from a deeper subduction zone. Thus the width of outcrop of the batholith in this sector may simply reflect the existence of this linear zone. Magmas may have been able to ascend during extensional phases, exploiting fractures formed during compressional phases. Partial melting within the plume at deeper levels continued to provide basaltic magmas in the form of the dyke swarm. The ring complexes, which were the last expression of plutonism within the batholith in the Lima segment, were also intruded along the axial zone, thus illustrating the longevity of this structure.

Fracture control of intrusive contacts

The fracture patterns which were exploited by the ascending granitic plutons are of two major classes: those which are Andean parallel and those which are transverse. The Andean parallel system provides the main component of elongation for all the major plutons and it is reasonable to suppose that plutons were emplaced within this linear system during periods of extension, enabling magmas generated in the axial zone to be emplaced in favourable locations. Similarly the swarm of mafic dykes, located mainly in the central zone, was probably also emplaced during extensional episodes. However, a very different pluton geometry was developed at the ends of the main linear plutons and here it is evident that cross fractures provided the main structural control. This is particularly evident in the valley of the Rio Huaura where, not only do the cross fractures and the outlines of the plutons converge towards the axial zone, but the plutons of the ring complexes were themselves emplaced precisely at the point of convergence. Bussell (1983) and Bussell & Pitcher (1985) have shown that a major dextral wrench fault, which bisects the Huaura ring complex, shows evidence for episodic movement during granite emplacement, indicated by the greater offsets of structures

predating the complex, than for those offsets within it.

Regional mapping has established that offsets of igneous contacts along these transverse fractures correspond to movements on sinistral and dextral transverse wrench faults (Fig. 3). The geometry of these structures shows that they were developed as a result of NE–SW compression. Conversely the NW–SE alignment of the major tonalite bodies and of the dyke swarms suggests

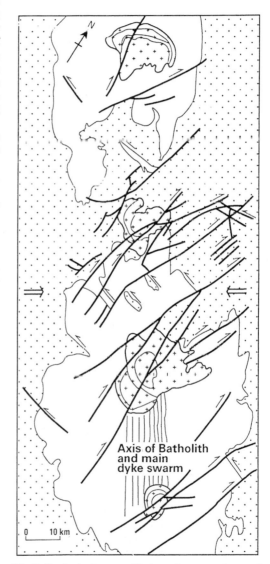

Fig. 3. Geological map to illustrate the concordance of linear pluton contacts with the sinistral–dextral regional wrench fault system. The ring complexes from south to north are Lumbre, Huaura, Paros and Fortaleza.

an extensional structural control for these components.

This major control of batholith geometry is reflected in every pluton where sufficient detailed work has been done. Contacts which at map scale are smooth are always jagged in detail. The granites were evidently emplaced by a process which exploited the preexisting fracture patterns in the country rocks resulting in the downward displacement of the roof material. This conclusion seems to be required because there are examples of undisturbed horizontal roof contacts for many plutons, which suggests that granite emplacement has resulted in mass exchange with the country rocks. In some cases, such as the Pativilca Pluton, huge blocks of the metavolcanic envelope can be seen frozen in the granite, while the granite itself is capped by volcanic rocks which pass continuously into the wall rocks at vertical contacts. In this case it is not possible to consider any structural adjustment of the envelope to accommodate the pluton; however, similar criteria are applicable to most plutons of the Batholith.

Towards the end of the emplacement history the ring complexes were emplaced in the central zone at regular intervals of 35–40 km corresponding to those areas where the transverse structures are best developed, and where they intersect the axial trend. The evidence suggests that the structural framework resulted from alternating periods of extension and compression. Actual emplacement probably took place during extensional periods, but exploited fractures which were formed during both extensional and compressional episodes. Magmatic stoping and block subsidence followed, modified by periodic reworking of older faults. This pattern of alternate compression and extension probably continued through the life of the Batholith.

The effects of contrasting systems of fracture control can be appreciated on maps of regional scale. However, it is a mechanism which operates on all scales, and detailed studies have shown that contacts which are smooth at map scale are actually quite jagged in detail with the intrusive magma exploiting different joint directions during emplacement. For example, the contact of the La Mina pluton at the western margin of the Huaura complex (Bussell & Pitcher 1985). These are very fine examples but similar phenomena have been observed during mapping at many places in the Batholith and it is probably true for most of the constituent plutons. This evidence, together with the general absence of deformational structures in both county rocks and granitoids, suggests that the principal mechanism for pluton emplacement was one of magmatic stoping.

The process of magmatic stoping must have been very efficient. The earliest plutons were emplaced in host rocks of the Casma Volcanic Group and the undisturbed roof and wall contacts indicate that the host rock material was stoped downwards. However, the later ring complexes and associated plutons were intruded into these earlier granites, which were already cold, and had been affected by the regional fault and fracture pattern. These earlier intrusives were in turn stoped downwards in the same manner, showing that the stoping mechanism happened twice in the same batholithic segment.

A generalized model of magmatic stoping incorporates a number of different processes which are not actually related, but which together combine to enable the transfer of low density material to a higher structural level, and of higher density material downwards: (1) first there is the fact of the pre-existing fracture pattern generated following the inversion of the Huarmey basin, but which continued to be active in alternating episodes of extension and compression throughout the life of the Batholith. Mylonites and breccia zones developed along these fractures which provided preferential lines of weakness for exploitation by rising granitoid magmas; (2) Myers (1975) has demonstrated that the earliest components of the Puscao plutons in the Huillapampa Quadrangle were tuffisites. These are attenuated gas charged magmas containing a high content of crystal and lithic fragments, mainly of the parent pluton, but also of other plutons and the country rocks. They penetrate mylonite zones isolating blocks of country rocks or of older plutons, which then sink through the magma thus providing space for new plutons. Generally the granite magma follows behind the tuffisitic precursor, clearing it away and leaving an apparently clean intrusive contact. This process was repeated many times throughout the emplacement history of the Batholith; (3) the process resulted in the quenching of the upper parts of the plutons with the earliest formed crystals enclosed in a finer grained base and produced a porphyritic rock. It continued through the plutonic crystallization history of plutons, resulting in rocks of progressively greater textural complexity. These are most commonly preserved in the ring dykes of the ring complexes (Bussell 1975).

The internal structure of plutons

Many of the plutons in the Batholith are zoned having a felsic core bordered by more mafic variants. However, some of the monzogranitic suites appear to have a different pattern of variation. One of the main components of the Huaura ring complex is the Puscao pluton which, together with the San Jeronimo unit, extends from the Rio Huaura northwards to the Quebrada Paros complex to form another ring complex similar to that of the Rio Huaura, being comprised of the same granite units.

The Puscao component of these complexes and the intervening area is grossly layered into three zones: a lower xenolithic zone, a normal middle zone and an upper layered zone of horizontal pegmatic and aplitic variants. The layered zone is best seen in the valley of the Rio Supe where it is about 400 m thick.

By contrast the Canas pluton, which is also located in the valley of the Rio Huaura and which is the youngest component of the complex, is texturally very homogeneous. For this reason it was selected by Taylor (1985) for a geochemical study. He established a grid on the mountainous outcrop and collected samples from the grid points for chemical analysis. His results showed that this apparently homogeneous pluton actually had an occult layered structure. At present these are the only known examples of layering in the Batholith. If this concept were found to be of a more general nature it could suggest that the Batholith as a whole may be relatively thin and that the problem of mass exchange by stoping may not be so insuperable.

The Incaic Orogeny: development of the Incaic erosion surface and migration of the magmatic arc

Plutonism along the batholith lineament in the region of the former Huarmey–Cancte basin stopped with the intrusion of the ring complexes at about 60 Ma. Some younger granites within the basin, such as the Pativilca granite with an age of 30 Ma do occur within the batholithic zone, but it was at this stage that the magmatic arc shifted eastwards where it gave rise to the plateau volcanics of the Calipuy group, together with associated, mainly tonalitic and grano-dioritic plutons.

With this eastward shift the whole character of the arc changed from a well defined linear zone of granites within an envelope of marine volcanics, to a very broad belt extending across the Western Cordillera and High Plateau on to the Eastern Cordillera. Its main products were the plateau volcanics of the Calipuy Group together with dispersed, isolated plutons of mainly tonalitic

and granodioritic composition, generally of rather small size.

The plateau volcanics are extremely variable, and sections measured in any traverse do not match with adjacent traverses. In general two modes of origin are possible: central volcanoes and fissure eruptions. At present insufficient systematic work has been done to establish which of these two alternatives is the most likely. However, these strata consist of basaltic and andesitic flows, agglomerates, breccias and tuffs, all of which could be interpreted as primary deposits from relatively small scale eruptions (Myers 1980; Atherton et al. 1985). They may have been erupted subaerially through extensional fissure-vent systems on a low relief surface, located inland and parallel to the trench.

The migration and broadening of the arc was preceded by strong folding of the Cretaceous and Lower Tertiary sedimentary formations of the inner arc and the high plateau followed by the development of a terrestrial erosion surface. This sequence of folding, uplift and erosion has been designated as the Incaic orogeny by Noble et al. (1979). Soler & Bonhomme (1990) have corre-lated the events of the Incaic orogeny with an increase in the rate of continent ocean conver-gence which began at about 50 Ma according to Pardo Casas & Molnar (1987) and Pilger (1983). They also correlated the eruption of the Calipuy volcanics and the intrusion of associated tonalitic and granodioritic plutons with the increased rate of convergence accompanied by a decrease in the dip of the subduction zone (Fig. 4).

Alternatively, Petford & Atherton (1996) have suggested that the magmatic arc moved east-wards as a result of Incaic uplift. This was considered to have induced a tensional situation in the continental slab, and ensured that the plumbing systems (fractures) moved eastwards, providing conduits for the rise of magmas over a wide area generated by decompressional partial melting in the mantle of a spinel–clinopyroxene–orthopyroxene–olivine lherzolite, which was not necessarily related to subduction.

The Incaic unconformity marks the uplift and erosion of the inner arc, domain, together with that of the former Huarmey–Canete basin, and the plutons of the Coastal Batholith emplaced by that time. These diverse components were unified and uplifted, and an erosion surface was devel-oped on the resulting composite terrain, the Incaic surface. The age of this surface is not well constrained since there is conflicting evidence for the age of the lowermost units of the overlying plateau volcanics of the Calipuy group. The oldest age of 53 Ma was obtained by Wilson (1975) and Cobbing et al. (1980) from Tapacocha

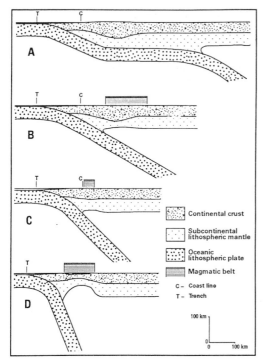

Fig. 4. Subduction geometry. A: present day. B: Eocene to early Pliocene, Eastern Stocks. C: Cenomanian to Campanian, main units of the Coastal Batholith. D: Albian (Casma volcanics). c, coast line; T, trench. (After Soler & Bonhomme 1990.)

150 km N of Lima, where the basal units rest discordantly upon folded strata of Lower Cretaceous age.

By contrast Noble et al. (1978, 1979) reported ages from 41 to 40 Ma for the lowermost units of the Calipuy group over a wide area ranging southwards from the latitude of Lima to Pisco. The ages for the basal units are remarkably consistent over this area and provide convincing evidence for the upper time limit of the Incaic erosion surface which was developed upon all units of early Palaeocene and earlier ages.

Noble et al. (1979) were concerned to establish the age of the lowermost units, and hence the latest age of the Incaic erosion surface. They did, however, take samples from c. 1000 m up the sequence and the youngest of these gave an age of 31.2 Ma. Thus the age range of the Calipuy group from the base to the highest dated unit was from 41.4–31.2 Ma. This contrasts with data obtained from further north where Wilson (1975) obtained an age of 53 Ma for the basal unit of Calipuy and 15 Ma from the highest units. Farrar & Noble (1976) also obtained an age of 15 Ma from the highest part of Calipuy. Thus the

possible age ranges for different sections of the Calipuy range from 53 to 15 Ma and from 41–31 Ma. There are many parts of the Calipuy outcrop where sections are greatly in excess of 1000 m and this may have a bearing on the problem of discrepancy in age.

The Pariacoto pluton, located at the eastern margin of the Coastal Batholith to the west of Huaraz, has an Rb/Sr isochron age of 49 Ma (Beckinsale et al. 1985). The pluton has intruded and thermally metamorphosed the lowermost units of the Calipuy group. This result, together with that of Wilson (1975) suggests that the lower units of the sequence may be older in this region than they are further south.

The available data indicates that the Incaic erosion surface has a minimum age of 40 Ma from Lima southwards, overlain by volcanic rocks with a thickness of 1000 m, whereas to the north of Lima it is as old as 53 Ma with a volcanic cover of 2000–3000 m.

The most southerly extent of the plateau volcanics normally attributed to the Calipuy group is only a little further south than the area covered by Noble et al. (1979) at the latitude of Pisco. From there southwards silicic tuffs predominate with ages ranging from 18 to 22 Ma showing them to be later than much of the Calipuy group.

The Mid-Tertiary plutonic inner arc

The Mid-Tertiary inner arc was a major feature of Andean geology producing great volumes of eruptive volcanics but rather smaller volumes of intrusive granitoids than the preceding Late Mesozoic mainly plutonic arc. It was also more unfocussed, having a greater lateral width and a wider dispersal of the plutonic component.

Plutons and small batholiths such as those of Acos and Surco with ages of c. 20 Ma (Cobbing et al. 1981; Beckinsale et al. 1985) were emplaced into the Calipuy volcanics and clearly they must have reached a very high level since they intruded a subaerial erosion surface covered only by the plateau volcanics. These and other intrusives are distributed over a wide zone corresponding approximately to the outcrop of the plateau volcanics, contrasting to the narrower, linear zone of the Coastal Batholith.

The granites of the 'inner arc' are essentially similar in their composition to those of the Coastal Batholith. They also consist predominantly of tonalites and granodiorites with subordinate monzogranites, but there is a lack of geochemical and isotopic information. Many of them are also texturally similar to the granitoids of the Coastal Batholith, but some are porphyritic, and in a few of these, textural evidence shows them to have been in an advanced stage of crystallization with the formation of a 'touching fabric' in which primocrysts were in contact and developed common grain boundaries before they were quenched, probably by decompression, with the development of a porphyritic–tuffisitic texture.

Final migration of the arc to the High Cordillera High Plateau region

The third arc in the region of the High Cordillera and High Plateau is less well defined in that the oldest ages of *c.* 30 Ma are similar to that of the Pativilca pluton, the youngest in the zone of the Coastal Batholith. Nevertheless, the preponderant age in this zone is 15 Ma or less. It is also marked by the presence of small, but widely scattered subaerial ignimbrite sheets that were erupted on to the surface of the plateau and which, in one case, flowed from the plateau down the palaeo canyon of the Rio Fortaleza, showing that uplift was well advanced at that time. This and other ignimbrite sheets at Yungay and Huaron all gave ages of *c.* 6 Ma (Cobbing *et al.* 1981).

This final Neogene arc most notably contains the Cordillera Blanca Batholith, but it mainly comprises a concentration of small, highly mineralized intrusives within a relatively narrow zone which includes the eastern part of the Western Cordillera and most of the High Plateau with a particular concentration of deposits along the junction between these two zones. Most of these intrusives are porphyritic, carrying megacrysts of sanidine, plagioclase, quartz, biotite and hornblende in a fine-grained micromonzonitic groundmass. Their SiO_2 content ranges from *c.* 60 to 67 wt% and they are of tonalitic to granodioritic composition (Cobbing *et al.* 1981). Very little in the way of modern petrological studies have been undertaken on these rocks which, in view of their economic importance, is surprising. Many of the small intrusions in this zone have ages ranging from 12 to 3 Ma (Soler & Bonhomme 1990). A number of these small stocks are mineralized and are an important component in the economic geology of the region. They were emplaced at a very high level and some of them are transitional into flow domes (Vidal pers. comm.). They represent the last episode of plutonism in Central Peru, which, from the mid-Cretaceous onwards, had provided a high proportion of the rock formations now forming the Western Cordillera.

The flow domes of the high plateau are often composite and may contain several circular structures, defined by flow foliations within each body. The foliations consist of a compositional flow banding on the scale of 1–5 cm, with each band being slightly different, having a greater or smaller proportion of megacrysts and other minerals with respect to adjacent bands. The banding is generally steep in the deeper parts of the structure and is funnel shaped, shallowing upwards and outwards. Hydrothermal alteration and mineralization is superimposed on the dome structure and in the area under discussion is of the acid–sulphate type. In Peru the best known example is at Julcani (C. Vidal pers. comm.) The general sequence of alteration is argillic–quartz alunite–vuggy silica rock–silicified rock with superimposed sulphides. This pattern or some variant of it characterizes most of the Peruvian examples. Where fully developed mineralization ranges from gold–silver in vuggy silica rock to lead zinc and copper in the sulphide zone.

The large batholith of the Cordillera Blanca consists of two facies, mafic diorites in the Carhuish stock and along the SW border of the Batholith which have U/Pb Zr ages of 13.7 Ma, while the main body of leucogranodiorite gives a U/Pb Zr age of 6.3 Ma (Mukasa 1986). Later work using $^{40}Ar/^{39}Ar$ confirms these values (Petford & Atherton 1992).

Petford & Atherton (1996) have suggested that the source region for the Cordillera Blanca granite and for the Yungay ignimbrite was a basaltic underplate forming the deep root of the high Andes. Whether this model can be applied to other intrusives in the high plateau zone remains to be established. Soler & Bonhomme (1990) did not specifically recognize a third arc but attributed subduction and magmatism at that time, PB4–PB7 of their notation, to high convergence rates of about 11 cm a^{-1}, and an almost perpendicular convergence direction along the Peru–Chile trench.

Petford & Atherton (1996) identified a number of geochemical factors which distinguished the granites of the Cordillera Blanca batholith from those of the Coastal Batholith, a selection of which are listed in Table 1. Although there is now a wealth of geochemical data for the Cordillera Blanca batholith there is very little for other granitic bodies in the western Cordillera High Plateau region. Nevertheless, the available data suggests that other plutons in that zone have similar geochemical characteristics. The data for a selection of ten high level stocks (Cobbing *et al.* 1981) show that these bodies all have higher levels of Al_2O_3, Na_2O and Sr than those of the

Table 1. *Main distinctive features between the Cordillera Blanca and Coastal Batholith*

	Cordillera Blanca	Coastal Batholith
Y (ppm)	<15	Average 17
La_n/Yb_n	>20	<20
Eu anomaly	No	Yes
Sr (ppm)	>300	Average 200
Rb/Sr	0.18–0.59	0.2–2.5
Sr/Y	>40	Average 12.4
Al_2O_3 (wt%)	>15	Average 14.6
Na_2O (wt%)	Average 4.3	Average 3.9

Data from Petford & Atherton (1996).

Coastal Batholith, in which respect they more closely resemble the Cordillera Blanca batholith.

Discussion and conclusions

Figure 5 shows the distribution of Tertiary volcanic formations and Cretaceous and Tertiary granitoids of the Coastal Batholith. The most fundamental event in the geological evolution of the region was the inversion of the marine volcanic and sedimentary basins in the Huarmey–Canete basin. This was an event of profound importance, which initiated a new system of plutonism and volcanicity for the remainder of the Mesozoic and Tertiary periods and which is still in place. It has been referred to as the 'magic switch' by Atherton (1990) and

the 'switchover' by Pitcher (1993). Soler & Bonhomme (1990) attributed this event to the opening of the Atlantic ocean towards the end of the Albian and the resulting increase in the rate of subduction at the Pacific margin. The event resulted in the subduction of formerly extensive forearc deposits, and of oceanic sediments (Soler & Bonhomme 1990). It also led to the generation of the granitic rocks of the Coastal Batholith and subsequently to the Inner Arc and High Plateau region.

During or shortly after basin inversion a conjugate set of ENE and WNW-trending strike slip faults developed, which continued to be episodically active during the period of pluton emplacement and after. It resulted in the imposition of a pattern of brittle deformation developed at all scales from major faults to the regional joint systems. These fracture patterns have controlled the geometry of the Batholith and of its constituent plutons and complexes.

The plutons of the Batholith were emplaced within the inverted Casma Basin in an episodic manner into an already brittle and fractured envelope over a time interval of 40 Ma. During this period the fractures were sporadically renewed. The axis of the former marginal basin provided the main conduit for the ascent of batholitic magmas, the largest of which were elongated along the Andean trending continental margin. The same structure continued to provide channelways for the mafic dykes up to the final

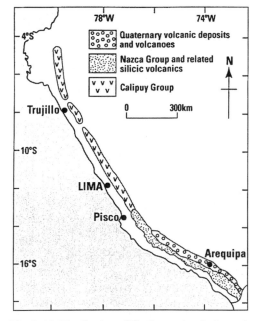

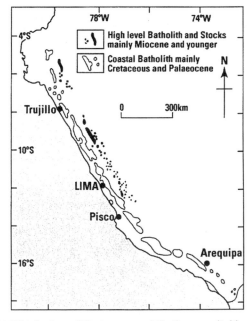

Fig. 5. (**a**) Distribution of Tertiary volcanic formations. (**b**) Distribution of Cretaceous and Tertiary granitoids.

episode of plutonic activity which was the emplacement of a chain of ring complexes at regular intervals in the central zone. This terminated magmatic activity in the Batholith lineament.

The continental margin was then uplifted and a surface of terrestrial erosion formed during the Palaeocene. Magmatic activity was displaced eastwards over a much broader area, from the eastern margin of the Batholith to the western edge of the High Plateau region, with some overlap in both cases. Its main product was a thick sequence of volcanic rocks which were deposited upon the erosion surface. Some of these later plutons, such as the Pativilca granite, were emplaced along the eastern margin of the Batholith, but most of the other plutons are widely dispersed and form a suite of generally small plutons of granodioritic composition and of broadly similar affinity to some components of the Coastal Batholith. These granites have not received the same degree of study as those of the Coastal Batholith and the Cordillera Blanca.

The last magmatic event was the emplacement of the Cordillera Blanca batholith and associated ignimbrite sheets in the High Plateau region at about six million years. This batholith was intruded along the Cordillera Blanca fault and its emplacement was facilitated by movement along the fault which is still active.

I thank A. Castro for inviting me to attend the meeting at Huelva in November 1997 and for suggesting an alternative emphasis for this contribution. Thanks to M. P. Atherton for his unstinted help in the preparation and review of this contribution and to M. A. Bussell for his helpful review.

References

AGUIRRE, L. & OFFLER, R. 1985. Burial metamorphism in the West Peruvian Trough; its relation to Andean magmatism and tectonics. *In*: PITCHER, W. S., ATHERTON, M. P., COBBING, E. J. & BECKINSALE, R. D. (eds) *Magmatism at a Plate Edge: the Peruvian Andes*. Blackie Halsted, 59–71.

ATHERTON, M. P. 1990. The Coastal Batholith of Peru: the product of rapid recycling of new crust formed within a rifted continental margin. *Geological Journal*, **25**, 337–349.

——, SANDERSON, L. M., WARDEN, V. & McCOURT, W. J. 1985. The volcanic cover: chemical composition and origin of the Calipuy Group. *In*: PITCHER, W. S., ATHERTON, M. P., COBBING, E. J. & BECKINSALE, R. D. (eds) *Magmatism at a Plate Edge: The Peruvian Andes*. Blackie Halsted, 273–284.

BECKINSALE, R. D., SANCHEZ-FERNANDEZ, A. W., BROOK, M., COBBING, E. J., TAYLOR, W. P. & MOORE, N. D. 1985. Rb–Sr whole rock isochron

and K-Ar determinations for the Coastal Batholith. *In*: PITCHER, W. S., ATHERTON, M. P., COBBING, E. J. & BECKINSALE, R. D. (eds) *Magmatism at a Plate Edge: The Peruvian Andes*. Blackie Halsted, 177–202.

BUSSELL, M. A. 1975. *The structural evolution of the Coastal Batholith in the provinces of Ancash and Lima. Central Peru*. PhD thesis, University of Liverpool.

—— 1983. Timing of tectonic and magmatic events in the Central Andes of Peru. *Journal of the Geological Society, London*, **140**, 279–286.

—— & PITCHER, W. S. 1985. The structural controls of batholith emplacement. *In*: PITCHER, W. S., ATHERTON, M. P., COBBING, E. J. & BECKINSALE, R. D. (eds) *Magmatism at a Plate Edge: The Peruvian Andes*. Blackie Halsted, 167–176.

COBBING, E. J., PITCHER, W. S., WILSON J. J., BALDOCK, J. W., TAYLOR, W. P., McCOURT, W. & SNELLING, N. J. 1981. *The geology of the Western Cordillera of Northern Peru*. Overseas Memoirs, Institute of Geological Sciences, **5**.

COSSIO, N. 1964. *Geologia de los cuadrangulos de Santiago de Chuco y Santa Rosa*, Boletin de la Comision de la Carta Geologica Nacional, Peru, **8**.

COUCH, R., WHITSETT, R. M., HUEHN, B. & BRICENO-GUARUPE, L. 1981. Structures of the continental margin of Peru and Chile. *In*: KULM, L. D., DYMOND, J., DASCH, E. J. & HUSSONG, D. M. (eds) *Nazca Plate: Crustal formation and Andean convergence*. Geological Society of America Memoirs, **154**, 703–726.

FARRAR, E. & NOBLE, D. C. 1976. Timing of late Tertiary deformation in the Andes of Peru. *Geological Society of America, Bulletin*, **87**, 1247–1250.

McCOURT, W. J. 1978. *The Geochemistry and Petrogenesis of the Coastal Batholith of Peru. Lima Segment*. PhD Thesis, University of Liverpool.

MUKASA, S. B. 1986. Zircon U-Pb ages of Super-units in the Coastal Batholith, Peru: Implications for magmatic and tectonic processes. *Geological Society of America Bulletin*, **97**, 241–254.

—— & TILTON, G. R. 1985. Zircon U–Pb ages of super-units in the Coastal Batholith, Peru: *In*: PITCHER, W. S., ATHERTON, M. P., COBBING, E. J. & BECKINSALE, R. D. (eds) *Magmatism at a Plate Edge: The Peruvian Andes*. Blackie Halsted, 203–207.

MYERS, J. S. 1975. Vertical crustal movements of the Andes in Peru. *Nature*, **254**, 672–674.

—— 1980. *Geologia de los Cuadrangulos de Huarmey y Huillapampa*. Servicio de Geologia y Mineria del Peru. Boletin, **33**.

NOBLE, D. C., McKEE, E. H. & MEGARD, F. F. 1978. Eocene uplift and unroofing of the Coastal Batholith near Lima, Central Peru. *Journal of Geology*, **86**, 403–405.

——, —— & —— 1979. Early Tertiary 'Incaic' tectonism, uplift and volcanic activity, Andes of Central Peru. *Geological Society of America Bulletin*, **90**, 903–907.

PARDO CASAS, F. & MOLNAR, P. 1987. Relative motion of the Nazca (Farallon) and South America plates since Late Cretaceous times. *Tectonics*, **6**, 233–248.

PETFORD, N. & ATHERTON, M. P. 1992. Granitoid emplacement and deformation along a major crustal lineament: the Cordillera Blanca, Peru. *Tectonophysics*, **205**, 171–185.

—— & ——1996. Na rich Partial Melts from Newly Underplated Basaltic Crust: The Cordillera Blanca, Peru. *Journal of Petrology*, **37**, 1491–1521.

PILGER, R. H. 1983. Kinematics of the South American subduction zone from global plate reconstructions. *In*: CABRE, R. (ed.) *Geodynamics of the eastern Pacific region, Caribbean and Scotia Arcs.* Geodynamics, **9**, 113–125.

PITCHER, W. S. 1993. *The nature and origin of granite.* Chapman & Hall, London.

—— & BUSSELL, M. A. 1985. Andean dyke swarms: andesite in synplutonic relationship with tonalite. *In*: PITCHER, W. S., ATHERTON, M. P., COBBING, E. J. & BECKINSALE, R. D. (eds) *Magmatism at a Plate Edge: The Peruvian Andes.* Blackie Halsted, 102–107.

——, ATHERTON, M. P., COBBING, E. J. & BECKINSALE, R. D. 1985. *Magmatism at a Plate Edge: The Peruvian Andes.* Blackie Halsted.

SOLER, P. & BONHOMME, M. G. 1990. Relation of magmatic activity to plate dynamics in central Peru from Late Cretaceous to present. *In*: KAY, S. M. & RAPELA, C. W. (eds) *Plutonism from Antarctica to Alaska.* Geological Society of America, Special Papers, **241**, 173–192.

TAYLOR, W. P. 1985. Three dimensional variation within granite plutons: a model for the crystallization of the Canas and Puscato plutons. *In*: PITCHER, W. S., ATHERTON, M. P., COBBING, E. J. & BECKINSALE, R. D. (eds) *Magmatism at a Plate Edge: The Peruvian Andes.* Blackie Halsted, 228–234.

WILSON, P. A. 1975. *K-Ar studies in Peru with special reference to the emplacement of the Coastal Batholith.* PhD Thesis, University of Liverpool.

Contrasts in morphogenesis and tectonic setting during contemporaneous emplacement of S- and I-type granitoids in the Eastern Lachlan Fold Belt, southeastern Australia

R. TRZEBSKI, P. LENNOX & D. PALMER

School of Geology, The University of New South Wales, Sydney NSW 2052, Australia
(e-mail: R.Trzebski@unsw.edu.au)

Abstract: Integrative gravity and structural modelling of Ordovician–Silurian granitoids in the Eastern Lachlan Fold Belt (southeastern Australia) revealed contrasts in emplacement mode and deformation style between coeval S- and I-type granites. The NNE–SSW directed contraction during the Benambran event of the Lachlan Orogen caused dextral movement along two major strike-slip faults (Carcoar Fault/Copperhannia Thrust) and simultaneous formation of both transtensional pull-apart and transpressional shear zones. The geometry and deformation style of the plutons and country rock, their spatial relationship at depth to adjacent faults and the structural history of both the granites and country rocks suggest a genetic linkage between magma emplacement and synmagmatic deformation. Synchronously, the Carcoar Granodiorite was emplaced into a transtensional pull-apart structure and the Barry Granodiorite and Sunset Hills Granite intruded transpressional shear zones. The I-type Carcoar and Barry granites are square to tabular, wedge-shaped bodies exhibiting a weak deformation; whereas the S-type Sunset Hills Granite is an elongated, tabular to sheet-like pluton showing a moderate deformation degree. The contrasts in 3D shape, emplacement mode and deformation style between the I- and S-type granites are due to differences in near-field stress regime, geometry of the emplacement sites, intrusion level with respect to thermal and rheological conditions, and in their response to deformation. This response is in part controlled by the proportion of resistant/non-resistant minerals in the granite and host rock. This study demonstrates that distinctive emplacement modes can operate simultaneously in different parts of a fault system under contrasting deformation conditions.

Granite intrusion and synmagmatic faulting are considered to interact during magma ascent and emplacement (Guineberteau *et al.* 1987; Hutton *et al.* 1990; Clemens & Mawer 1992; Brown *et al.* 1995). Faults facilitate magma transport through the crust and provide space for magma accumulation (Hollister & Crawford 1986; Vigneresse 1995; Trzebski *et al.* 1997; Lacroix *et al.* 1998). The fault geometry, kinematics and position within the stress field essentially influence the emplacement mode and the morphogenesis of the granite. Strike-slip faults aid magma transfer and intrusion along pull-apart domains (Guineberteau *et al.* 1987; Morand 1992; Lennox *et al.* 1998), transtensional fractures (Tobisch & Cruden 1995; Lacroix *et al.* 1998), shear-zone terminations (Hutton 1988) or P-shear bridges (Tikoff & Teyssier 1992). Vigneresse (1995) discriminated between extension, shear faulting and overlapping shear faulting as the main regimes of regional deformation which are related to the distinctive shapes of the intrusive

bodies. The gravity method is the most effective approach to determine the three-dimensional geometry of granites and their position at depth (Vigneresse 1990). The integration of gravity and structural data reveals the basal morphology, depth, volume, orientation and number of root zones of granites, and thus provides critical criteria to interpret the deformation style during magma emplacement.

Palaeozoic granitoids constitute nearly one third of the crystalline rocks in the Lachlan Fold Belt (Powell 1984). Intrusion of large volumes of granitic material in the crust leads to high rates of heat transfer, material movement and displacement, chemical mixing and assimilation and disruption of the near field stress regime. Although considerable data sets exist on the mineralogy, geochemistry and mineralization of the granites (Chappell & White 1974), their structuring and subsurface shape are poorly understood.

This paper focuses on two coeval granite groups in the Eastern Lachlan Fold Belt with

From: CASTRO, A., FERNÁNDEZ, C. & VIGNERESSE, J. L. (eds) *Understanding Granites: Integrating New and Classical Techniques.* Geological Society, London, Special Publications, **168**, 123–140. 1-86239-058-4/99/$15.00 © The Geological Society of London 1999.

fairly contrasting mineralogical, geochemical and structural features. Combining gravity and structural data enables us to present the three-dimensional geometry of the plutons and deduce an emplacement model. This study shows that locally simultaneous magma emplacement can generate distinctive intrusion shapes in plutons related to their position in the stress field. We provide evidence that one granite intruded a transtensional pull-apart, whereas the other granites used transpressional shear faults during emplacement.

Geological setting

The geology of the Lachlan Fold Belt and its evolution in terms of Circum-Pacific plate tectonics of the Tasmanides have been described by numerous authors (Powell 1984; Collins & Vernon 1992; Coney 1992; Gray *et al.* 1997). The Lachlan Fold Belt is a composite terrane that amalgamated during the Palaeozoic and Early Mesozoic due to the convergence of the Australian craton and the Pacific plate. One main characteristic of the Lachlan Fold Belt is

the large percentage of Siluro-Devonian granites which intruded Ordovician rocks as north–south elongated belts (Powell 1984). The traditional differentiation into peraluminous S-type and metaluminous I-type granitoids is mainly based on their mineralogy and geochemistry; however, it also reflects their distinctive deformation style and response to deformation (Vernon & Flood 1988). Generally, S-type granites are more deformed and penetratively foliated than the I-type granites due to the proportions of resistant (e.g. plagioclase, hornblende) and non-resistant (e.g. quartz, biotite) minerals.

The Carcoar, Barry and Sunset Hills granites occur in the Molong–Wyangala–Jerangle–Kuark Zone which is situated in the northern part of the Eastern Lachlan Fold Belt (Scheibner 1998; Fig. 1). The Molong–Wyangala Zone in the region of interest consists of metasedimentary and volcaniclastic rocks that were metamorphosed from prehnite–pumpellyite to greenschist facies and deformed during the Benambran Orogeny in the Late Ordovician/Early Silurian (Smith 1969). This event was followed by extensional and transcurrent deformation from

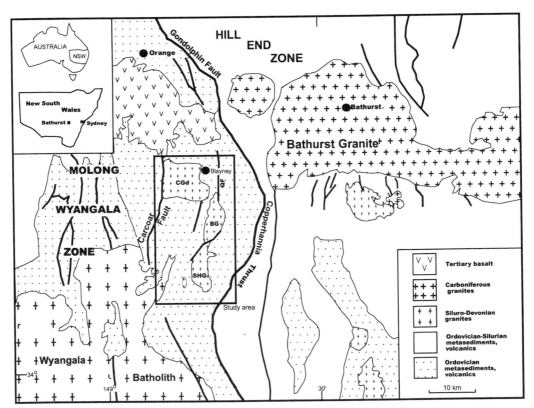

Fig. 1. Simplified regional geology of the study area. CGd, Carcoar Granodiorite; BG, Barry Granodiorite; SHG, Sunset Hills Granite; QF, Quarry Fault; NSW, New South Wales.

the mid-Silurian to the mid-Devonian, which led to the formation of sedimentary basins (e.g. Hill End Zone) and emplacement of other major granitoids (Collins & Vernon 1992; Glen 1998).

The Carcoar Granodiorite is a roughly equant (10 × 10 km) pluton bounded by the Carcoar Fault in the west and the Quarry Fault in the east (Fig. 2). It is composed of fine- to medium-, occasionally coarse-grained hornblende-biotite granodiorite to tonalite, similar to other I-type granites in the area, and minor intrusions of diorite (Long Hill Diorite), monzonite, aplite and pegmatite. The more mafic phase is represented by the Long Hill Diorite located in the northwest. The granodiorite consists of 45% plagioclase feldspar, 10% alkali-feldspar, 25% quartz, 10% biotite and 10% hornblende, with accessory apatite, zircon, pyroxene and ilmenite (Wilkins & Smart 1997). Alteration is characterized by secondary chlorite, clinozoisite, epidote and sericite. Two types of enclaves occur at various locations throughout the pluton. The more abundant enclaves consist of mafic, hornblende–plagioclase, microgranitic xenoliths with tabular to well-rounded shapes and are prolific in the southern part of the pluton. The second randomly distributed enclave is basaltic in composition and angular in shape. Sparse basaltic dykes intruded the Carcoar Granodiorite along N–S- to NE–SW-striking fractures. Recent $^{40}Ar/^{39}Ar$ isotope dating of magmatic hornblende indicates cooling through the 600 °C isotherm of the Carcoar Granodiorite at 416.2 ± 1.1 Ma, approximately during the Bowning event of the Lachlan Orogen (Perkins et al. 1995; Gray et al. 1997). Dating by the $^{40}Ar/^{39}Ar$ isotope technique of contact metamorphic or alteration biotite and amphibole show ages of 423.6 ± 1.0–424.4 ± 1.0 Ma, which presumably indicate the cooling of the country rocks in the Mid- to Late Silurian (Perkins et al. 1995).

The Barry Granodiorite is a meridionally elongated body (5 × 12 km) extending parallel to the Quarry Fault (Fig. 2). It consists mainly of a hornblende and biotite granodiorite of tonalitic composition with minor microtonalite enclaves (Lennox et al. 1998). The granodiorite contains 30% quartz, 30–40% plagioclase, 5–20% hornblende/biotite and up to 10% alkali-feldspars and thus mineralogically overlaps with that of the Carcoar Granodiorite; although the Barry Granodiorite is more mafic. Microgranitoid enclaves occur in five exposures and are sometimes aligned parallel to the foliation in the granite. K/Ar and $^{40}Ar/^{39}Ar$ isotope dating of biotite and hornblende gives ages of 408.2 ± 11.1/410.2 ± 11.6 Ma and 413 ± 4 Ma which are the minimum ages for the emplace-

ment of the Barry Granodiorite (Lennox et al. 1998).

The Sunset Hills Granite crops out as a north–south elongated body (6 × 13 km) roughly bounded in part by the Quarry Fault in the west and the Copperhannia Thrust in the east (Fig. 2). The Sunset Hills Granite has often been correlated with the Wyangala Batholith in the west due to genetic and compositional affinities of the predominantly S-type granites (Wyborn & Henderson 1996). However, the geochemical characteristics are not properly constrained as the pluton shows an I-type geochemical character, but an S-type aluminium saturation index. It is mainly composed of a biotite granite intruded in places by aplite and muscovite-leucogranite dykes. The granite contains 40–50% quartz, 30–40% plagioclase, 10–20% biotite and less than 10% muscovite (Lennox et al. 1998). Metasedimentary xenoliths of up to 5 m in length and 0.3–0.5 m in width crop out along the eastern and southern margin of the pluton. The xenoliths which represent stoped blocks of the host rocks (Adaminaby Group), show a foliation which is oblique to the foliation in the granite indicating deformation prior to granite emplacement (Lennox et al. 1998). Radiometric dating of biotite yields ages ranging between 363.1 ± 1.8 Ma (Rb/Sr), 371 ± 2 Ma ($^{40}Ar/^{39}Ar$) and 378 ± 8.5 Ma (K/Ar), which reflect partial resetting of magmatic biotite due to deformation at lower greenschist facies conditions during the Early Carboniferous Kanimblan event (Lennox et al. 1998).

Gravity modelling and the geometry of the plutons

Gravity is most efficient in determining the subsurface shape of granitic bodies, and combined with structural data, provides important constraints on the spatial relationship between plutons and deformation structures at depth (Guineberteau et al. 1987; Vigneresse 1995; Trzebski et al. 1997; Brown & Solar 1998). The accuracy of the gravity method is closely related to the resolution of the gravity field, which is given by the average station spacing. The Australian National Gravity Network has an average resolution of 0.25 points per km^2 in southeastern Australia, which is insufficient for detailed studies at a local scale.

A total of 1046 new gravity observations were recorded covering the Carcoar, Barry and Sunset Hills granites, which provided an average station spacing of 500 m. The measurements were taken, using a La Coste–Romberg gravimeter (Model

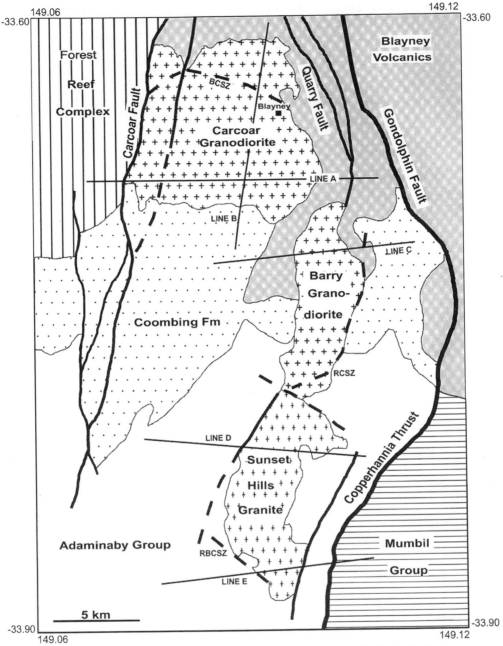

Fig. 2. Local geology of the study area. BCSZ, Browns Creek shear zone; RCSZ, Reedy Creek shear zone; RBCSZ, Rocky Bridge Creek shear zone. Lines A, B, C, D and F indicate the locality of the cross-sections in Fig. 4.

G) with an accuracy ± 0.05 mGal. The position and elevation control of the gravity stations were obtained, using GPS equipment (Leica SR299) with a post-processing component, giving an accuracy in the horizontal of <0.01 m and in the vertical of <0.05 m.

Bouguer anomalies were calculated against the International Gravity Formula 1980 and referred to the Australian National Gravity Network. Inner zone terrain corrections (to Hammer zone G) were estimated from 1:25 000 scale topographic maps enlarged to a scale of

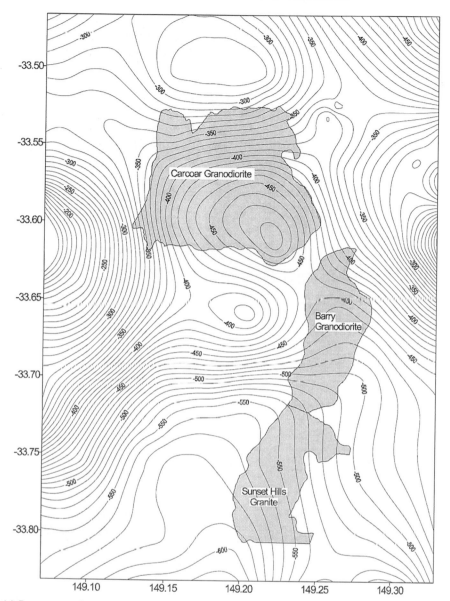

Fig. 3. (a) Bouguer gravity map of the study area; contour interval 5 m s^{-2}.

1:10 000. Outer zone terrain corrections (to a radius of 23.5 km) were computed automatically from a digital terrain model with a resolution of 250 m, kindly provided by the Australian Geological Survey Organisation. The Bouguer gravity data were gridded at a 500 m spacing using the minimum curvature method based on the triangulation and interpolation algorithm after Renka (1984).

The Bouguer anomaly map, which shows the new data, merged with reprocessed data from previous published surveys in surrounding areas, clearly reflecting two distinctive negative anomalies with maximum magnitudes of 18 mGal in the north and 8 mGal in the south (Fig. 3a). Although both anomalies entirely cover the exposed plutons, a correlation between the anomalies and the granites is not well constrained as the local, near-surface gravity variations are obscured by the regional, deep-crustal gravity variations. The northern anomaly is slightly elongated in the northwest–southeast

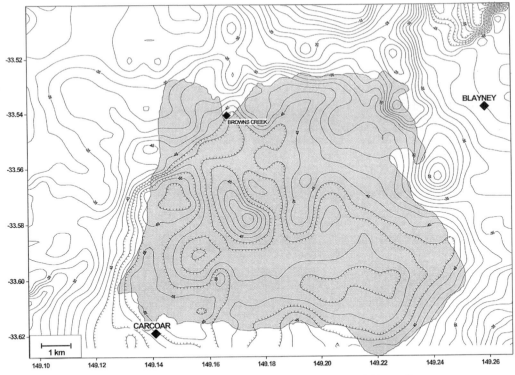

Fig. 3. (b) Residual gravity map of the Carcoar Granodiorite; contour interval 10 m s^{-2}.

direction and suggests a subsurface linkage between the Carcoar and Barry granites at its southeastern end. The southern anomaly, partly covering the Barry and Sunset Hills granites, is largely affected by the Wyangala Batholith, located in the southwest of the study area (Fig. 1). The positive anomalies surrounding the granites, mainly correspond to the large-scale volcanic and volcaniclastic complexes, which have significantly higher densities. The Bouguer gravity gradients show the general, NNW–SSE to N–S trend of the structural grain in the Lachlan Fold Belt.

In order to interpret the three-dimensional shape of the granites at depth, their anomalies were isolated by modelling, thus removing the effect of the surrounding geological formations and defining the trend of the background field across the region. The Bouguer anomalies were separated to calculate the residual gravity, using computer programs based on the method of Talwani & Ewing (1960). The residual gravity map of the Carcoar Granodiorite is characterized by a suite of negative anomalies, mainly aligned in the meridional and latitudinal directions along the pluton margins (Fig. 3b). The maximum magnitudes of *c*. 12 mGal occur in the

southwest and along the southern margin of the pluton. A north–south-striking, positive anomaly divides the pluton along its centre in two segments. Steep gravity gradients occurring along the western and eastern margins of the pluton correspond to the Carcoar Fault and the Quarry Fault, respectively (Figs 1 and 3b). The gradient lineaments are deflected in NE–SW and NW–SE (Browns Creek Shear Zone) directions (Fig. 2), which suggests vertical and lateral displacement of the pluton along the associated faults (Fig. 3b). Although the east–west-striking gradients along the northern margin of the Carcoar Granodiorite also suggest a fault, no-field evidence for such a structure has been found. In contrast, the southern margin is marked by smooth gradients, indicating a subsurface continuation of the pluton to the south (Fig. 3b).

The residual gravity field of the Barry Granodiorite shows two minor, negative anomalies in the north, which are separated by a positive anomaly along SW–NE-striking gradients (Fig. 3c). The positive anomaly is correlated to a small diorite body with a relatively high density of 2.88×10^3 kg m^{-3}. The maximum magnitudes of 8 mGal occur in the north and decrease to approximately 3 mGal in the south. The eastern

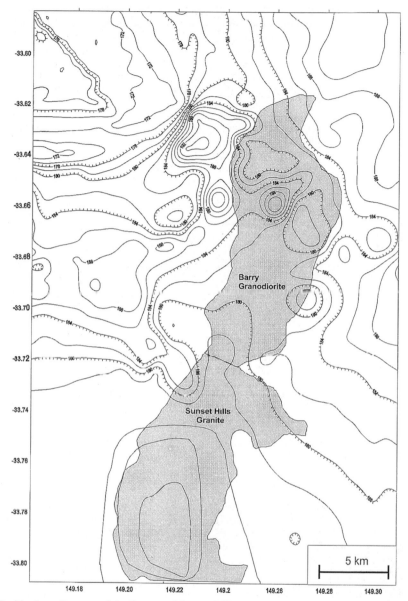

Fig. 3. (c) Residual gravity map of the Barry and Sunset Hills granites; contour interval 2 m s^{-2}.

margin of the pluton is marked by relative smooth, north–south-striking gradients, corresponding with the Quarry Fault (Fig. 3c). The Sunset Hills Granite exerts a weak negative anomaly with a magnitude of <4 mGal, which is characterized by smooth gradients along the pluton margins (Fig. 3c). In general, the residual gravity field of the Barry and Sunset Hills granites and surrounding host rocks is dominated by NW–SE and NE–SW gradients, which reflects the change of the general structural trend

to the south. No attempt was made to interpret the small-scale, positive anomalies over the host-rocks, which mainly correspond to the Tertiary basalt flows.

Three techniques were applied for the computation of models for the granites and related structural features: (1) 2.5D density modelling with the Gravmag computer program (Rasmussen & Pederson 1979), (2) Linsser filtering (Linsser 1967) and (3) inverse modelling (Tombs 1976). The 2.5D density modelling is based on forward modelling with the option of interactive geometry and density variation, until a satisfactory fit is obtained between the observed and calculated gravity data. The program calculates the gravity response of the model and displays the observed and calculated fields. In total 13 traverses were calculated with the Gravmag computer program across the granites in the study area. Five representative model lines are shown in Fig. 4. Linsser filtering generates a depth-selective tomography of the gravity field, using variable density contrasts to determine the geometry, depth extension, and orientation of the source bodies and related structures (Linsser 1967; Conrad *et al.* 1996; Trzebski *et al.* 1997). The filter is particularly useful for contouring the granite-related gravity gradients at selective depth levels, thus revealing the variations of the pluton shape with depth. This technique also helps to correlate the gravity gradients with faults, and to interpret their spatial relationship with plutons. In comparison with conventional methods of gravity analysis (e.g. field transformation, high-/low-pass filtering), Linsser filtering allows a quantitative analysis of the gravity field at selected depth levels and provides parameters to construct the geometry of the source bodies. Inverse modelling is based on the calculation of a theoretical gravity field due to a three-dimensional model, consisting of vertical square prisms, and the iterative adjustment of the top and basal surface until a satisfactory fit with the residual anomaly is obtained (Tombs 1976). It has been used to model the basal morphology of the plutons and to produce the depth isoline maps (Fig. 6).

The densities of the lithologies encountered in the study area were determined on 17 representative samples of both the granites and country rocks. The density contrast (Dr) between the granitoids (granodiorite: 2.65×10^3 kg m^{-3}, biotite granite: 2.62×10^3 kg m^{-3}) and the host-rocks (Blayney Volcanics: 2.88×10^3 kg m^{-3}, Forest Reef Complex: 2.82×10^3 kg m^{-3}, Coombing Formation: 2.76×10^3 kg m^{-3} and Adaminaby Group: 2.72×10^3 kg m^{-3}) ranges between 0.1 and 0.26×10^3 kg m^{-3} and is thus sufficient for the gravity modelling.

The Carcoar Granodiorite was emplaced into rocks of the Late Ordovician Cabonne Group; the Forest Reef Complex to the west and Blayney Volcanics to the north and east, and Early Ordovician Coombing Formation to the south (Fig. 2). The geometry of the pluton as revealed by the 2.5D density modelling shows two symmetrical steep- and flat-sided keels, striking approximately north–south, and separated by a shallow central section (Fig. 4a). Both steep

flanks are bounded by steeply dipping faults; the Carcoar Fault in the west and the Quarry Fault in the east (Fig. 4a). The maximum depth along the marginal root-zones is about 7 km and less than 3 km along the central rise (see also Fig. 6). The Carcoar Granodiorite gradually deepens from north to south to form a broad, basal low along the southern margin of the pluton (Fig. 4b). The total volume of the pluton has been calculated to be about 600 km^3. The Linsser filtering shows predominant north–south- and east–west-striking gradients, which correspond to the faults (Carcoar Fault, Quarry Fault) and fractures in the Carcoar Granodiorite (Fig. 5a). These gradients are in turn cross-cut by the NE–SW- and NW–SE-striking gradients, which appear to be associated with the fault system along the Carcoar and Browns Creek shear zones. These shear zones are responsible for both dextral and sinistral displacement of the pluton following the movement along the meridional and latitudinal faults. The basal morphology of the Carcoar Granodiorite shows two marginal root-zones, which strike approximately in a north–south direction.

The Barry Granodiorite forms an elongated, wedge-shaped body with moderate to steep flanks in the east and flat flanks in the west (Fig. 4c). The maximum depth reaches about 3 km along the narrow, NNE–SSW-striking root zone. In the north–south direction the Barry Granodiorite shows two segments slightly uplifted to a depth of < 3 km in its central section (see also Fig. 6). North–south-trending gravity gradients, that correspond to the general structural grain of the Barry Granodiorite, dominate among the linear features produced by Linsser filtering (Fig. 5b). Moderate gradients extending along the eastern margin of the pluton reflect the contact between the granite and host-rocks along the extension of the Quarry Fault. Sporadic northeast–southwest striking gradients are present and appear to sinistrally displace the north–south gradients. Similarly aligned lineaments occurring in the centre of the pluton may be associated with the fault system of the Rocky Creek Shear Zone (Lennox *et al.* 1998). The basal morphology shows a NNE–SSW-trending root-zone, which parallels the extension of the Quarry Fault along its eastern flank (Fig. 6). The volume of the pluton was calculated to be about 180 km^3.

The Sunset Hills Granite, which intruded the Adaminaby Group, shows a shallow (< 3 km), wedge-shaped body with a steep flank against the Quarry Fault in the northwest. In the southern direction its shape gradually changes to a thin, sheet-like body, moderately dipping towards the west to southwest (Fig. 4d, e). The pluton

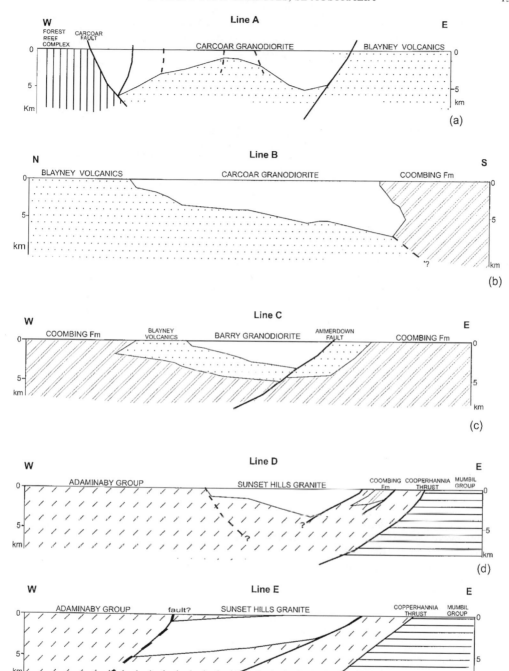

Fig. 4. Representative cross-sections across the Carcoar (**a**, **b**), Barry (**c**) and Sunset Hills (**d**, **e**) granites as derived from the 2.5D density models of the gravity data. The localities of the lines are indicated in Fig. 2.

thickens from north (<2 km) to south (*c.* 5 km) and appears to be more extensive beneath the Adaminaby Group in the southwest (Figs 4e and 6). Linsser filtering reveals weak north–south-striking gradients on the western margin in the north and in the southern extension along the

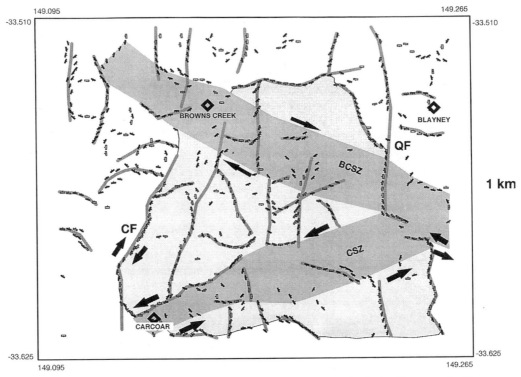

Fig. 5. (a) The analysis of the gravity gradients with Linsser filtering. **(a)** The display illustrates the Linsser indications (tick marks) corresponding to the residual gravity gradients of the Carcoar Granodiorite (light dotted) anomaly. The lines show the interpretation of the Linsser indications in terms of faults (lines), shear zones (dark dotted) and their motion. BCSZ, Browns Creek shear Zone; CSZ, Carcoar shear zone; CF, Carcoar Fault; QF, Quarry Fault.

eastern margin of the pluton (Fig. 5b). In the north, the western margin of the Sunset Hills Granite is also marked by northeast–southwest-striking gradients, which correspond to the Quarry Fault. Northwest–southeast-striking gradients occur at the northern and southern margins of the Sunset Hills Granite and correspond with the dextral system of the Rocky Bridge Creek Shear Zone (Lennox *et al.* 1998). The volume of the pluton under the outcrop area was calculated to be about 150 km³, which is a minimum value as considerable portions of the pluton are located beneath the host-rocks, i.e. outside the related negative anomaly in the residual gravity field.

Due to the nature of the gravity method, uncertainties in the input data, and limitations of the applied techniques, the models presented in this study necessarily contain a degree of ambiguity. The 2.5D density modelling and the inverse modelling are limited by the fact that the model density of the rocks occurring underneath the granites may differ from the assumed value, and result in too high or too low depths of the

pluton. However, the maximum variations in density ($\pm 0.05 \times 10^3$ kg m^{-3}) changes the thickness of the pluton by only 500 m and thus range within the accuracy of the gravity data. Furthermore, the inverse modelling after Tombs (1976) is restricted to negative anomalies and to the outcrop area of the plutons that yield minimum values for the calculations. In order to reduce the density-related ambiguity of the Linsser method, several tests were performed by varying the density contrast by $\pm 0.05 \times 10^3$ kg m^{-3}. The results showed an alteration in both the lateral subsurface distribution of the plutons and the dip of their flanks by less than 15%.

Structural framework

The already meridionally folded and foliated Adaminaby Group in the south, Coombing Formation in the middle and Blayney Volcanics $\pm$ Forest Reef Complex in the north were intruded by the granites (Fig. 2). All the granites were foliated syntectonically and developed weak (Carcoar, Barry) to moderately intense (Sunset

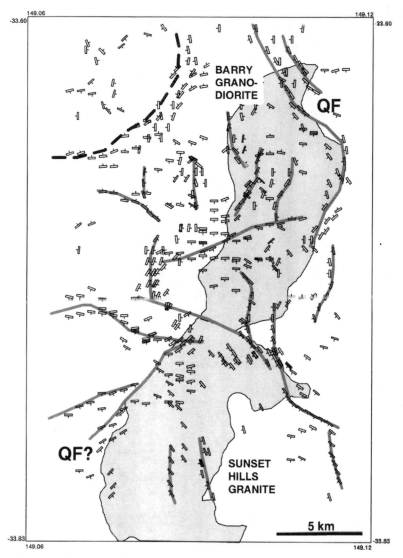

Fig. 5. (b) The Linsser analysis for the Barry Granodiorite (light dotted) and the Sunset Hills Granite (dark dotted). QF, Quarry Fault.

Hills), meridionally striking foliations, defined by aligned biotite books, deformed and cracked feldspars, recrystallized quartz, which have been cross-cut by later C–S structures, (conjugate?) mylonite and shear zones. The kinematics from these C–S structures are consistent with the regionally observed pattern of dextral movement on northeast-trending segments of the Copperhannia Thrust and sinistral movement on the NNW–SSE-trending segment of the Gondolphin Fault (Fig. 1). Late movement along the north-west–southeast-trending shear zones caused dextral, kinking or bending of the regionally

developed granite and country rock foliation, consistent with the effect of the Early Carboniferous Kanimblan event.

Metamorphic isogrades cross this area obliquely from northeast to southwest between the Carcoar and other granites, and define a transition from the actinolite zone (quartz–albite–chlorite–epidote–actinolite–carbonate) to biotite zone (quartz–albite–chlorite–epidote–biotite–actinolite–carbonate) in the south (Smith 1969). There is a change from brittle structures (faults, joints) to more ductile structures (mylonites, C–S structures) from north to south,

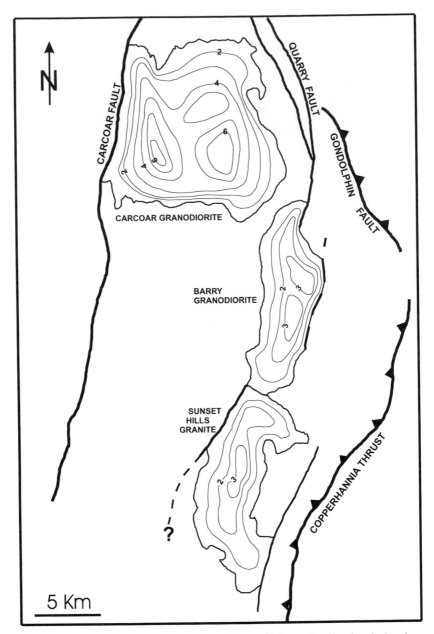

Fig. 6. Depth isoline map in km of the Carcoar, Barry and Sunset Hills granites showing the basal morphology of the plutons.

representing an increase in the depth of crust cropping out at the surface. The contact metamorphic aureole around each granite, except the Carcoar Granodiorite, is very narrow (<1000 m) and development of metamorphic spottings (sericite after cordierite) minimal, suggesting rapid cooling for these high-level granites.

Foliations of strictly magmatic origin are sometimes identifiable. Solid-state deformation is weak to moderate and as such aligned igneous minerals (hornblende), and sometimes enclaves or xenoliths commonly define a weakly developed magmatic foliation. All three plutons show evidence of solid-state deformation or recrystallization. Although the intensity of deformation is generally weak, especially for the Carcoar and Barry granites, it is locally more strongly developed, where regional faults or shear zones occur, such as on the northwestern edge of the Sunset Hills Granite or the western margin of the Carcoar Granodiorite (Fig. 7). Solid-state deformation consists of undulose extinction in quartz, recrystallization of quartz, growth of new pale brown biotite, especially in the Barry and Sunset Hills granites, cracking and distortion of the multiply twinned, invariably sericitized plagioclase and, particularly in the Carcoar and Barry granites, the breakdown of plagioclase to epidote and sericite. Biotite is commonly chloritized in the Sunset Hills Granite and the feldspars destroyed.

Where the intensity of deformation is higher the granites show C and S type fabrics with S, representing the dominant foliation in the granite. Occasionally, millimetre-wide mylonites or sub-millimetre-scale foliation form a mesh-like pattern oriented obliquely to the dominant granite foliation. Analysis of mylonite zones from the western margin of the Sunset Hills Granite indicates that the NE–SW-trending zones underwent dextral oblique-slip movement, with the northwest block moving up relative to the southeast block (Lennox et al. 1998). From exposures of C surfaces and mylonites in the Sunset Hills Granite, those trending 050–070° commonly have dextral displacement; whereas those trending 160–180° commonly have sinistral displacement. If these C surfaces and mylonites were conjugates, then the resolved contraction direction would be approximately 115° consistent with an east–west contraction. The dominance of meridional foliations in both the granites and country rocks is consistent with this east–west shortening during various phases of the Lachlan Orogeny.

Three distinct changes in the orientation of the dominant foliation within the country rock or granites are used to define three ductile shear zones. The NW–SE-trending, 500–800 m wide Rocky Bridge Creek Shear Zone (RBCSZ), in part marks the southwestern margin of the Sunset Hills Granite. A change in orientation of the dominant foliation from north–south-striking in the south to NE–SW-striking in the north occurs in both the Adaminaby Group and Sunset Hills Granite (Fig. 7). The 500–800 m wide Reedy Creek Shear Zone (RCSZ) contains deformed granite, is oriented northeast–southwest, parallel to a topographic low and possibly marks the zone, along which the Barry Granodiorite was displaced eastward during east–west contraction, causing dragging of the dominant

foliation. Alternatively, gravity studies indicate this shear zone may be oriented NW–SE (Fig. 5b), parallel to the other two major shear zones in this area. The 2.5–3 km wide Browns Creek Shear Zone has been defined by disruption to the NNE–SSW-trending foliation in the southern Carcoar Granodiorite and country rock and meridional-trending foliation in the Blayney Volcanics north of the Carcoar Granodiorite (Fig. 7). Gravity studies are consistent with the presence of such a zone (Fig. 5a).

Pluton emplacement and regional deformation

The similarity in geometry and spatial relationship between the granites and adjacent faults, and the parallel nature of both the magmatic and solid-state fabrics in the granites and host-rocks suggest that magma emplacement was controlled by regional deformation. Based on the 3D shape of the plutons and their spatial relationship with faults at depth, and structural features in granites and wall-rocks, we propose a model for emplacement of the Carcoar, Barry and Sunset Hills granites and their tectonic setting that features two contrasting emplacement modes and structural styles, operating simultaneously in different parts of the same strike-slip fault system.

The gravity modelling revealed two distinctive types of pluton shapes, which imply granite emplacement controlled by two deformation regimes. Although the geometrical and structural features of the granites are not entirely consistent with the characteristics of emplacement models after Vigneresse (1995), we suggest a bridging solution for a synchronous granite emplacement, alternating between extension and shear faulting. The 3D geometry, basal morphology and volume of the Carcoar Granodiorite, and its spatial relationship at depth with adjacent faults suggest that the pluton was emplaced into a transtensional pull-apart structure during wrench-faulting. The root-zones parallel the Carcoar and Quarry faults, extending along the western and eastern margins of the pluton, and thus suggest interaction between emplacement and faulting. However, the absence, or only weak development of, magmatic and solid-state fabrics in the Carcoar Granodiorite do not support a fault-controlled emplacement during shear deformation. The steep lineations, near-vertical foliations and minor mylonite zones only developed along the Carcoar and Quarry faults indicate minor dextral shearing along the pluton margins (Fig. 7). The overall absence or weak development of magmatic and solid-state fabrics in the

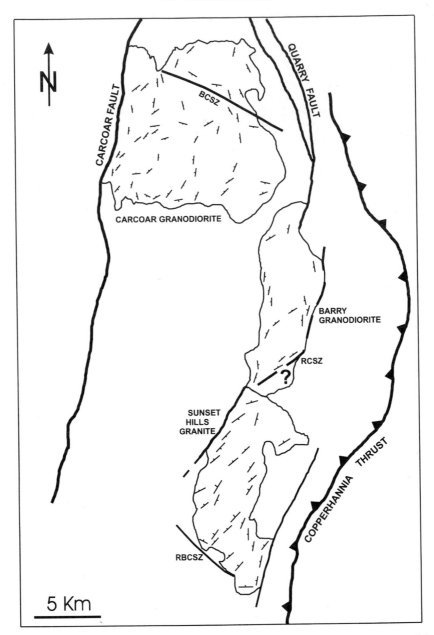

Fig. 7. Map showing the distribution and orientation of tectonic foliation in the Carcoar, Barry and Sunset Hills granites. RCSZ, Reedy Creek shear zone; RBCSZ, Rocky Bridge Creek shear zone.

Carcoar Granodiorite may in part be caused by the relatively high content of resistant minerals (e.g. hornblende, plagioclase), which can considerably affect the response to deformation of I-type granites (Vernon & Flood 1988). The magma intruded the brittle host-rocks, metamorphosed under prehnite-pumpellyite facies conditions at a depth of 6 ± 2.6 km (A. Müller pers. comm.). The emplacement of the Carcoar Granodiorite corresponds well with the model of granite emplacement into overlapping shear structures (Vigneresse 1995), except for the number of root zones. Our model shows two root zones, which presumably operated as channels for magma ascent along the adjacent faults.

The Barry Granodiorite was emplaced under similar conditions in the dextral overlapping shear structure of the Quarry Fault zone (Fig. 8b). In contrast to the Carcoar Granodiorite, the Barry Granodiorite has a narrow, elongate shape with a single root-zone, which parallels the Quarry Fault (Fig. 8c). It shows weak to moderately developed, magmatic and solid-state fabrics, presumably formed due to more intense shear deformation, which was likely enhanced by the narrow shape of the pluton and the shorter distance between the anastomosing shear planes of the Quarry Fault zone.

In contrast to the other plutons, the Sunset Hills Granite intruded as sheets in en échelon dilational areas of dextral shear zones along the Quarry and Copperhannia Thrust faults (Fig. 8b). The pluton formed an elongate, fault-hosted, steep to moderately dipping sheet, emplaced during transpressional shear faulting. The strongly developed magmatic and solid-state fabrics in the Sunset Hills Granite parallel the ductile structures in the host rocks, which were deformed under upper greenshist facies conditions. The difference in deformation style of the Sunset Hills Granite in comparison with the Carcoar and Barry granites is due to either the deeper emplacement levels, and/or higher deformation degree due to its higher content of non-resistant minerals, e.g. quartz and biotite (Vernon & Flood 1988). The similarity between the foliation in the host rocks and the syn-intrusive fabrics in the granite suggests that the emplacement and fabric development were controlled by the deformation of the country rocks.

In the Eastern Lachlan Fold Belt the Siluro-Devonian magmatism is associated with the regional contractional deformation during the Benambran to Bindian event of the Lachlan Orogen (Gray et al. 1997; Glen 1998). The dominant structural features are crustal-scale, meridional thrust and/or strike-slip faults, which formed due to a series of compressional events (Glen & Watkins 1994). The polyphase and multifaceted movements along the Carcoar, Gondolphin and Quarry faults, and the Copperhannia Thrust (Fig. 8c–e), and their along strike continuation controlled the structurally induced emplacement of the Carcoar, Barry and Sunset Hills granites (Lennox et al. 1998). Based on structural and radiometric data of the Carcoar, Barry and Sunset Hills granites, we postulate three, syn- to post-intrusive deformation stages, which were active during the Lachlan Orogeny in the Late Ordovician to Early Carboniferous.

The first stage has been constrained by recent SHRIMP isotope dating on zircons of the granites in the study area, which gives an age of c. 435 Ma (R. Armstrong pers. comm.), which we interpret as the emplacement time. We suggest that this date corresponds with the Benambran event, which caused southwest–northeast-oriented compression in the Eastern Lachlan Fold Belt (Gray et al. 1997; Glen 1998). The Benambran event caused dextral movement along the regional strike-slip faults, which, in turn, provided space for magma emplacement into transtensional and transpressional structures (Fig. 8c). Deflection and/or segmentation along transcurrent faults led to formation of releasing bends and stepovers, which favour the development of transtensional/transpressional linking bridges between the offset strike-slip faults (Segall & Pollard 1980; Woodcock & Fischer 1986; Fig. 8b). The north–south striking faulting was simultaneous with the formation of subsidiary east–west shear zones, which are clear in the residual gravity gradients (Fig. 4a, b).

The second stage corresponds to regional compression during the Early Devonian Bowning event, which resulted in east–west thrusting along the major thrust faults (Fig. 8d). On a local scale, the Bowning event caused dextral strike-slip faulting along the Carcoar Shear Zone in the Carcoar Granodiorite and the Reedy Creek Shear Zone, and reorientation of existing foliations in the Barry Granodiorite. During the second stage, the hornblende/biotite Rb/Sr isotope system of the Carcoar and Barry granites was reset at c. 415 Ma (Lennox et al. 1998). The third stage of deformation is characterized by north–south compression during the Early Carboniferous Kanimblan event (Powell et al. 1985; Lennox et al. 1998). This stage produced dextral dip- and strike-slip faulting along the Rocky Bridge Creek and Browns Creek shear zones (Fig. 8e). Rb/Sr and $^{40}Ar/^{39}Ar$ isotope dating on biotite of the Sunset Hills Granite gives ages around 375 Ma that document the uplift of the pluton to the present outcrop level due to the north–south-oriented thrusting (Lennox et al. 1998). The lack of the 375 Ma age in the Carcoar and Barry granodiorites may be due either to the absence of biotite and/or higher position of both plutons in the crust during the Kanimblan event.

Conclusions

Integrative modelling of gravity and structural data revealed the 3D geometry of the granites, and their spatial relationship with adjacent faults. The models developed for the Carcoar, Barry and Sunset Hills granites in the Eastern Lachlan fold Belt, combined with structural data of the granites and host-rocks, and radiometric ages essentially helped to interpret the emplace-

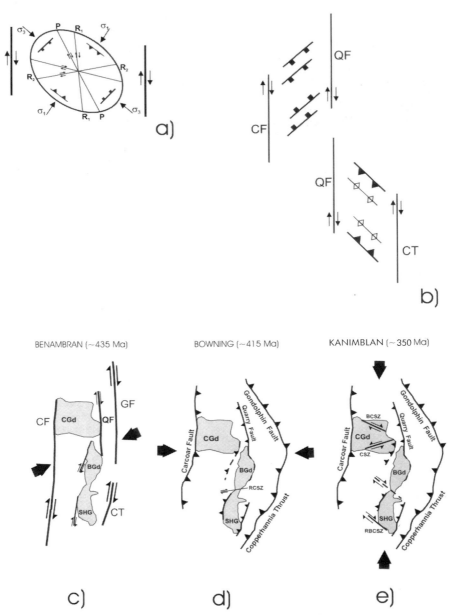

Fig. 8. Schematic illustration of the tectonic setting along the Carcoar Fault and the Copperhannia Thrust during emplacement of the Carcoar, Barry and Sunset Hills granites. (**a**) Strain ellipsoid during dextral shear movement. (**b**) (left/top) Formation of transtensional cavities during dextral shear faulting (pull-apart) as suggested for the emplacement of the Carcoar Granodiorite (CGd); and (right/bottom) formation of transpressional cavities during dextral shear faulting as suggested for the emplacement of the Barry (BGd) and Sunset Hills (SHG) granites. (**c**) Simplified tectonic model for emplacement of the Carcoar, Barry and Sunset Hills granites during dextral, strike-slip faulting between the Carcoar, Quarry and Gondolphin faults due to a southwest–northeast oriented contraction of the Benambran event (*c.* 435 Ma). The granites alternatively intruded transtensional and transpressional sites formed between the strike-slip faults. (**d**) East–west oriented contraction during the Bowning event (*c.* 415 Ma) caused thrusting on the bounding faults and probable dextral strike-slip movement along the Reedy Creek Shear Zone displacing the Barry Granodiorite to the east. (**e**) Dextral shear faulting along the Browns Creek Shear Zone (BCSZ), Rocky Bridge Creek Shear Zone (RBCSZ) and a presumed shear fault between the Barry and Sunset Hills granites due to a north–south contraction during the Kanimblan event (*c.* 350 Ma).

ment mode and tectonic setting during granite emplacement. The modelling demonstrated that, particularly in areas with relatively poor outcrop conditions, the application of the gravity for three-dimensional mapping of granites and associated structures is an adequate means to support field observations.

The Carcoar, Barry and Sunset Hills granites intruded structural cavities along major transcurrent faults; the magma emplacement occurred contemporaneously at different sections of the fault system. Due to a north–south- to north-east–southwest-directed contraction during the Benambran event in the Lachlan Orogen, deflection and/or segmentation along the transcurrent faults, locally formed releasing bends and oblique-slip stepovers, which operated as transtensional pull-apart, whilst elsewhere transpressive shear zones operated along the strike-slip faults. The Carcoar Granodiorite intruded a dextral, transtensional pull-apart structure, formed between the Carcoar Fault and the Copperhannia Thrust/Quarry Fault; whereas the Barry Granodiorite and the Sunset Hills Granite intruded transpressional dilational jogs along the dextral shear zone of the Copperhannia Thrust/Quarry Fault.

The two distinctive emplacement modes are well illustrated by the morphogenesis of the plutons and the deformation style of the granites. The 3D shape of the plutons, their relationship at depth with adjacent faults, and their deformation are consistent with the models of Vigneresse (1995), which suggest that the emplacement of the Carcoar Granodiorite was extension-related and the emplacement of the Barry Granodiorite and Sunset Hills Granite was rather shear-related.

The difference in geometry of the two plutons and deformation style of the granites and wall rock, together with the radiometric ages, suggest that the Barry Granodiorite intruded a transpressive dilational jog at the southeastern termination of the Carcoar Granodiorite pull-apart structure, and the Sunset Hills Granite was emplaced during synchronous or slightly subsequent oblique upthrust along the Copperhannia Thrust forming gently-dipping sheets. The Sunset Hills Granite was emplaced in deeper and more ductile rocks, whereas the Barry and Carcoar granites were emplaced at higher crustal levels in a more brittle environment. The I-type (Carcoar/Barry) and S-type (Sunset Hills) granites differ in their geometry and degree of deformation. The Carcoar and Barry granites are square to tabular, blocky bodies in profile; whereas the Sunset Hills Granite is elongated tabular to sheet-like in shape. These granites have a weakly developed magmatic foliation, but a weakly (Carcoar/Barry) to moderately developed (Sunset Hills) tectonic foliation. The Carcoar and Barry granites exhibit more brittle structures (faults, joints, brecciated joints); whereas the Sunset Hills Granite contains more ductile structures (S–C fabrics, mylonites).

This study was funded by the Deutsche Forschungsgemeinschaft (Tr 402/2-1) and Hargraves Resources NL Pty Ltd. We thank J. Graham, R. Cotton, C. Wilkins, I. Cooper, G. Smart, S. Paterson and S. Siegesmund for their assistance and for interesting discussions.

References

BROWN, M. & SOLAR, G. S. 1998. Granite ascent and emplacement during contractional deformation in convergent orogens. *Journal of Structural Geology*, **20**, 1365–1393.

——, RUSHMER, T. & SAWYER, E. W. 1995. Segregation of melts from crustal protoliths: mechanisms and consequences. *Journal of Geophysical Research*, **100**, 15 551–15 564.

CHAPPELL, B. W. & WHITE, A. J. R. 1974. Two contrasting granite types. *Pacific Geology*, **8**, 173–174.

CLEMENS, J. D. & MAWER, C. K. 1992. Granitic magma transport by fracture propagation. *Tectonophysics*, **204**, 339–360.

COLLINS, W. J. & VERNON, R. H. 1992. Palaeozoic arc growth, deformation and migration across the Lachlan Fold Belt, southeastern Australia. *Tectonophysics*, **214**, 381–400.

CONEY, P. J. 1992. The Lachlan Belt of eastern Australia and Circum-Pacific tectonic evolution. *Tectonophysics*, **214**, 1–26.

CONRAD, W., BEHR, H. J. & TRZEBSKI, R. 1996. Die Linsser-Filterung der Süddeutschen Großscholle und ihre Interpretation. *Zeitschrift Geologischer Wissenschaften*, **26(4)**, 20–32.

GLEN, R. A. 1998. The Eastern Belt of the Lachlan Orogen. *In*: FINLAYSON, D. M. & JONES, L. E. A. (eds) *Mineral Systems and the crust-upper mantle of southeast Australia*. Australian Geological Survey Organisation Record, **2**, 80–82.

—— & WATKINS, J. J. 1994. The Orange 1:100.000 sheet: a preliminary account of stratigraphy, structure and tectonics, and implications for mineralisation. *Quarterly Notes Geological Survey NSW*, **95**, 1–17.

GRAY, D. R., FOSTER, D. A. & BUCHER, M. 1997. Recognition and definition of orogenic events in the Lachlan Fold Belt. *Australian Journal of Earth Sciences*, **44**, 489–501.

GUINEBERTEAU, B., BOUCHEZ, J. L. & VIGNERESSE, J. L. 1987. The Mortagne granite pluton (France) emplaced by pull-apart along a shear-zone: structural and gravimetric arguments and regional implications. *Geological Society of America, Bulletin*, **99**, 763–770.

HOLLISTER, L. S. & CRAWFORD, M. L. 1986. Melt-enhanced deformation: A major tectonic process. *Geology*, **14**, 558–561.

HUTTON, D. H. W. 1988. Granite emplacement mechanisms and tectonic controls: inferences from deformation studies. *Transactions of the Royal Society Edinburgh: Earth Sciences,* **79**, 245–255.

——, DEMPSTER, T. J., BROWN, P. E. & BECKER, S. D. 1990. A new mechanism of granite emplacement: intrusion in active extensional shear zones. *Nature,* **343**, 452–455.

LACROIX, S., SAWYER, E. W. & CHOWN, E. H. 1998. Pluton emplacement within extensional transfer zone during strike-slip faulting: an example from the late Archean Abitibi Greenstone Belt. *Journal of Structural Geology,* **20**, 43–59.

LENNOX, P. G., FOWLER, T. J. & FOSTER, D. 1998. The Barry Granodiorite and Sunset Hills Granite: Wyangala-style intrusion at the margin of a regional ductile shear zone. *Australian Journal of Earth Sciences,* **45**, 849–863.

LINSSER, H. 1967. Investigation of tectonics by gravity detailing. *Geophysical Prospecting,* **15**, 480–515.

MORAND, V. J. 1992. Pluton emplacement in a strike-slip fault zone: the Doctors Flat Pluton, Victoria, Australia. *Journal of Structural Geology,* **14**, 205–213.

PERKINS, C., WALSHE, J. L. & MORRISON, G. 1995. Metallogenic episodes of the Tasman Fold Belt system, Eastern Australia. *Economic Geology,* **90**, 1443–1466.

POWELL, C. McA. 1984. Ordovician to earliest Silurian: marginal sea and island arc. *In*: VEEVERS, J. J. (ed.), *Phanerozoic Earth History of Australia.* Oxford University Press, Oxford, 29–309.

——, COLE, J. P. & CUDAHY, T. J. 1985. Megakinking in the Lachlan Fold Belt, Australia. *Journal of Structural Geology,* **7**, 281–300.

RASMUSSEN, R. & PEDERSON, L. B. 1979. End corrections in potential field modelling. *Geophysical Prospecting,* **27**, 749–760.

RENKA, R. J. 1984. Algorithm 624, triangulation and interpolation at arbitrarily distributed points in the plane. *ACM Transformation Mathematics Software,* **10**, 440–442.

SCHEIBNER, E. & BASDEN, H. (eds) 1998. *Geology of New South Wales — Synthesis. Volume 2, Geological Evolution.* Geological Survey of New South Wales, Memoirs in Geology, **13**.

SEGALL, P. & POLLARD, D. D. 1980. Mechanics of discontinuous faults. *Journal of Geophysical Research,* **85**, 4337–4350.

SMITH, R. E. 1969. Zones of progressive regional burial metamorphism in the part of the Tasman Geosyncline, Eastern Australia. *Journal of Petrology,* **10**, 144–163.

TALWANI, M. & EWING, M. 1960. Rapid computation of gravitational attraction of three-dimensional bodies of arbitrary shape. *Geophysics,* **25**, 203–225.

TIKOFF, B. & TEYSSIER, C. 1992. Crustal scale, en échelon 'P-shear' tension bridges: A possible solution to the batholithic room problem. *Geology,* **20**, 927–930.

TOBISCH, O. T. & CRUDEN, A. R. 1995. Fracture controlled magma conduits in an obliquely convergent continental magmatic arc. *Geology,* **23**, 941–944.

TOMBS, J. M. C. 1976. *A package of three-dimensional gravity programs.* Applied Geophysics Unit Computer Program Report No. 19, Institute of Geological Sciences (unpubl.).

TRZEBSKI, R., BEHR, H. J. & CONRAD, W. 1997. Subsurface distribution and tectonic setting of the late-Variscan granites in the northwestern Bohemian Massif. *Geologische Rundschau,* **86**, 64–78.

VERNON, R. H. & FLOOD, R. H. 1988. Contrasting deformation of S- and I-type granitoids in the Lachlan Fold Belt, eastern Australia. *Tectonophysics,* **147**, 127–143.

VIGNERESSE, J. L. 1990. Use and misuse of geophysical data to determine the shape at depth of granitic intrusions. *Geological Journal,* **25**, 249–260.

—— 1995. Control of granite emplacement by regional deformation. *Tectonophysics,* **249**, 173–186.

WILKINS, C. & SMART, G. 1997. Browns Creek gold-copper deposit. *In*: BERKMAN, D. A. & MACKENZIE, D. H. (eds) *Geology of Australian and Papua New Guinean Mineral Deposits.* The Australasian Institute of Mining and Metallurgy: Melbourne, 575–580.

WOODCOCK, N. H. & FISCHER, M. 1986. Strike-slip duplexes. *Journal of Structural Geology,* **8**, 725–735.

WYBORN, D. & HENDERSON, G. A. M. 1996. *Blayney 1 : 100 000 Sheet area.* Australian Geological Survey Organisation Record, **56**.

Structure and geophysics of the Gåsborn granite, central Sweden: an example of fracture-fed asymmetric pluton emplacement

ALEXANDER R. CRUDEN[1], HÅKAN SJÖSTRÖM[2] & SVEN AARO[3]

[1]*Department of Geology, University of Toronto, Mississauga, Ontario L5L 1C6, Canada (e-mail: cruden@credit.erin.utoronto.ca)*

[2]*Department of Earth Sciences, Uppsala University, Villavägen 16, S-752 36 Uppsala, Sweden*

[3]*Geological Survey of Sweden, Box 670, SE-751 28 Uppsala, Sweden*

Abstract: The emplacement mechanisms of the Palaeoproterozoic Gåsborn granite, a satellite of the Transscandinavian Igneous Belt (TIB), are investigated using an integrated structural and geophysical approach. The pluton is discordant to *c.* 1.89 Ga folded supracrustal rocks that were deformed and metamorphosed at *c.* 1.85–1.80 Ga during the Svecokarelian Orogeny. Emplacement occurred at a depth of *c.* 10 km, within a regime of late Svecokarelian dextral transpression. Deformation of the pluton during cooling resulted in the formation of a variably developed foliation in the granite and deflection of less competent wall-rock units around its western and eastern contacts. Later E–W Sveconorwegian shortening resulted in the formation of shear zones that affect one of the pluton margins and may contribute to a component of the observed wall-rock distortion. The granite is situated above strong NNW-trending linear magnetic and negative gravity anomalies, which are interpreted to correspond to an important early Svecokarelian shear zone. The geophysical data indicate that the pluton is markedly asymmetric and modelling of the residual gravity field suggests that it consists of a deep root zone in the west and a thin sill-like body, which makes up most of the east and south parts of the body. Emplacement of the sill-like part occurred by lateral flow of magma from the root zone accommodated by downwarping of the underlying units. Intrusion of the thicker, discordant west part may have been accommodated by a combination of roof lifting and floor depression, aided by displacement on an active shear zone.

Emplacement studies of granites normally involve structural analysis of the wall-rocks (e.g. Pitcher & Berger 1972; Castro 1986; Paterson *et al.* 1991) in combination with measurements of fabrics developed within the pluton (e.g. Marre 1986; Hutton 1988; Bouchez *et al.* 1990; Cruden & Launeau 1994). Geochronological and geochemical work can also be incorporated to determine the time dimension, where the magma came from, the emplacement depth, and how it evolved (e.g. Pitcher 1993; Brown 1994; Vigneresse & Bouchez 1997). However, a critical factor in determining how space is created for an individual intrusion is constraining its depth extent and 3D geometry. With the exception of a few unique studies of plutons exposed in areas of extreme vertical relief and fortuitous exposure (e.g. Bridgewater *et al.* 1984; Myers 1975; Coleman *et al.* 1995; Scaillet *et al.* 1995; Grocott *et al.* 1998), the only practical

method available is to carry out a gravity survey (e.g. Bott & Smithson 1967; Bothner 1974; Sweeny 1976; Wikström & Aaro 1986; Vigneresse 1990). Careful mapping of wall-rock strains and displacements and internal fabrics of the pluton can then be used to place reasonable constraints on 2D or 3D models of the residual gravity field (e.g. Brun *et al.* 1990; Cruden & Aaro 1992; Vigneresse & Bouchez 1997; Dehls *et al.* 1998). Knowledge of the third dimension is often critical in order to differentiate between otherwise equivocal models for the emplacement of the pluton.

As an example of the above approach, we present results of a structural and geophysical study of the Gåsborn granite, a small, roughly circular Paleoproterozoic pluton in the Baltic Shield, Sweden. It is moderately exposed in densely forested terrain of low relief ($\leqslant 200$ m). Detailed mapping, high-resolution aeromagnetic

From: CASTRO, A., FERNÁNDEZ, C. & VIGNERESSE, J. L. (eds) *Understanding Granites: Integrating New and Classical Techniques.* Geological Society, London, Special Publications, **168**, 141–160. 1-86239-058-4/99/$15.00 © The Geological Society of London 1999.

data and a precise gravity survey reveal that the pluton is a markedly asymmetric tabular body. It was emplaced into previously folded metasedimentary and metavolcanic rocks, and is inferred to have been fed by a NNW-trending conduit related to a regional basement structure. Although ascent of the magma may have been tectonically controlled, its emplacement appears to have been largely influenced by variations in the mechanical properties of its wall rocks and their preintrusive structure. However, subsequent cooling of the pluton within a Svecokarelian transpressive strain field and later brittle–ductile (Sveconorwegian) and brittle faulting events profoundly affected its external and internal structure.

Geological setting

Regional geology

The Gåsborn granite is located in the western part of Bergslagen, a *c.* 1.89 Ga volcano-sedimentary belt within the Svecokarelian orogen of central Sweden (Fig. 1) (Welin 1987; Lundström 1990; Stephens *et al.* 1994, 1997; Allen *et al.* 1996). The pluton is considered to be a satellite to the Transscandinavian Igneous Belt (TIB), a N–S-trending, 1.85–1.65 Ga belt of granitoids and related mafic and felsic volcanic rocks (Wilson 1982; Gaál & Gorbatchev 1987; Jarl & Johansson 1988; Stephens *et al.* 1997). West Bergslagen and much of the TIB have been affected by tectonothermal over-printing related to the 1.1–0.9 Ga Sveconorwegian orogen exposed in SW Sweden (Fig. 1) (Wahlgren & Stephens 1990, 1996; Wahlgren *et al.* 1994).

West Bergslagen is characterized by a series of NNW-trending belts of *c.* 1.89 Ga felsic meta-volcanic rocks, alternating with belts of clastic meta-sediments, meta-pyroclastics and minor marbles that are exposed in synclinal structures. Geochemical, ore-genetic and sedimentological studies suggest that the supracrustal rocks were formed in a magmatic arc setting, probably inboard of an active continental margin in an environment comparable to a continental back-arc region (e.g. Allen *et al.* 1996). Subsequent deformation and peak regional metamorphism during the Svecokarelian orogeny occurred between *c.* 1.85 and 1.80 Ga (Wahlgren & Stephens 1996; Stephens *et al.* 1997). Recent regional studies indicate that peak metamorphism generally outlasted the main deformation (Allen *et al.* 1996; Bergman & Sjöström 1996; Sjöström & Bergman 1998) and may overlap with *c.* 1.78 Ga regional contact metamorphism

associated with emplacement of the TIB (Anderson 1997). In high-grade metamorphic areas to the NE and SE of W Bergslagen as well as in parts of Finland, late Svecokarelian low-pressure metamorphism, possibly related to Svecokarelian orogenic collapse, is a typical feature (Korja *et al.* 1993; Väisänen *et al.* 1994; Korja & Heikkinen 1995; Stephens & Wahlgren 1995; Bergman & Sjöström 1996; Sjöström & Bergman 1998).

The outcrop pattern of supracrustal rocks of the Bergslagen Supracrustal Series (BSS) in W Bergslagen (Fig. 1) is interpreted either to be due to structural control of sedimentation in early Svecokarelian grabens (Oen *et al.* 1982; Baker *et al.* 1988) or to represent deformed relics of an originally more extensive sedimentary succession (Allen *et al.* 1996). In the latter interpretation extensional, sediment-filled regional basins formed during and after the waning stages of volcanism. Structural interpretations have suggested folding and faulting of regionally continuous stratigraphy (Stålhös 1981, 1984, 1991). In the study area, the major structural features are the Saxå and Grythyttan synclines, which are occupied by meta-shales and meta-pyroclastic rocks (Torrvarpen Formation; Fig. 1). The intervening domain exposes meta-rhyolites and meta-andesites of the underlying Älgen Formation in a faulted anticlinal structure. The nature of the contacts between the Älgen and Torr-varpen formations are not known with certainty. They have been interpreted as probable synde-positional normal faults (Baker *et al.* 1988) that have been reactivated (inverted) by subsequent Svecokarelian and Sveconorwegian deformation events. Alternatively they have been interpreted as primary features subsequently affected by local faulting (Allen *et al.* 1996). In the Gåsborn area the Älgen and Torrvarpen Formation contact is interpreted as a Svecokarelian shear zone (see below). A regional foliation (S2) in the supracrustal rocks is axial planar to tight folds with steeply W dipping axial planes and gently plunging axes. Tight earlier folds with axial planar foliation (S1) parallel to bedding (S0) occur locally, and are of uncertain significance. Thus, the regional foliation and the related NNW-trending regional and minor folds are D2 structures. Peak regional meta-morphic middle-greenschist facies conditions were synchronous with or subsequent to the D2 event.

The Gåsborn granite is located *c.* 30 km to the west of the inferred location of the Sveconorwegian Frontal Deformation Zone (SFDZ) in W Bergslagen (Fig. 1) (Wahlgren *et al.* 1994). The SFDZ and a set of approximately N–S-trending

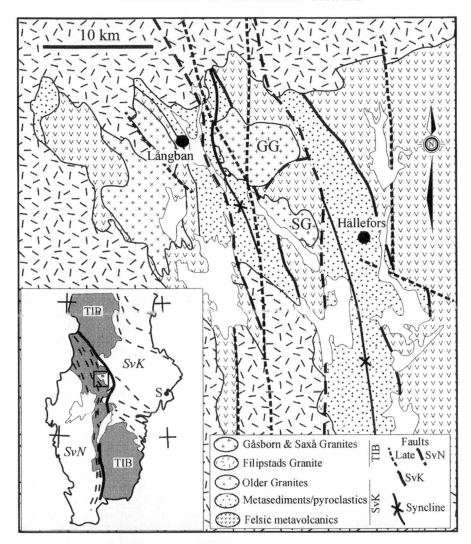

Fig. 1. Regional geology of the W Bergslagen District and S Sweden. Inset: geology of S Sweden (after Stephens *et al.* 1994), grey is Transcandinavian Igneous Belt (TIB); SvN, Sveconorwegian; SvK, Svecokarelian; S, Stockholm, heavy line, eastern limit of Sveconorwegian deformation (SFDZ); dotted lines, major Sveconorwegian and older shear zones. Box indicates location of W Bergslagen District. Main map shows geology of W Berslagen (after Baker 1985; Björk 1985; Lundegårdh 1987*b*; Lundström 1995*b*), GG and SG, Gåsborn and Saxå Granites.

Sveconorwegian shear zones to the west have affected the TIB and younger (1.75–1.55 Ga) mainly intrusive rocks, dividing the Sveconorwegian orogen into three tectonic segments (Berthelsen 1980; Stephens *et al.* 1994). Generally, Sveconorwegian deformation becomes increasingly penetrative from E to W across the TIB. Several older WNW–ESE, NW–SE and NNW–SSE trending shear zones occur to the east of the SFDZ. Early activity along some of these zones has been interpreted to have occurred at *c.* 1.80 Ga (Bergman & Sjöström 1994;

Högdahl *et al.* 1996; Stephens & Wahlgren 1996; Beunk *et al.* 1996). Most of these shear zones contain evidence of reactivation. One such example occurs along the eastern margin of the TIB to the north of Bergslagen, where mylonites were developed in a *c.* 1.70 Ga granitoid during dextral transpressive shearing (Bergman & Sjöström 1994; Sjöström & Bergman 1995, 1996). In south-central Sweden, deformation under retrograde conditions along shear zones is poorly constrained to the interval 1.78–1.56 Ga (Stephens *et al.* 1997).

The TIB is thought to have been emplaced in a continental magmatic arc setting (Wilson 1982; Gaál & Gorbatchev 1987). Since deformation within shear zones in Svecokarelian rocks east of the SFDZ overlaps temporally with the emplacement ages of the TIB (i.e. 1.85–1.65 Ga) and the formation of 1.85–1.75 Ga anatectic granites (Öhlander & Romer 1996), intimate relationships between shear zones and magma ascent are likely (Korja & Heikkinen 1995; Sjöström & Bergman 1998). Consequently, ascent of TIB magmas west of the SFDZ was probably also controlled by Svecokarelian structures, but these have been obscured by the Sveconorwegian overprint. To date, metamorphic minerals in shear zones west of the SFDZ only record Sveconorwegian overprinting ages (1.009–0.965 Ga for hornblende, 0.930 to 0.905 Ga for muscovite; Page *et al.* 1996).

The *c.* 1.79–1.76 Ga Filipstad granite suite (Jarl & Johansson 1988; Persson & Ripa 1993; Sundbladh *et al.* 1993) is the dominant phase of the TIB that intrudes supracrustal rocks of W Bergslagen. Although the general trend of the TIB is NNW, it is locally strongly discordant to the structural grain of W Bergslagen, which occurs within a large embayment on the eastern margin of the TIB (Fig. 1). The Gåsborn granite occurs in the northern part of this embayment.

Sveconorwegian overprinting in W Bergslagen is indicated by the development of moderately to steep SW-dipping ductile reverse shear zones that offset contacts of the TIB and the older supracrustal units. Rb–Sr dating of metamorphic biotites associated with Sveconorwegian shear zones in the Filipstad granite suite rocks SW of Långban (Fig. 1) yields a minimum age of 904 ± 31 Ma for this overprinting event (Verschure *et al.* 1987).

Regional geophysics

West Bergslagen, and the Gåsborn granite are located on the flank of a major NNW-trending gravity trough corresponding to the mass deficiency associated with the TIB (Fig. 2). The regional Bouguer gravity decreases from *c.* −40 to *c.* −50 mGal from NE to SW across the area. The Gåsborn granite sits above a well-defined linear gravity low, flanked by gravity highs corresponding to the Saxå and Grythyttan synclines. The area north of the Gåsborn granite underlain by the TIB is characterized by relatively high Bouguer gravity values, which suggests that this eastern extension of the Filipstad granite suite is relatively thin (Fig. 2). Gravity studies of TIB granitoids to the south have established that thicknesses of 2–5 km for

individual plutons are common (Wikström & Aaro 1986).

A strong NNW-trending structural grain within W Bergslagen and the TIB is evident in regional aeromagnetic data (Fig. 3). The magnetic signatures are characterized by homogeneous areas, which outline various rock units and structures (e.g. Gåsborn granite and Grythyttan syncline), and regions with strong NNW-trending anomaly patterns. The NNW-trending magnetic grain is likely due to the cumulative effects of Svecokarelian and Sveconorwegian structures as well as late Proterozoic diabase dykes.

Geology of the Gåsborn granite

The Gåsborn granite is a *c.* 6 × 6 km, moderately exposed, semi-circular pluton (Fig. 4; Björk 1985, 1986). The west part of the pluton intrudes rocks of the Torrvarpen Formation on the east limb of the Saxå syncline, and the east part of the granite intrudes west-dipping units of the Älgen Formation. The pinching out of lithological units along the contacts between the Älgen and Torrvarpen formations, particularly to the north and south of the Gåsborn granite (Fig. 4), indicates that they are separated by a tectonic contact, either an early normal fault (Baker *et al.* 1985) or a post-depositional Svecokarelian shear zone. The minimum age of this structure is constrained by the fact that it is truncated by the *c.* 1.78 Ga Gåsborn pluton. The Saxå granite, a smaller NW-trending intrusion of similar composition and inferred age, is exposed *c.* 1 km S of the Gåsborn granite. The relationship between these two plutons has not been investigated in detail, but it is likely that they are connected at depth, since they both lie within a major NW-trending gravity low (Fig. 2), and their associated magnetic anomalies are connected (Fig. 3). The Saxå granite at least partly straddles the contact between the Älgen and Torrvarpen formations, suggesting that its emplacement may also be related to this structure (Figs 1 and 4).

Wall-rock structure and metamorphism

The west margin of the pluton is strongly discordant, cutting across the axial trace of the Saxå syncline and across several stratigraphic contacts in the NW. Layer parallel, metre- to kilometre-scale sheet injections of granite into supracrustals are developed locally (e.g. the large apophysis in the SW; Fig. 4) close to the contact. Where not affected by shear zones, exposed contacts are sharp with no apparent increase in wall-rock strain related to the intrusion. Primary

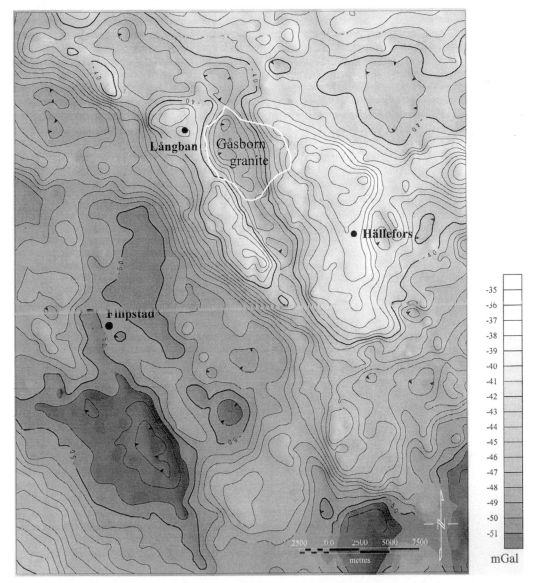

Fig. 2. Regional Bouguer gravity anomaly map of the W Bergslagen District. Contour interval is 1 mGal. Measurements by the Geological Survey and National Land Survey of Sweden (IGSN71 1943–1981).

features like chilled margins (reduction in grain size and locally porphyritic texture) occur in granite adjacent to the wall rocks. The Saxå Syncline appears to be deflected, tightened and overturned to the east, close to the western margin of the pluton (Fig. 4). This map-scale deformation may be related to post-intrusive Svecokarelian deformation and is discussed in a later section.

In most localities, contact metamorphic effects adjacent to the western margin are minimal,

except for olivine growth in dolomitic marble north of the granite, and a *c.* 1 × 0.4 km, NW-trending area of spotted hornfels in grey shales in the core of the Saxå Syncline, located between the main body of the pluton and the large NNW-trending apophysis (Fig. 4). In this locality, over a distance of 400 m approaching the contact, metamorphic assemblages change from pyrophyllite–biotite ± andalusite, to andalusite–almandine–biotite, to andalusite–cordierite, to sillimanite–K-feldspar. These parageneses,

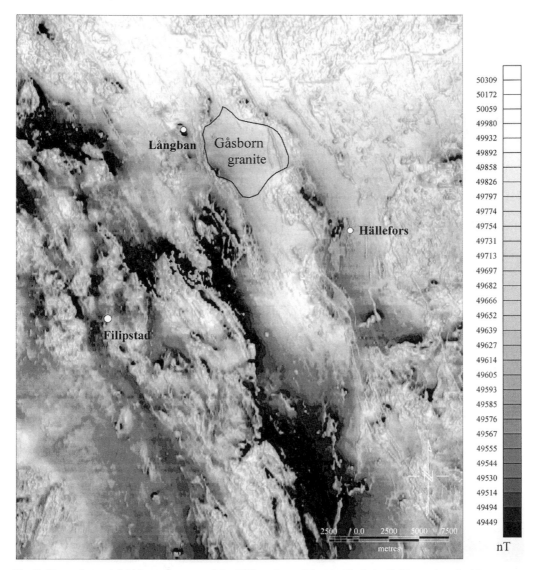

Fig. 3. Magnetic total field intensity map of the W Bergslagen District. Shaded relief is produced with an illumination from the ENE (declination 075°/inclination 32°). Measurements by the Geological Survey of Sweden (1973).

together with preliminary garnet–biotite geo-thermometry indicate an isobaric temperature increase from *c.* 430° to *c.* 680 °C at the contact (J. Arnbom unpublished data 1983). The depth of emplacement of the Gåsborn granite is bracketed by these reactions to between 2 and 3 kbar. Texturally, development of the hornfelses is associated with a grain-size coarsening and loss of the S2 foliation. Cordierite aggregates vary from displaying random shapes and orientations to a strong planar NNW-trending vertical alignment. This could be interpreted as an emplacement-related strain fabric, however the adjacent rocks of the Gåsborn granite in this locality also contain a very strong, locally protomylonitic, foliation developed under medium grade conditions (see below) that is parallel to the cordierite fabric. These relation-ships suggest that the pluton and its contact metamorphic aureole were locally deformed during post-crystallization cooling. The age of this deformation is interpreted to ˙be late

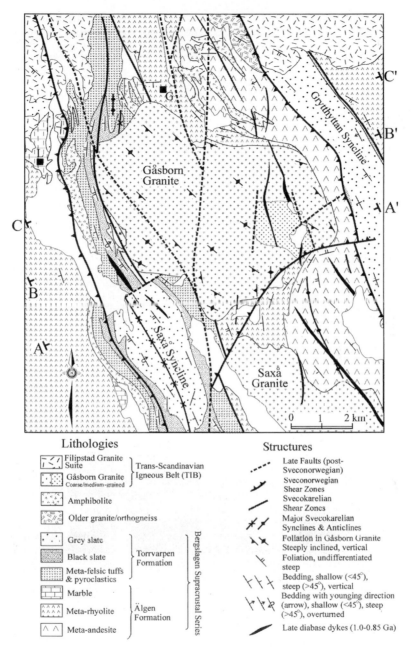

Lithologies

<table>
<tr><td>Filipstad Granite Suite</td><td rowspan="2">Trans-Scandinavian Igneous Belt (TIB)</td></tr>
<tr><td>Gåsborn Granite Coarse/medium-grained</td></tr>
</table>

Amphibolite

Older granite/orthogneiss

Grey slate

Black slate — Torrvarpen Formation

Meta-felsic tuffs & pyroclastics

Marble

Meta-rhyolite — Älgen Formation

Meta-andesite

Bergslagen Supracrustal Series

Structures

- - - - Late Faults (post-Sveconorwegian)

Sveconorwegian Shear Zones

Svecokarelian Shear Zones

Major Svecokarelian Synclines & Anticlines

Foliation in Gåsborn Granite Steeply inclined, vertical

Foliation, undifferentiated steep

Bedding, shallow (<45°), steep (>45°), vertical

Bedding with younging direction (arrow), shallow (<45°), steep (>45°), overturned

Late diabase dykes (1.0–0.85 Ga)

Fig. 4. Geological map of the Gåsborn granite (after Björk 1985; Baker 1985; Baker *et al.* 1988; Lundegård 1987*b*; Lundström 1995*b*, and this study). L, Långban; G, Gåsborn. Note that the Gåsborn granite is discordant to wall-rock structure along its western margin but also that the axial trace of the Saxå syncline is distorted. Along the eastern margin there is considerable wall-rock deflection. These observations, combined with likely rheological differences between the pelitic rocks in the west and the siliceous rocks in the east are the basis for the evaluation of emplacement-related distortion v. post-solidification Svecokarelian deformation. Svecokarelian shear zones occur along the western margin of the pluton (coinciding with the axial trace of the Saxå syncline) and along the contact separating the Älgen and Torrvarpen Formations north and south of the pluton. This shear zone is the inferred conduit for Gåsborn granite. Sveconorwegian shear zones occur both within the Saxå and Grythyttan synclines and as a curvilinear transfer structure SE of the pluton. The locations of cross sections (Fig. 8) are indicated with tick marks.

Svecokarelian and is associated with regional dextral transpression.

The eastern part of the Gåsborn granite intrudes more competent felsic meta-volcanics of the Älgen Formation, exposed in an upright anticline SE of the pluton. Apart from local growth of anthophyllite close to the contact, supracrustal rocks on this side of the pluton have not developed any observed contact meta-morphic assemblages or textures. The contact here is both discordant and related to a significant deflection of the axial trace of the anticline and the meta-rhyolite/meta-andesite contact to the east (Fig. 4). Another important observation is that the wall rocks dip between 30° to 80° towards the pluton contact. These units also young towards the contact, indicating that they pass under the granite without any emplace-ment-related overturning. Exposures of the NNW-trending easternmost contact verify this relationship, displaying a *c.* 45° W dipping intrusive contact characterized by *c.* 1 m wide granite sheets with chilled margins intruding parallel to S0 + S1 in felsic meta-volcanics. The roles of pre-, syn- and post-emplacement deformation in producing the structural pattern around the pluton are discussed in a later section.

The pluton

With the exception of a small area of porphyritic granodiorite in the SE, the Gåsborn granite is composed predominantly of coarse-grained K-feldspar megacrystic biotite granite, which is compositionally similar to the main phase of the Filipstad granite suite of the TIB (Fig. 4). Meta-supracrustal xenoliths and fine-grained grano-dioritic enclaves are rarely observed, small (<50 cm) and most often occur close to the contacts. The granodiorite phase in the SE displays a fine- to medium-grained porphyritic texture with millemetre- to centimetre-sized euhedral K-feldspar phenocrysts, similar to those in the surrounding granite. The granodio-rite is also similar to fine-grained enclaves that occur within the granite, suggesting that both phases are co-magmatic. Contacts of the grano-diorite with the surrounding granite are not well exposed, but are rectilinear at the map-scale suggesting control by faults of unknown age. The fine- grain size of the granodiorite, together with its possible correlation to enclaves in the main granite, suggests that it may represent an early pulse of the Gåsborn granite. It is interpreted to have been emplaced and crystallized close to the roof of the pluton, and subsequently to have foundered into the underlying granite. This interpretation suggests that the erosion level of

the eastern portion of the pluton may be close to the eroded roof and that the original intrusion may have been horizontally stratified, with an upper lid of porphyritic granodiorite.

Most outcrops of the granite contain a weak to strong planar preferred orientation of biotite, quartz aggregates and elliptical to tabular, *c.* 3 × 2 cm K-feldspar megacrysts. The foliation maintains a constant NW strike and steep SW dip throughout the pluton, except close to the NE contact where steep E–W orientations are observed (Figs 4 and 5a). Rare mineral linea-tions, defined by biotite, quartz aggregates and K-feldspar alignment, plunge down dip (Fig. 5a). In thin section, biotite grains are bent and internally kinked. Large quartz grains display prismatic subgrain boundaries and some samples contain both prismatic and basal-plane subgrain boundaries (chessboard pattern; Kruhl 1996). Dynamically recrystallized matrix quartz grains display polygonal granoblastic texture. Plagio-clase grains are fractured, but also contain bent and kinked twins and are locally recrystallized. K-feldspar grains frequently contain flame perthite and they often have recrystallized tails. Formation of muscovite and chlorite, saussuri-tization of plagioclase and development of flame perthites indicate that the foliation developed under upper greenschist facies conditions (e.g. Pryer & Robin 1995). This is consistent with the overall microstructure of the granite, which is dominated by medium-grade type features (e.g. Passchier & Trouw 1996). The local occurrence of chessboard extinction in quartz indicates that deformation of the granite began under high temperature subsolidus con-ditions (Kruhl 1996). Hence, foliation develop-ment the Gåsborn granite, and locally in its wall rocks appears to have occurred as the granite cooled from its solidus down to regional conditions (e.g. Gapais 1989).

Protomylonites with quartz ribbons and mortar textures are observed within and adjacent to centimetre- to metre-wide shear zones that traverse the granite. Similar mineralogy within these rocks and the adjacent foliated granites suggest that shear zone development and the more pervasive, homogeneous foliation develop-ment were broadly coeval. Local deviations of the NW trend of this fabric may be due to preservation of an earlier emplacement-related alignment of K-feldspars and biotites in the pluton.

In order to place limits on the degree of bulk strain associated with post-emplacement over-printing, strain analysis of K-feldspar megacryst populations was carried out using the normalized centre-to-centre Fry analysis method (Fry 1979;

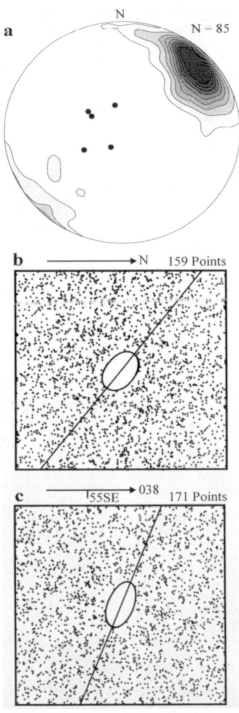

Fig. 5. (a) Equal-area, lower-hemisphere projection of poles to foliations in the Gåsborn granite. 85 poles, Gaussian contoured with $K = 100$ and an extra fine grid. Black dots represent mineral aggregate lineations. **(b and c)** Examples of normalized centre-to-centre Fry plots of K-feldspar centres from a locality in the Gåsborn granite. **(b)** Horizontal surface, ellipse axial ratio = 1.54. **(c)** Steep SE-dipping surface, ellipse axial ratio = 2.0. The foliation in the outcrop is oriented 128°/75°, consistent with these sectional strain ellipses.

Erslev 1988). Oriented photographs were taken of horizontal outcrop surfaces and steeply inclined faces oriented sub-perpendicular to the foliation at 27 sites distributed throughout the pluton. K-feldspar megacryst centres were digitized from the photographs and normalized centre-to-centre plots were produced using the algorithm of Erslev (1988; Fig. 5b and c). Results were then evaluated in terms of the definition of the resulting ellipse and correlation of the ellipse orientations with the foliation measured at each outcrop. This process resulted in eight sites with well-defined strain ellipses on surfaces subparallel to the YZ (horizontal) and XZ (steeply inclined) planes of the local fabric ellipsoid (e.g. Fig. 5). Strain ratios (R) varied from $R_{xz} = 2.0$–1.6, $R_{yz} = 1.3$–1.6, giving estimated values for $R_{xy} = 1.0$–1.4. These results are also in agreement with analyses from sites were only one surface could be analysed.

The fabric observed in the Gåsborn granite can therefore be associated with an average strain ellipsoid with axial ratios $R_{xz} \approx 1.8$, $R_{yz} \approx 1.4$ and $R_{xy} \approx 1.3$, a steeply W-dipping XY-plane and a down-dip X-axis. The pluton has undergone a post-emplacement late Svecokarelian bulk general flattening strain with an approximately down-dip stretching component. Although no independent markers are available to determine absolute displacements associated with this strain, if we assume a c. 10% stretch in the Y-direction (NW–SE), this gives a horizontal shortening of c. 22% and a vertical stretch of c. 40%. However, these strains must be regarded as an upper bound because Fry analysis of K-feldspar centres has not been calibrated against an independent strain gauge, and the total strain is a composite of primary and secondary fabrics. Furthermore, they represent analyses of sites in which the fabric was strong enough to produce well-defined ellipses using the Fry method; many other sites had fabrics too weak to produce a significant result.

Several shear zones of different ages result in map-scale displacements of the pluton's contacts (Fig. 4). The western margin of the pluton is bound by a NNW- to N-trending, steeply W-dipping ductile shear zone that contains S–C fabrics and steep stretching lineations. The displacement on this shear zone is interpreted to be dominantly W-side-up with a component of dextral transcurrent shear compatible with the kinematics of late Svecokarelian deformation along the eastern margin of the TIB further north (Sjöström & Bergman 1996). The mineralogy and microstructure of the protomylonitic granites in this shear zone formed under similar conditions to the main foliation within the

pluton (see above), hence it may have been active during and after emplacement.

The curvilinear ENE-trending shear zone that bounds the SE part of the pluton is interpreted to be a Sveconorwegian structure. This is because it contains a low grade mineral assemblage, displays oblique W-side-up shear sense indicators, and it appears to be a transfer structure associated with the NNW-trending Sveconorwegian reverse shear zone that bounds the W limb of the Grytthyttan syncline (Lundström 1995*a*, *b*). The NW- and N-trending faults that cut the pluton have uncertain age and kinematics. However, their map pattern suggests that they are normal faults, possibly associated with the extensional event that produced the 1.0–0.85 Ga diabase dykes that traverse the region.

Geophysics and 3D shape of the Gåsborn granite

Aeromagnetic data

High-resolution aeromagnetic data acquired by the Geological Survey of Sweden, with a N–S line spacing of 200 m and flight altitude of 30 m, reveals several important aspects of the 3D shape of the Gåsborn granite and the structure of its wall-rocks (Fig. 3, see also detailed aeromagnetic map in Björk 1986). The strong linear anomaly on the western side of the pluton corresponds to pyrite-rich meta-shales of the Torrvarpen Formation and clearly shows the displacement of the Saxå syncline. Likewise, deflection of high susceptibility units of the Älgen Formation is observed around the NE and SE contacts of the granite. The aeromagnetic anomaly associated with the pluton is asymmetric with a lower intensity, relatively homogeneous western half and a higher intensity eastern half with a NE–SW gradient (Fig. 3). This pattern is consistent with a thicker NW–SE-trending body on the eastern side that shallows to the east over higher susceptibility volcanic rocks of the Älgen Formation. The magnetic susceptibility and q-values of the Gåsborn granite are homogeneous over the whole pluton (90–200×10^{-6} SI), with no difference between the eastern and western parts.

Gravity survey

A detailed gravity survey of the Gåsborn granite was carried out to constrain better the 3D shape of the pluton. A total of 555 measurements were made of the granite and its surroundings with a Worden gravimeter. Most measurements were taken along the extensive road network in the area and elevations were constrained using benchmarks and levelling. The precision of these elevations is estimated to be better than 0.5 m, except for a few stations where lake levels and topographic contours on 1:10,000 scale maps were used, in which case the error is higher (5 m). Rock samples were collected from outcrops near some of the gravity stations within the granite and its surroundings and their densities were determined in the laboratory.

Reduction of the levelling and gravity data was performed using standard procedures with a reference density of 2670 kg m^{-3}. Terrain corrections were applied to all stations. The accuracy of the resulting Bouguer anomalies is estimated to be better than 0.2 mGal. Regional gravity data measured by the Geological Survey of Sweden and by the National Land Survey, which were used to supplement the current survey and to determine the regional gravity field, have a somewhat lower precision of 0.4 mGal. Regional gravity anomalies with a station spacing of c. 20 km were removed from the detailed Bouguer gravity field by visual graphical smoothing (e.g. Gupta & Ramani 1980). The resulting residual gravity anomaly map (Fig. 6) is used to interpret the 3D shape of the pluton below.

Residual gravity of the Gåsborn granite and gravity model

The pluton is associated with a markedly asymmetric negative residual gravity anomaly (Fig. 6). The western margin of the pluton roughly coincides with the 5 mGal contour, with residual gravity values falling rapidly towards a NNW-trending 1 to 3 km wide trough. The thickest part of the granite lies at the NW end of this trough and is marked by the 0 mGal contour. At the southern end of the trough, gravity contours display a well defined NNW- to NW-trending linear low that is either due to a linear root or to the effects of the gravity field of the Saxå granite. The eastern part of the pluton has little effect on a strong NW-trending gradient associated with the Grythyttan syncline and the W-dipping Älgen Formation. This suggests that a large part of the eastern half of the pluton is very thin, becoming thicker in the vicinity of the 2 mGal contour. The residual gravity data are therefore consistent with the aeromagnetic data (Fig. 3) and indicate that the Gåsborn granite is characterised by a thicker, rather flat bottomed western part underlain by a NNW-trending root zone of variable depth and

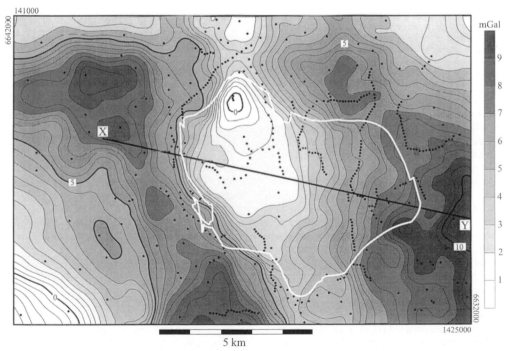

Fig. 6. Residual gravity anomaly map of the Gåsborn granite and surroundings. Dots are gravity stations, white line is the boundary of the pluton, X–Y indicates location of gravity model profile (Fig. 7). Contour interval is 0.5 mGal, shading interval is 1.0 mGal.

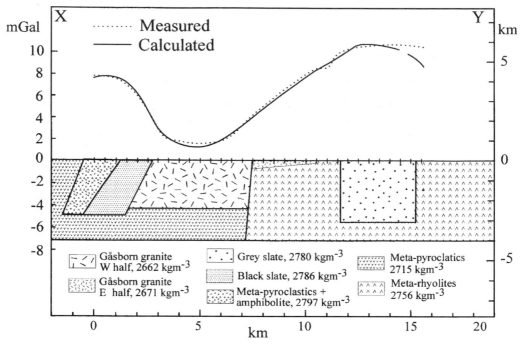

Fig. 7. Measured and calculated residual gravity anomalies along profile X–Y in Fig. 6. Symbols and units are based on those in Fig. 4. Density values assigned to prisms are based on measurements of samples collected in the field.

width, and a very shallow, inward dipping flap of granite in the east.

In gravity studies of granites it is common practice to derive a 3D model of the residual gravity anomaly (e.g. Vigneresse 1990, 1995; Dehls *et al.* 1998). In cases where this is most successful, the margin of the pluton conforms closely to one of the residual gravity contours, and positive anomalies in the surroundings are simple or absent. Strong positive anomalies associated with the Saxå and Grythyttan synclines as well as the complex nature of the negative anomaly associated with the granite (Fig. 6) make derivation of a 3D gravity model of the pluton difficult and prone to large errors. Therefore, in order to provide estimates on the thickness variations of the Gåsborn granite we have confined our analysis to a simple 2.5D model.

Results of a WNW ESE 2.5D model profile across the pluton are shown in Fig. 7. Density values used in the model are based on measurements of 57 samples of the Gåsborn granite and more than 400 samples of the surrounding rocks. The geometry of rock bodies is highly idealised after the geological map (Fig. 4). The positive anomalies associated with the Saxå and Grythyttan synclines are modelled reasonably as dense subvertical bodies extending to depths of *c.* 3 km.

Supracrustal units adjacent to these structures are assigned densities representative for a mixture of amphibolites and meta-pyroclastics, and Älgen Formation meta-volcanics, respectively. A good fit between the calculated and modelled negative anomaly associated with the Gåsborn granite is achieved by a *c.* 2 km thick granite body in the west with a steeply outward dipping contact and a 100–300 m thick tabular unit of granite in the east (Fig. 7). A slightly higher mean density was used for the east part of the granite based on results of the density measurements.

Note that the shapes used in the model are not intended to simulate nature exactly, rather to place constraints on the thicknesses of rock bodies responsible for the gravity anomalies. Refinements to the model could include modification of the flat base of the western body to have a deeper root and a thinner western part, and allowing for a more gradual thickening of the eastern body towards the root. In the model of the western part, a 2 mGal difference corresponds to a thickness of granite of about 1000 m (Fig. 7). Applying this to the residual anomaly map gives an estimated thickness of the deepest part of the root (i.e. <0 mGal) of *c.* 3 km, and the shallowest part of the western area is *c.* 1 km thick in the SW. Assigning a lower density

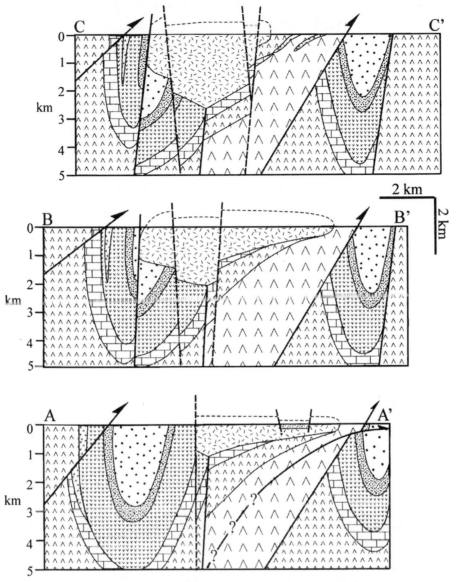

Fig. 8. Cross sections A–A′ to C–C′ (see Fig. 4 for locations and legend) through the Gåsborn granite. The sections are interpretations based on geological mapping of the pluton, constrained at depth by local aeromagnetic data and the gravity data shown in Figs 6 and 7. The original asymmetric geometry of the pluton is still distinct in spite of local overprinting by Sveconorwegian reverse shear zones and steep, post-Sveconorwegian normal faults. A major Svecokarelian shear zone occurs below the pluton (the interpreted conduit in all three profiles). A second, dextral, W-side-up Svecokarelian shear zone is parallel to the axial plane of the Saxå syncline (profiles B–B′ and C–C′).

(i.e. 2662 kg m^{-3} rather than 2671 kg m^{-3}) to the thin flap in the east would increase its model thickness by c. 100 m. Alternatively, using an average value of 2666 kg m^{-3} would result in a thinner western part and thicker eastern part, but does not significantly alter the overall shape of the profile in Fig. 7.

3D structure of the Gåsborn granite

Results of the gravity survey and structural mapping of the pluton's wall rocks are integrated in three NE–SW cross-sections (Fig. 8; profile lines indicated in Fig. 4). The sections show the tightening and rotation of the Saxå syncline noted previously. Shallowing and downward deflection of the meta-rhyolite/meta-andesite contact beneath the eastern flap of granite is inferred by projection of contacts into the section lines. The 3D shape of the granite is based on the magnetic and gravity anomaly data as well as geological mapping. The dips and throws on shear zones and faults that cut the granite are poorly constrained estimates based on the map pattern and gravity data.

Interpretation

Significance of Svecokarelian and Sveconorwegian overprinting

The effects of penetrative late Svecokarelian ductile strains and Sveconorwegian shear zone displacements on the structure of W Bergslagen, although currently poorly constrained, are likely to be significant. We have attempted to account for these effects in the Gåsborn granite qualitatively so that they are not incorrectly attributed to emplacement.

The shape of the pluton in plan view, after removal of lateral displacements associated with shear zones, is roughly circular, although in detail the pluton contacts are quite rectilinear (Fig. 9). Gaps in the restoration are due to material removed out of the map plane by vertical displacements. Development of the NW-trending foliation in the pluton is consistent with its behaviour as a relatively competent inclusion within a general dextral shear regime (Fig. 9). The angle between the 135° foliation and the mean 155° trend of Svecokarelian shear zones is c. 20° in plan view, consistent with a dextral transpressive regime. This is in agreement with the regional kinematics during the late stages of Svecokarelian deformation along the eastern margin of the TIB (Bergman & Sjöström 1996) and also supports the proposal that the TIB, including the Gåsborn granite, was emplaced within a dextral transpressive regime (Stephens & Wahlgren 1996; Sjöström & Bergman 1996). Less competent wall rocks would have experienced greater strains with a flattening plane lying closer to the NNW Svecokarelian trend (Fig. 9). Because the competence contrast between the granite and its surroundings is unknown, the magnitude of Svecokarelian strain in the wall-rocks cannot be estimated with certainty. However, an order of magnitude difference in viscosity could account for the observed deflection of the trace of the Saxå syncline around the western margin. A much higher viscosity contrast and/or a greater regional strain is required to remove the curvature of the

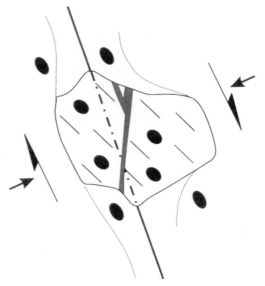

Fig. 9. Sketch of the kinematics of regional Svecokarelian deformation and restored shape of the Gåsborn granite after removal of lateral offsets on faults. Dark grey indicates missing area due to displacements on faults out of the map plane. Ellipses in the pluton are the average sectional YZ ellipses associated with the foliation (dashed lines), as determined by Fry analysis of K-feldspars. Ellipses in the wall-rocks are the inferred regional strain increment due to Svecokarelian transpression. Heavy black solid (observed) and dashed (projected) line is the tectonic contact between the Torrvarpen and Älgen Formations, interpreted as the conduit for the Gåsborn granite magma. Thin lines represent the traces of the distorted Saxå syncline on the west side and an important stratigraphic contact on the east side of the pluton, respectively.

meta-rhyolite/meta-andesite contact on the eastern margin. Given the siliceous composition of the rocks in the Älgen Formation this seems unlikely, suggesting that a significant part of the observed wall-rock deflection east of the pluton is emplacement related, whereas the distortion of the Saxå syncline in pelitic rocks of the Torrvarpen Formation is most likely a result of post-emplacement, Svecokarelian deformation.

Sveconorwegian sinistral transpression in W Bergslagen resulted in the formation of steep to moderately W-dipping reverse shear zones and at least one NE-trending oblique reverse transfer structure that effects the SE margin of the Gåsborn granite. With the exception of foliation development in these shear zones, Sveconorwegian overprinting does not appear to have resulted in the formation of penetrative fabrics in the study area. However, we cannot rule out the possibility that some of the apparent deflection of stratigraphic markers adjacent to the NE and SE contacts of the Gåsborn granite is due to Sveconorwegian strain.

Emplacement of the Gåsborn granite

The pre-deformational geometry of the Gåsborn granite is that of an asymmetric tabular sheet that thickens from E to W. The western half of the pluton is underlain by a NNW-trending root zone that displays a marked variation in width and depth extent along strike, being deepest at the northern end of the structure. The trace of the root zone approximately follows the projection of an important older Svecokarelian shear zone beneath the pluton (i.e. the faulted contact between the Älgen and Torrvarpen formations; Fig. 4). Emplacement of the thicker western part was discordant to supracrustal units of the Saxå syncline and does not appear to have imposed any significant strain on these units, once post-emplacement Svecokarelian strain effects have been considered. Most of the western margin of the pluton is bound by a Svecokarelian shear zone that may have been active during emplacement of the granite. Space creation must therefore have been accomplished by removing material out of the present level of exposure. Potential mechanisms include stoping (e.g. Marsh 1982; Paterson et al. 1996), roof lifting and/or floor depression (e.g. Corry 1988; Petford 1996; Cruden 1998; Grocott et al. 1998). Of these options, stoping is the least likely since wall-rock xenoliths are rarely observed in the granite and the 3D shape inferred from the gravity data is incompatible with a magma body that has eaten its way into place, which would have resulted in vertical sides and uniform thickness. There are no observations available that can differentiate the relative roles of roof lifting and floor depression in creating space for the western part of the pluton. However, the emplacement depth of c. 10 km inferred from contact metamorphic assemblages is on the high side for emplacement of laccoliths, which tend to occur at depths <3 km (Corry 1988). The inclination of the pluton floor towards the root of the granite, as suggested by the shape of the residual gravity anomaly (Fig. 6), is compatible with the cantilever mechanism of floor depression, in which space is made for the magma by rotation of the floor of the pluton toward the conduit (Cruden 1998; Dehls et al. 1988). A component of vertical downward translation of the floor may have also been accommodated by displacement on the shear zone along the steeply west-dipping western margin of the pluton, which has a component of pluton-side-down, post-emplacement shear. An alternative possibility is that some of the space for the magma could have been made by roof lifting on oblique reverse shear zones active during syn-emplacement regional transpression, a situation modelled elegantly by Benn et al. (1998).

The thickness of the eastern part of the pluton was probably not much greater than the current estimate of ⩽300 m, judging from the presence of the down-dropped medium-grained granodiorite body, which is interpreted to have crystallized close to the roof of the pluton. Emplacement of this sub horizontal sheet was also discordant to the wall rocks in map view and is transgressive to the stratigraphy in cross section (Fig. 8). However, arguments presented above suggest that emplacement of the eastern sheet was responsible for some of the symmetrical ductile deflection of marker horizons on its N and S sides (Fig. 9). It may have also resulted in a rotation of beds on the western limb of the anticline to shallower dips (Fig. 8). This was most likely achieved by the drag effect of the magma body as it spread laterally from the root zone to the east, and/or by downward depression of rocks beneath the sheet as it inflated. The first mechanism should also be accompanied by significant shear strains in the wall rocks adjacent to the sheet. Since these are not observed we favour the latter mechanism for producing the observed structure adjacent to the eastern part of the pluton.

The NNW-trending root zone of the Gåsborn granite lies within a regional gravity structure parallel to the structural grain in W Bergslagen that connects with the TIB to the north and the Saxå granite to the south (Figs 2 and 3). This structure may therefore have fed parts of the TIB, and the Saxå and Gåsborn granites. The trace of the root zone within the Gåsborn granite also appears to be approximately coincident with the inferred tectonic contact between the Älgen and Torrvarpen formations. It is likely that this structure was repeatedly active during the Svecokarelian orogeny and the emplacement of the TIB and may have acted as a locus for magma transport for the Gåsborn and Saxå granites.

An emplacement model for the Gåsborn granite based on the above observations and arguments is presented in Fig. 10. The initial phase of intrusion involved the establishment of a magma conduit, possibly localised on the tectonic contact between the Älgen and Torrvarpen formations, and emplacement of a thin sill at a depth of c. 10 km (Fig. 10a). Upward propagation of magma was stopped by an unknown sub-horizontal mechanical barrier that was at a high angle to steeply inclined S0 in the wall rocks. Candidates include the brittle–

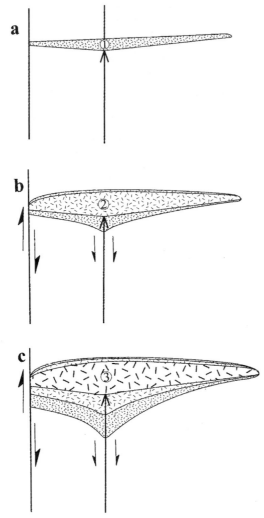

Fig. 10. Schematic NE–SW cross-section, appoximately equivalent to B–B' in Fig. 8, illustrating one possible model to account for the emplacement of the Gåsborn granite. (**a**) Initial ascent controlled by an older Svecokarelian shear zone and subsequent sill emplacement into the folded Bergslagen supracrustal series (stage 1 magma, equivalent to the medium-grained granodiorite phase of the pluton). (**b**) Inflation of the pluton is accommodated by downwarping of rocks beneath the eastern tongue, and floor depression accommodated by displacement on an active Svecokarelian fault on the west side (stage 2 magma). (**c**) Continued vertical inflation, involving both downward ± upward displacements (stage 3 magma) results in the formation of an asymmetric pluton with dimensions equivalent to the Gåsborn granite. Subsequent Svocokarelian ductile strains, Svoconorwegian brittle-ductile shear zone displacemements, and late Proterozoic brittle normal faults resulted in the final geometry of the pluton (Figs 4 and 8).

ductile transition (e.g. Brown 1994) or a sub-horizontal, freely-slipping fracture (e.g. Clemens & Mawer 1992). Horizontal sill propagation and lateral flow of overpressured magma was more effective in the Älgen Formation compared to the meta-shales and meta-pyroclastics of the Saxå syncline. This was probably a consequence of the more brittle behaviour of the siliceous meta-volcanics, which favoured propagation of the crack tip at the edge of the sill. Tip-propagation into meta-shales in the west on the other hand was either arrested quickly due to the relatively high ductility of these rocks or to intersection of the propagating crack tip with an active N–S-trending shear zone.

After initial sill-emplacement the pluton was inflated to its final thickness, at which time the magma driving pressure became equal to the lithostatic load (e.g., Pollard & Johnston 1973; Hogan *et al.* 1998). Inflation of the eastern part was accommodated mainly by tilting of the floor towards the conduit and ductile deflection of the underlying rocks (Fig. 10b). Vertical growth of the western part likely occurred by a combination of roof lifting and floor depression on pre-intrusive wall-rock discontinuities, possibly assisted by syn-emplacement transpression, and tilting of the floor (e.g. Dehls *et al.* 1998; Benn *et al.* 1998). The most important of these structures was the N-trending shear zone that bounds the western margin of the pluton, but the rectilinear character of many of the other pluton contacts supports the notion that other, minor discontinuities in the wall-rock were utilized during the space-making process. A similar scenario has been proposed for similar sized sub-circular Proterozoic Rapakivi granites in the Ketilidian orogen, S Greenland, whose 3D shape has been determined by field mapping as opposed to gravity (Grocott *et al.* 1998).

The final shape and outcrop pattern of the Gåsborn granite is a consequence of post-emplacement late Svecokarelian deformation, and to a lesser extent Sveconorwegian shear zones and late Proterozoic normal faults (Fig. 4). Svecokarelian deformation resulted in NE–SW shortening of the pluton and its wall rocks and continued oblique reverse dextral displacement on steep SW side-up ductile shear zones (Fig. 10c). Much of the map-scale deflection of units on the western and eastern sides of the pluton is attributed to this deformation event rather than to emplacement of the granite. Note that although the Gåsborn granite is syntectonic with respect to late-Svecokarelian dextral transpression, there is a marked contrast in the contribution of regional tectonics to the space-making process for the pluton and its subsequent ductile deformation during cooling to regional greenschist facies conditions. This is a consequence of the relative rates of pluton emplacement versus regional deformation. The time required for sill-emplacement and vertical growth of a pluton the size of the Gåsborn granite could have been very rapid (tens to tens of thousands of years), depending on the magma supply rate (e.g. Petford 1996; Cruden 1998). In contrast regional ductile Svecokarelian ductile strains were likely accumulated over a time frame of millions of years, far longer than the duration of pluton emplacement, crystallisation and cooling (e.g. Paterson & Tobisch 1992).

Conclusions

An integrated structural and geophysical approach has led to a model for the 3D geometry of the Gåsborn granite and allowed the separation of syn- and post-emplacement strains. In the absence of extreme relief or deep drilling, this approach is considered to be the most effective method for the evaluation of pluton emplacement mechanisms. In the case of the Gåsborn granite, we propose that it was emplaced via a NNW-trending magma conduit coincident with an early regional structure formed in the Svecokarelian orogeny. This structure and many related parallel ones may have supplied the bulk of magma to the TIB. Growth of the Gåsborn granite to its current thickness occurred by a combination of roof lifting and floor depression possibly assisted by syn-emplacement regional dextral transpression. The observed asymmetry of the pluton is a consequence of the different mechanical properties of the wall rocks on the western and eastern sides. Many sub-circular plutons, traditionally attributed to diapirism or stoping, may have similar emplacement histories.

Initial research on the Gåsborn granite by H.S. and S.A. was carried out during the project 'Granit och gnejsplutonernas dynamiska utveckling i det Svenska urberget' (NFR grant G2392-106-109, STU grant 79-5643), directed by H. Ramberg. Later work by A.R.C. was supported by a Linnéus travel stipend and Uppsala University. Additional contributions were made by J. O. Arnbom, H. Skogby and L. Hansen on the metamorphic and igneous petrology of the pluton. Comments on a draft of the manuscript by C.-H. Wahlgren, and reviews by K. Benn and A. Berger helped to clarify our presentation considerably.

References

ALLEN, R. L., LUNDSTRÖM, I., RIPA, M., SIMEONOV, A. & CHRISTOFFERSON, H. 1996. Facies Analysis of a 1.9 Ga, Continental Margin, Back-Arc, Felsic Caldera Province with Diverse Zn–Pb–Ag–(Cu–Au) Sulfide and Fe Oxide Deposits, Bergslagen Region Sweden. *Economic Geology*, **91**, 979–1008.

ANDERSON, U. B. 1997. *The late Sveocofennian, high-grade contact and regional metamorphism in southwestern Bergslagen (central southern Sweden)*. Unpublished report, Swedish Geological Survey.

BAKER, J. H. 1985. *The petrology and geochemistry of 1.8–1.9 Ga granitic magmatism and related subseafloor hydrothermal alteration and ore-forming processes, W. Bergslagen, Sweden*. PhD thesis, University of Amsterdam, *GUA Papers of Geology*, Series 1, No. 21.

——, BAKER, J. H., HELLINGWERF, R. H. & OEN, I. S. 1988. Structure, stratigraphy and ore-forming processes in Bergslagen; implications for the development of the Svecofennian of the Baltic Shield. *Geologie en Mijnbouw*, **67**, 121–138.

BENN, K., ODONNE, F. & DE SAINT BLANQUAT, M. 1998. Pluton emplacement during transpression in brittle crust: New views from analogue experiments. *Geology*, **26**, 1079–1082.

BERGMAN, S. & SJÖSTRÖM, H, 1994, *The Storsjön-Edsbyn Deformation Zone, central Sweden*. Unpublished report, Swedish Geological Survey.

—— & —— 1996. Metamorphic and tectonic evolution in south central Sweden. *Geologiska Föreningens i Stockholm Förhandlingar*, **118**, Jubilee Issue, A8-9.

BERTHELSEN, A. 1980. Towards a palinspastic analysis of the Baltic Shield. *In*: COGNÉ, J. & SLANSKY, M. (eds) *Geology of Europe from Precambrian to Post-Hercynian Sedimentary Basins*. International Geological Congress, Colloquium **C6**, Paris, 5–21.

BEUNK, F. F., PAGE, L. M., WIJBRANS, J. R. & BARLING, J. 1996. Deformational, metamorphic and geochronological constraints from the Loftahammar—Linköping Deformational Zone (LLDZ) in SE Sweden: implications for the development of the Svecofennian Orogen. *Geologiska Föreningens i Stockholm Förhandlingar*, **118**, Jubilee Issue, A9.

BJÖRK, L. 1985. *Berggrundskartan 11E NV*. SGU Af, **147**.

—— 1986. *Beskrivning till berggrundskartan Filipstad NV (with English summary)*. SGU Af, **147**.

BOTT, M. H. P. & SMITHSON, S. B. 1967. Gravity investigations of subsurface shape and mass distributions of granite batholiths. *Geological Society of America Bulletin*, **78**, 859–878.

BOTHNER, W. A. 1974. Gravity study of the Exeter pluton, southeastern New Hampshire. *Geological Society of America Bulletin*, **85**, 51–56.

BOUCHEZ, J.-L., GLEIZES, G., DJOUADI, T. & ROCHETTE, P. 1990. Microstructure and magnetic susceptibility applied to emplacement kinematics of granites: the example of the Foix pluton (French Pyrenees). *Tectonophysics*, **184**, 157–171.

BRIDGWATER, D., SUTTON, J. & WATTERSON, J. 1974. Crustal downfolding associated with igneous activity. *Tectonophysics*, **21**, 57–77.

BROWN, M. 1994. The generation, segregation, ascent and emplacement of granite magma: the migmatite-to-crustally-derived granite connection in thickened orogens. *Earth Science Reviews*, **36**, 83–130.

BRUN, J. P., GAPAIS, D. & COGNE, J. P. 1990. The Flamanville granite (northwest France): an unequivocal example of a syntectonically expanding pluton. *Geological Journal*, **25**, 271–286.

CASTRO, A. 1986. Structural pattern and ascent model in the Central Extramadura batholith, Hercynian belt, Spain. *Journal of Structural Geology*, **8**, 633–645.

CLEMENS, J. D. & MAWER, C. K. 1992. Granitic magma transport by fracture propagation. *Tectonophysics*, **204**, 339–360.

COLEMAN, D. S., GLAZNER, A. F., MILLER, J. S., BRADFORD, K. J., FROST, T. P., JOYE, J. L. & BACHL, C. A. 1995. Exposure of a Late Cretaceous layered mafic–felsic magma system in the central Sierra Nevada batholith, California. *Contributions to Mineralogy and Petrology*, **120**, 129–136.

CORRY, C. E. 1988. *Laccoliths: Mechanics of emplacement and growth*. Geological Society of America, Special Papers, **220**.

CRUDEN, A. R. 1998. On the emplacement of tabular granites. *Journal of the Geological Society*, **155**, 853–862.

—— & AARO, S. 1992. The Ljugaren granite massif, Dalarna, central Sweden. *Geologiska Föreningens i Stockholm Förhandlingar*, **114**, 209–225.

—— & LAUNEAU, P. 1994. Structure, magnetic fabric and emplacement of the Archean Lebel Stock, S.W. Abitibi Greenstone belt. *Journal of Structural Geology*, **16**, 677–691.

DEHLS, J. D., CRUDEN, A. R. & VIGNERESSE, J. L. 1998. Fracture control of late-Archean pluton emplacement in the Northern Slave Province, Canada. *Journal of Structural Geology*, **20**, 1145–1154.

ERSLEV, E. A. 1988. Normalized center-to-center strain analysis of packed aggregates. *Journal of Structural Geology*, **10**, 201–209.

FRY, N. 1979. Random point distributions and strain measurement in rocks. *Tectonophysics*, **60**, 806–807.

GAÁL, G. & GORBATSHEV, R. 1987. An outline of the Precambrian evolution of the Baltic Shield. *Precambrian Research*, **35**, 15–52.

GAPAIS, D. 1989. Shear structures within deformed granites: mechanical and thermal interactions. *Geology*, **17**, 1144–1147.

GROCOTT, J., GARDE, A., CHADWICK, B., CRUDEN, A. R. & SWAGER, C. 1998. Emplacement of Rapakivi granite and syenite by floor depression and roof uplift in the Paleoproterozoic Ketilidian orogen, South Greenland. *Journal of the Geological Society, London*, **156**, 15–24.

GUPTA, V. K. & RAMANI, N. 1980. Some aspects of regional residual separation of gravity anomalies in a Precambrian terrain. *Geophysics*, **45**, 1412–1426.

HOGAN, J. P., PRICE, J. D. & GILBERT, M. C. 1998. Magma traps and driving pressure: consequences for pluton shape and emplacement in an extensional regime. *Journal of Structural Geology*, **20**, 1155–1168.

HÖGDAHL, K., BERGMAN, S. & SJÖSTRÖM, H. 1996. Geochronology and tectonic evolution of the Hagsta Gneiss Zone, east Central Sweden. *In*: KORHONEN, T. and LINDBERG, B. (eds) *Abstracts of oral and poster presentations 22nd Nordic Geological Winter Meeting*, Åbo, Finland, 73.

HUTTON, D. W. H. 1988. Granite emplacement mechanisms and tectonic controls: inferences from deformation studies. *Transactions of the Royal Society of Edinburgh*, **79**, 245–255.

JARL, L.-G. & JOHANSSON, L. 1988. U.Pb zircon ages of granitoids from the Småland–Värmland granite-porphyry belt, southern and central Sweden. *Geoloogiska Föreningens i Stockholm Förhandlingar*, **110**, 21–28.

KORJA, A. & HEIKKINEN, P. J. 1995. Proterozoic extensional tectonics of the central Fennoscandian shield: Results from the Baltic and Bothnian echoes from the Lithosphere experiment. *Tectonics*, **14**, 504–517.

——, KORJA, T., LUOSTO, U. & HEIKKINEN, P. J. 1993. Seismic and geoelectric evidence for collisional and extensional events in the Fennoscandian Shield—implications for Precambrian crustal evolution. *Tectonophysics*, **219**, 129–152.

KRUHL, J. H. 1996. Prism- and basal-plane parallel subgrain boundaries in quartz: a microstructural geothermobarometer. *Journal of Metamorphic Geology*, **14**, 581–589.

LUNDEGÅRDH, P. H. 1987a. Beskrivning till berggrundskartan Filipstad SV (with English summary). SGU Af, **157**.

—— 1987b. Berggrundskartan 11F Filipstad SV. SGU Af, **157**.

LUNDSTRÖM, I. 1988. Regional interrelationships in the Proterozoic of Bergslagen and southeastern central Sweden. *Geologie en Mijnbouw*, **62**, 157–164.

—— 1990. Proterozoic crustal evolution in Bergslagen, south-central Sweden—a brief review. *Geologiska Föreningens i Stockholm Förhandlingar*, **112**, 186–188.

—— 1995a. Beskrivning till berggrundskartorna Filipstad SO and NO (with English summary). SGU Af, **177**, **185**

LUNDSTRÖM, I. 1995b. Berggrundskartan 11F Filipstad NO. SGU Af, **185**.

MARRE, J. 1986. *The structural analysis of granitic rocks*. North Oxford Academic, Oxford.

MARSH, B. D. 1982. On the mechanics of igneous diapirism, stoping and zone melting. *American Journal of Science*, **282**, 808–855.

MYERS, J. S. 1975. Cauldron subsidence and fluidization: mechanisms of intrusion of the coastal batholith of Peru into its own volcanic ejecta. *Geological Society of America Bulletin*, **86**, 1209–1220.

OEN, I. S., HELMERS, H., VERSCHURE, R. H. & WIKLANDER, U. 1982. Ore deposition in a Proterozoic incipient rift zone environment: A tentative model for the Filipstad–Grythyttan–

Hjulsjö region, Bergslagen, Sweden. *Geologische Rundschau*, **71**, 182–194.

ÖHLANDER, B. & ROMER, R. L. 1996. Zircon ages occurring along the Central Swedish Gravity Low. *Geologiska Föreningens i Stockholm Förhandlingar*, **118**, 217–225.

PASSCHIER, C. W. & TROUW, R. A. J. 1996. *Microtectonics*. Springer-Verlag, Berlin.

PAGE, L. M., STEPHENS, M. B. & WAHLGREN, C.-H. 1996. 40Ar/39Ar geochronological constraints on the tectonothermal evolution of the Eastern Segment of the Sveconorwegian Orogen, south-central Sweden. *In*: BREWER, T. S. (ed.) *Precambrian Crustal Evolution in the North Atlantic Region. Geological Society, London, Special Publications*, **112**, 315–330.

PATERSON, S. R. & TOBISCH, O. T. 1992. Rates of processes in magmatic arcs: implications for the timing and nature of pluton emplacement and wall rock deformation. *Journal of Structural Geology*, **14**, 291–300.

——, FOWLER, T. K. JR & MILLER, R. B. 1996. Pluton emplacement in arcs: a crustal-scale exchange process. *Transactions of the Royal Society of Edinburgh: Earth Sciences*, **87**, 115–123.

——, VERNON, R. H. & FOWLER, T. K. 1991. Aureole Tectonics. *In*: KERRICK, D. M. (ed.) *Contact Metamorphism*. Mineralogical Society of America Reviews in Mineralogy, **26**, 673–722.

PERSSON, P.-O. & RIPA, M. 1993. U–Pb zircon dating of a Järna-Type granite in western Bergslagen, south-central Sweden. *In*: LUNDQVIST, T. (ed.) *Radiometric dating results, SGU C*, **823**, 41–45.

PETFORD, N. 1996 Dykes or diapirs? *Transactions of the Royal Society of Edinburgh: Earth Sciences*, **87**, 105–114.

PITCHER, W. S. 1993. *The nature and origin of granite*. Blackie Academic, London.

—— & BERGER, A. R. 1972. *The geology of Donegal: a study of granite emplacement and unroofing*. John Wiley & Sons, New York.

POLLARD, D. D. & JOHNSON, A. M. 1973. Mechanics of growth of some laccolithic intrusions in the Henry Mountains, Utah, II. Bending and failure of overburden layers and sill formation. *Tectonophysics*, **18**, 311–354.

PRYER, L. & ROBIN, P.-Y.F. 1995. Retrograde metamorphic reactions in deforming granites and the origin of flame perthites. *Journal of Metamorphic Geology*, **13**, 645–658.

SCAILLET, B., PÊCHER, A., ROCHETTE, P. & CHAMPENOIS, M. 1995. The Gangotri granite (Garhwal Himalaya): Laccolithic emplacement in an extending collisional belt. *Journal of Geophysical Research*, **100 (B1)**, 585–607.

SJÖSTRÖM, H. & BERGMAN, S. 1995. Deformation zones in central Sweden—relation to the Svecokarelian orogen. *In*: KORHONEN, T. & LINDBERG, B. (eds) *Abstracts of oral and poster presentations 22nd Nordic Geological Winter Meeting*, Åbo, Finland.

—— & —— 1996. Regional structures in central Sweden. *Geologiska Föreningens i Stockholm Förhandlingar*, **118**, Jubilee Issue, A25.

—— & —— 1998. Svecofennian Metamorphic and Tectonic Evolution of East Central Sweden. *Unpublished Report, Swedish Geological Survey*, 50 pp.

STÅLHÖS, G. 1981. A tectonic model for the Sveokarelian folding in east central Sweden. *Geologiska Föreningens i Stockholm Förhandlingar*, **103**, 33–46.

—— 1984. Svecokarelian folding and interfering macrostructures in eastern Central Sweden. *In*: KRÖNER, A. & GREILING, R. (eds) *Precambrian Tectonics Illustrated*. Schweizerbartsche Verlagsbuchhandlung, 369–379.

—— 1991. *Beskrivning till berggrundskartorna Östhammar NV, NO, SV, SO*. SGU Af, **161**, **166**, **169**, **172**.

STEPHENS, M. B. & WAHLGREN, C.-H. 1995. Thermal and mechanical responses to bilateral collision in the Svecokarelian orogen. *In*: KORHONEN, T. & LINDBERG, B. (eds) *Abstracts of oral and poster presentations 22nd Nordic Geological Winter Meeting*, Åbo, Finland, 203.

—— & —— 1996. Post 1.85 Ga tectonic evolution of the Svecokarelian orogen with special reference to central and SE Sweden. *Geologiska Föreningens i Stockholm Förhandlingar*, **118**, Jubilee Issue, A26.

——, —— & Weihed, P. 1994. *Geological Map of Sweden*. SGU Ba, **52**.

——, —— & —— 1997. Sweden. *In*: MOORES, E. M. & FAIRBRIDGE, R. W. (eds) *Encyclopedia of European and Asian regional geology*. Chapman & Hall, London, 690–704.

SUNDBLADH, C., AHL, M. & SCHÖBERG, H. 1993. Age and geochemistry of granites associated with Mo-mineralizations in western Bergslagen, Sweden. *Precambrian Research*, **64**, 319–335.

SWEENEY, J. F. 1976. Subsurface distribution of granitic rocks, south-central Maine. *Geological Society of America Bulletin*, **87**, 241–249.

VÄISÄNEN, M., HÖLTTÄ, P., RASTAS, J., KORJA, A. & HEIKKINEN, P. 1994. Deformation, metamorphism and the deep structure of the crust in the Turkku area, southwestern Finland. *In*: PAJUNEN, M. (ed.) *High temperature-low pressure metamorphism and deep crustal structures*. Geological Survey of Finland, **Guides**, **37**, 35–41.

VERSCHURE, R. H., OEN, I. S. & ANDRIESSEN, P. A. M. 1987. Isotopic age-determinations in Bergslagen, Sweden: VIII. Sveconorwegian Rb–Sr resetting and anomalous radiogenic argon in the Gothian Transcandinavian Småland–Värmland Granitic Belt and bordering parts of the Svecokarelian Bergslagen Region. *Geologie en Mijnbouw*, **66**, 111–120.

VIGNERESSE, J. L. 1990. Use and misuse of geophysical data to determine the shape at depth of granitic intrusions. *Geological Journal*, **25**, 249–260.

—— 1995. Crustal regime of deformation and ascent of granitic magma. *Tectonophysics*, **249**, 187–202.

—— & BOUCHEZ, J. L. 1997. Successive Granitic Magma Batches During Pluton Emplacement: the Case of Cabeza de Araya (Spain). *Journal of Petrology*, **38**, 1767–1776.

WAHLGREN, C.-H & STEPHENS, M. B. 1990. Post-Svecofennian plastic and brittle-plastic shear zones in westernmost Bergslagen, Sweden. *Geologiska Föreningens i Stockholm Förhandlingar*, **112**, 204–205.

—— & —— 1996. Polyphase tectonometamorphic reworking of the Transscandinavian Igneous Belt, east of Lake Vänern—regional tectonic implications for southern Sweden. *Geologiska Föreningens i Stockholm Förhandlingar*, **118**, Jubilee Issue, A27–A28.

——, CRUDEN, A. R. & STEPHENS, M. B. 1994. Kinematics of a major fan-like structure in the eastern part of the Sveconorwegian Orogen, Baltic Shield, south-central Sweden. *Precambrian Research*, **70**, 67–91.

WELIN, E. 1987. The depositional evolution of the Svecofennian supracrustal sequence in Finland and Sweden. *Precambrian Research*, **35**, 95–113.

WIKSTRÖM, A. & AARO, S. 1986. *The Finspång augen gneiss massif.* SGU, **C813**.

WILSON, M. R. 1982. Magma types and the tectonic evolution of the Swedish Proterozoic. *Geologische Rundschau*, **71**, 120–129.

Emplacement of the Joshua Flat–Beer Creek Pluton (White Inyo Mountains, California): a story of multiple material transfer processes

CARLO DIETL

Ruprecht-Karls-Universität Heidelberg, Geologisch-Paläontologisches Institut,
Im Neuenheimer Feld 234, D-69120 Heidelberg, Germany
(e-mail: f16@ix.urz.uni-heidelberg.de)

Abstract: The Joshua Flat–Beer Creek Pluton (JBP) in the White Inyo Mountains, California, is part of the Inyo batholith, which intruded a Neoproterozoic to Lower Cambrian metasedimentary sequence at about 180–160 Ma ago. Contact metamorphism around the JBP reached hornblende-hornfels- to lower amphibolite-facies conditions. The intrusion consists of three distinct phases. Field relations suggest an intrusion sequence Marble Canyon Diorite–Joshua Flat Monzonite–Beer Creek Granodiorite as nested diapirs. In the Marble Canyon Diorite as well as the Joshua Flat Monzonite the subsequent intrusions led to brecciation and stoping, but also to mingling and mixing between the Joshua Flat Monzonite and the Beer Creek Granodiorite. Fabrics in the contact aureole document several emplacement mechanisms such as stoping and dyking, ductile downward flow, partial melting and magma chamber expansion. The relative importance of the different emplacement mechanisms through time is as follows: the Marble Canyon Diorite probably intruded as dyke/sill, whereas stoping and dyking, ductile downward flow together with assimilation acted during the emplacement of the Joshua Flat Monzonite. Magma chamber expansion represents only the latest stage of intrusion during the emplacement of the Beer Creek Granodiorite into the already existing magma chamber of the JBP. AMS, quartz c-axis and strain measurements support the field observations.

Since the question about the origin of granitoids was decided for the 'plutonistic' faction in the 1950s, geoscientists argue about the main problem of pluton emplacement: which processes make space for a rising molten batch of rock? As one result of the debate two major schools were established, one preferring dyking the other diapirism as the main intrusion mode.

Magmatic overpressure acting in individual dykes together with an extensional regional stress regime and a final forceful expansion of the magma chamber are proposed by 'dykers' to be the major emplacement process for constructing an extended magma chamber. In contrast, to 'diapirists' buoyancy and vertical material transport are the most important features for the ascent and emplacement of plutons.

As already stated by Paterson & Fowler (1993) most emplacement processes do not really make space for a rising pluton, but lead to a replacement and redistribution of the wall rocks due to horizontal and vertical material-transfer processes. But, according to Buddington (1959), a combination of several material-transfer processes active at different times and places during the ascent and emplacement can enhance and accelerate the intrusion of a pluton.

Field observations in the aureole and the pluton provide information regarding the relative importance of individual material-transfer processes and the overall emplacement mode: dyking or diapirism. Clemens *et al.* (1997) suggest several field relations, although very simplified, which indicate if a pluton is a diapir or not. These are in the aureole: (1) narrow high temperature shear zones; (2) steep lineations; (3) rim synclines; (4) pluton-side-up kinematics; and in the pluton: (1) steeply plunging, high temperature, possibly radial, lineations; (2) margin parallel foliation. On the other hand dyking is indicated by (1) numerous dykes, (2) a pluton composed of sheeted complexes, (3) extensional fault zones bounding the pluton.

The JBP and its wall rocks are well exposed, which makes the JBP a perfect object to study material transfer and, possibly, space-making

From: CASTRO, A., FERNÁNDEZ, C. & VIGNERESSE, J. L. (eds) *Understanding Granites:*
Integrating New and Classical Techniques. Geological Society, London, Special Publications,
168, 161–176. 1-86239-058-4/99/$15.00 © The Geological Society of London 1999.

processes, their relative importance and, finally, to evaluate the ruling emplacement mode in this case study. In order to do that, detailed petrographic and structural mapping at the SW margin of this pluton in the Deep Springs Valley (DSV) area was complemented by laboratory studies such as AMS, quartz c-axis and strain measurements.

Regional setting

The JBP is situated at the southern end of the Inyo batholith in the White Inyo Mountains (Fig. 1), which is generally regarded as a precursor of the Sierra Nevada batholith. The alkaline to calc-alkaline Soldier Pass Intrusive Suite (Bateman 1992), consisting of granitoids with monzonitic, dioritic, granodioritic and granitic composition, occupies the southern part of the batholith. The Marble Canyon Diorite, Joshua Flat Monzonite and Beer Creek Granodiorite, making up the JBP, are part of this suite. According to U–Pb dating (Sylvester et al. 1978; Gillespie 1979; Stern et al. 1981) the Soldier Pass Intrusive Suite intruded about 180–160 Ma ago into a Late Proterozoic to Cambrian sequence of carbonate and siliciclastic rocks, which was deposited at the western, then passive, continental margin of North America (Nelson et al. 1991). This shelf sequence was deformed during the Antler (Late Devonian to Mississippian) and Sonoma orogeneses (Late Permian to Early Triassic) into N–S-trending (Antler) and NE–SW-trending (Sonoma) folds (Dunne et al. 1978; Bateman 1992), respectively, and metamorphosed under greenschist-facies conditions (Ernst 1996). The intrusion of the Inyo batholith coincides with the deformational events summarized as Nevadan orogenesis (Jurassic through Cretaceous; Dunne et al. 1978; Bateman 1992), although no structures related to this regional deformation phase were observed in and around the contact aureole of the JBP. The Nevadan orogenesis is characterized by thrust tectonics (Dunne et al. 1978), which can be distinguished from deformations related to the intrusion leading to tight or isoclinal folds (parallel to the contact with the plutons), boudinage and thinned strata around the plutons of the Inyo batholith (Stein & Paterson 1996). According to P–T data from aureole rocks, the intrusion took place at a depth of about 10 km (Ernst 1996). In the thermal aureoles of the intrusive bodies contact metamorphism reached hornblende-hornfels up to lower amphibolite-facies conditions (Ernst 1996). According to U–Pb (Gillespie 1979; Hanson et al. 1987) and K–Ar and ^{40}Ar/^{39}Ar dating (McKee &

Conrad 1996), further plutons intruded into the batholith during two later periods of igneous activity ranging from 145 to 140 Ma and from 100 to 80 Ma. The Cenozoic era (especially the last 5 Ma) in the White Inyo Mountains is marked by extensional tectonics typical for the Basin and Range Province, accompanied by a bimodal basaltic–rhyolitic volcanism (Nelson et al. 1991).

Field and microscopic observations

Petrography of the pluton

Mapping and thin section investigations (Figs 2 and 3) prove the JBP to consist of three distinct intrusions: the Marble Canyon Diorite, the Joshua Flat Monzonite and the Beer Creek Granodiorite. The Marble Canyon Diorite is a medium- to coarse-grained, isotropic honblende-rich diorite with biotite, titanite and magnetite as accessories. The Joshua Flat Monzonite, a medium-grained monzonite to quartzmonzonite, is characterized by a strong magmatic foliation made up by hornblende (typically with pyroxene cores) and the high amount of titanite, visible even with an unaided eye. The Beer Creek Granodiorite, a medium- to coarse-grained, biotite- and hornblende-bearing granodiorite contains 1–3 cm long alkali feldspar phenocrysts, which make up a weak magmatic foliation together with dark, fine grained, monzonitic enclaves.

Field relationships and structures inside the JBP

The Joshua Flat Monzonite and the Beer Creek Granodiorite make up c. 90% of the JBP following a concentric pattern with the Joshua Flat Monzonite at the rim of the pluton and the Beer Creek Granodiorite in its center. In the entire pluton the Marble Canyon Diorite is abundant as xenoliths in the Joshua Flat Monzonite as well as in the Beer Creek Granodiorite (Fig. 3a). Moreover, stoped blocks of the Joshua Flat Monzonite appear in the Beer Creek Granodiorite all over the working area. In addition to these monzonitic xenoliths, characterized by sharp edges and an internal magmatic foliation discordant with the external magmatic foliation in the Beer Creek Granodiorite, monzonitic enclaves with smooth edges and internal magmatic fabrics concordant with the external fabrics, interpreted to be derived from the Joshua Flat Monzonite, occur in the Beer Creek Granodiorite. A third kind of contact

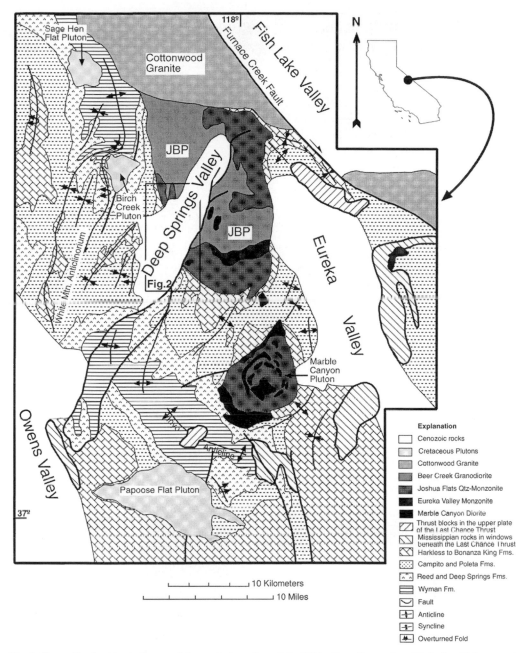

Fig. 1. Generalized geological map of the central portion of the White Inyo Range (modified after Nelson *et al.* 1991); rectangle refers to Fig. 2; arrow on the map of California shows the location of the White Inyo Range.

between the Joshua Flat Monzonite and the Beer Creek Granodiorite is characterized by mixing and mingling of both magmas. Mingling led to schlieren of Joshua Flat Monzonite in Beer Creek Granodiorite and vice versa, whereas sheets of an igneous material (1–100 m wide),

characterized by porphyritic alkali feldspar as well as hornblende with pyroxene cores lying between the Joshua Flat Monzonite and the Beer Creek Granodiorite are interpreted as a result of magma mixing between the monzonite and the granodiorite (Fig. 3d). The magmatic foliation in

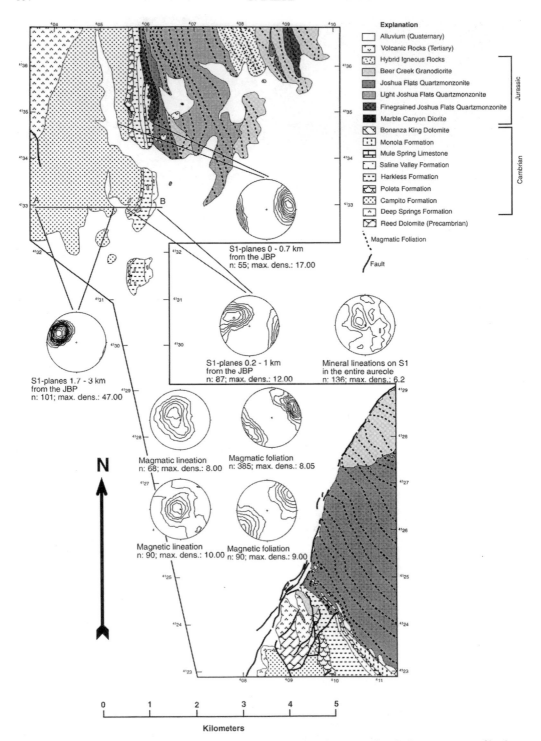

Fig. 2. Schematic geological map of the field area at the SW rim of the JBP; line A–B represents profile along which samples for the strain and quartz c-axis measurements were taken. Included are also stereoplots of the magmatic and magnetic fabrics in the JBP, as well as stereoplots showing the orientation of S1 planes in several parts of the mapping area NW of DSV.

Fig. 3. Field and microscopic observations in and around the JBP. (a) Brecciated blocks of the Marble Canyon Diorite in the Joshua Flat Monzonite. (b) Asymmetrical shear fold and boudins in a marble of the Deep Springs Formation indicating ductile downward flow (left NNE, right SSW). (c) Pluton–wall rock contact near Birch Creek with overturned strata (left W, right E). (d) Diffuse contact between the fine-grained Joshua Flat Monzonite and the Beer Creek Granodiorite both still in the magmatic state: alkali feldspar being transferred from the Beer Creek Granodiorite into the Joshua Flat Monzonite. (e) Intensively folded stoped block with smooth edges, indicating partial melting at its rims. (f) Reaction fabric between biotite, sillimanite, andalusite, plagioclase and quartz on one side and cordierite, alkali feldspar on the other side (long edge: 2 mm, PPL).

the JBP is oriented concentrically along the steep contact to the wall rocks; therefore it strikes NNW in the mapping area (Fig. 2), where it cross-cuts even intraplutonic contacts. Stoped blocks of aureole material are common not only at the rim of the JBP, but also in its centre. Many of these xenoliths show fabrics typical for intense ductile deformation, such as isoclinal folds (Fig. 3e) and chocolate-tablet boudins.

Petrography of the aureole rocks

The JBP intruded a sequence of carbonatic and psammopelitic rocks which had suffered only a weak regional metamorphism in greenschist facies with biotite as a kind of index mineral. This greenschist facies is overprinted by a later contact metamorphism up to lower amphibolite-facies conditions. Three contact metamorphic zones could be distinguished according to the occurrence of index minerals in the metapelites. The andalusite–cordierite zone (1.8–0.5 km distance from the pluton) is characterized by the mineral reaction chlorite + muscovite + ilmenite $\rightarrow$ biotite + cordierite + andalusite + H_2O. The transition into the next zone, the sillimanite zone (0.5–0.25 km) is indicated by the overgrowth of biotite and coexisting andalusite by sillimanite. Although the deduced poly-morphic transition from andalusite to sillimanite is the only observable mineral reaction, additional reactions might have also taken place at the andalusite–cordierite zone–sillimanite zone boundary. In the immediate contact area (<0.25 km) a cordierite–alkali-feldspar zone is characterized by leucocratic veins, prob-ably the products of dehydration melting due to the mineral reaction sillimanite/andalusite + bio-tite + quartz $\rightarrow$ cordierite + alkali feldspar + melt + H_2O (Fig. 3f). At least two reactions are dehydration reactions. Therefore water must have been present during and after the emplace-ment of the JBP, probably not only from dehydration reactions, but also water from the pluton, released during the continuous crystal-lization of the melt. The influence of fluids is visible all around and inside the pluton: The retrograde chloritization of andalusite and bio-tite as well as the extensive pinitization of cordierite in pelitic rocks, skarns, copper mineralizations at the contact between the JBP and marbles and numerous epidote veins, all observed in the innermost aureole, prove that fluids were present during the ascent and emplacement of the JBP.

Structures and field relations in the contact aureole

The most prominent fabric in the country rocks is an E- to NE-striking cleavage (S1), which developed parallel to the bedding planes, prob-ably as a pressure-solution cleavage. Regional deformation associated with the greenschist-facies regional metamorphism deformed this fabric into open E- to NE-trending folds. Inside the mapped part of the contact aureole along the western rim of the JBP the S1 cleavage is continuously rotated into a contact parallel direction. The dip of the stratigraphic units continuously steepens towards the contact, they even may be overturned next to the pluton (Figs 2 and 3c). The same observations can be made at the northern and southern rim of the JBP (Stein & Paterson 1996). Thanks to the detailed stratigraphic work of Nelson (1962, 1966) it was possible to correlate the strongly meta-morphosed and deformed layers in the contact aureole with only regionally deformed strata outside the aureole. The strata become younger when approaching the pluton. Preserved sedi-mentary structures, such as ripple marks and cross bedding, indicate that the stratigraphic younging is directed toward the pluton. There-fore, the aureole of the JBP can be regarded as a funnel-like structure (Fig. 1). Moreover, a comparison of the thickness of the same strati-graphic units outside and inside the aureole, using Nelson's stratigraphic column (Nelson 1962), shows that the stratigraphic units have been thinned from 30% in the outer part of the aureole up to 80% in the inner part of the aureole. The Poleta formation has a thickness of c. 400 m outside the aureole but just 100 m inside; i.e. it has been thinned by 75%.

Discordant contacts along large sections with the pluton are common. This feature corres-ponds very well with the numerous stoped blocks observed in the JBP. The structural inventory of the inner aureole (up to 500 m away from the JBP) is characterized by ductile deformational structures. Chocolate-tablet boudins confirm the amount of strain deduced from the strata's thickness and are very often associated with small scale asymmetrical shear folds with hori-zontal fold axes and axial planes dipping toward the pluton (Fig. 3b). Moreover, isoclinal, cylind-rical folds in the centimetre- to metre-scale, often with moderately to steeply plunging fold axes parallel to mineral stretching lineations, can be found. The axial planes of these folds strike NNE and dip steeply to the WSW or ENE. The moderate to steep mineral lineations lie on the S1 planes and show a similar trend as those: they steepen continuously toward the pluton, more-over, they increase in number. The mineral stretching lineations consist partly of elongated black spots, interpreted as the remnants of the typical contact metamorphic mineral cordierite. In the cordierite–alkali-feldpar zone mylonite zones can be found, where sillimanite forms S–C fabrics, which indicate upward movement of the stratigraphic hanging wall, i.e. the pluton side. Most other kinematic indicators in shear zones of the inner aureole, as σ- and δ-porphyroclasts and asymmetrical shear folds, show the same transport direction: pluton-side up relative to the rest of the aureole (Fig. 3b). In the outer part of the aureole (0.5–1.8 km away from the JBP) ductile structures that can be related with the emplacement of the JBP are missing and only extensional kink bands can be found. More-over, monzonitic and granodioritic dykes at all scales—millimetres to metres wide—were observed in the inner part of the aureole.

Measurements of the anisotropy of the magnetic susceptibility (AMS)

According to Tarling & Hrouda (1993) the AMS is a good tool to measure weak fabrics, e.g. magmatic foliations and lineations as well as the intensity of magmatic fabrics very easily and quickly. Here only a short description of the important parameters for the AMS ellipsoid shall be given. Hrouda (1982) proposed the following parameters to describe the AMS ellipsoid:
foliation factor $F = k_y/k_z$;
lineation factor $L = k_x/k_y$;
anisotropy factor $P = k_x/k_z$;
corrected anisotropy factor $P' = \exp\{2(\ln k_x - \ln k_m)^2 + 2(\ln k_y - \ln k_m)^2 + 2(\ln k_z - \ln k_m)^2\}^{1/2}$
with $= k_m$ (mean susceptibility) $= (k_x + k_y + k_z)/3$; form factor $T = (\ln F - \ln L)/(\ln F + \ln L)$.

P' describes the degree of ellipticity/anisotropy of the AMS ellipsoid. Flattening AMS ellipsoids are characterized by T values >0, whereas constrictional AMS ellipsoids have T values <0. The most useful graphical depiction to describe and interpret AMS measurements is the Jelinek diagram (Hrouda 1982), where P' is plotted against T.

Measuring conditions

Measurements were carried out with a Kappabridge KLY-2. The measuring conditions were as follows: temperature $= 20\,°C$, magnetic field $= 300\,A\,m^{-1}$ (weak magnetic field). Cylinders with a height of 2.08 cm and a diameter of 2.54 cm were used as samples. Their magnetic susceptibility is measured in 15 positions and the AMS ellipsoid calculated with the ANISO14g computer program. Measurements were done at 90 samples: 53 from the Joshua Flat Monzonite and 37 from the Beer Creek Granodiorite). Each sample consists of 4 to 10 cylinders. Only in one case a single cylinder was measured, due to the small size of the hand specimen.

Magnetic susceptibility and AMS in the JBP

Lithological units of the JBP are mostly ferromagnetic with susceptibilities between $20\,000$ and $30\,000 \times 10^{-6}$ SI. Only 10% of all samples lie below $10\,000 \times 10^{-6}$ SI and are, therefore, paramagnetic. Consequently, the main carrier of the magnetic susceptibility is magne-tite, with biotite and hornblende as minor contributors. The orientation of the magnetic lineations and foliations are fairly consistent for all samples (Fig. 2) and therefore for all lithologies. For both the poles of the magnetic foliation and the magnetic lineation eigenvalues and eigenvectors were calculated (Scheidegger 1965). The eigenvector of the largest normalized eigenvalue represents the average magnetic foliation and the average magnetic lineation, respectively. In case the largest normalized eigenvalue is higher than 0.5 the corresponding eigenvector can be regarded as well defined. Accordingly the average magnetic foliation is oriented 237/90 (dip direction/dip angle) with an eigenvalue of 0.6688. The mean magnetic linea-tion has the orientation 300/84 (trend/plunge), the eigenvalue is 0.5644. Therefore, both lie parallel to the magmatic foliations and linea-tions respectively (Fig. 2) and imply the con-temporaneous formation of both magmatic and magnetic fabrics.

The anisotropy of the AMS ellipsoids (Fig. 4) measured in the Joshua Flat Monzonite lies between 2.5% ($P' = 1.025$) and 59.3% ($P' = 1.593$) with an average value $P' = 1.212 \pm 0.114$ and is much stronger than the anisotropy of those in the Beer Creek Granodiorite, where P' ranges between 1.049 and 1.285 (average $P' = 1.127 \pm 0.056$). The broad range in P' for the AMS ellipsoids of the Joshua Flat Monzonite has two reasons: of the ten ellipsoids with $P' < 1.1$ six are para-magnetic, the two ellipsoids with $P' > 1.4$ were measured in samples that had suffered ductile subsolidus deformation. The samples from the Beer Creek Granodiorite are much more con-sistent with respect to their magnetology and structural inventory: only one sample is para-magnetic, none showed signs of subsolidus deformation. Most AMS ellipsoids have an oblate shape. Only 14 of all samples are prolate.

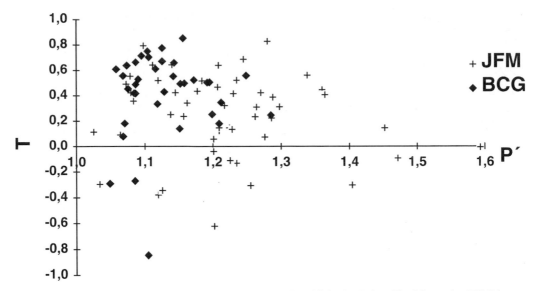

Fig. 4. Jelinek diagram for AMS results from the JBP. The ellipsoids in the Joshua Flat Monzonite (JFM) have a much stronger anisotropy (indicating stronger deformation) than those measured in the Beer Creek Granodiorite (BCG). Ellipsoids in both lithologies are generally oblate, probably due to flattening deformation.

The average T for the Joshua Flat Monzonite is 0.258 ± 0.321 and 0.414 ± 0.328 for the Beer Creek Granodiorite.

Investigations regarding the viscosity contrast between the carrier of susceptibility (magnetite) and the magma according to Hrouda & Lanza (1989) suggest the magnetic foliation, and therefore also the magmatic foliation, to have developed late in the magmatic state of the JBP: Hrouda & Lanza established a relation between the anisotropy factor P', the strain, the viscosity ratio magnetite:magma, the pluton's radius and the intrusion depth. They introduced a graphical solution for this question. As already stated, P' in the Beer Creek Granodiorite lies between 1.049 and 1.285 equivalent to 4.9–28.5% anisotropy and between 1.025 (=2.5% anisotropy) and 1.593 (=59.3% anisotropy) in the Joshua Flat Monzonite. The intrusion depth is about 10 km and the pluton's radius amounts to 15–20 km. Plotting these values in the diagram of Hrouda & Lanza shows that the viscosity ratio between the carrier of susceptibility, magnetite, and its matrix, the magma, is 20:1 to 10:1. That is, both magmas were already very viscous when the magnetic and, thus, the magmatic foliation were developed, i.e. the most important fabric in the JBP originated late in the near solidus magmatic state of the pluton, confirming the field observations.

Strain measurements in the aureole of the JBP

Samples and measurement method

The strain/shape orientation analyses were carried out with a VIDS image analysis equipment and the program VIDS V, using the R_f/Φ-method (Ramsay 1967). The computer program RS PLOT, which is based on the geological strain analysis of Lisle (1985), was used for data processing and their evaluations. The triaxial strain ellipsoids were calculated by the computer program TRISEC. Strain measurements were carried out at six samples collected along one profile W of the JBP (Fig. 2). Samples were taken at 500 m, 550 m, 600 m, 1700 m, 2000 and 2200 m away from the pluton. The samples are part of the Harkless and Campito Formations and have pelitic composition. Black spots, in thin sections identified as partly altered cordierite porphyroblasts, were used as strain markers.

Shape and orientation of the strain/fabric ellipsoids

All but one of the measured strain ellipsoids have a prolate shape indicating a more linear than planar character of the cordierite strain fabrics. Neither R_{xy} nor R_{yz} are higher than 2.1 and the

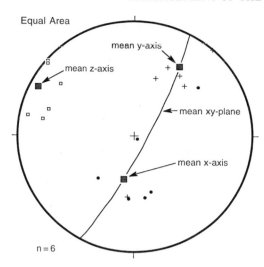

Fig. 5. Stereoplot for the six strain measurements, including the orientation of the xy plane and the three principle axes of the average strain ellipsoid.

stereonet with reference to the lineation and foliation measured at the individual samples. The N–S direction of the stereonets corresponds to the pole of the foliation and the E–W direction to the lineation. To make the c-axis fabrics (Fig. 6) clearer the lineations are shown either as lines or, if a shear sense could be deduced from the c-axis patterns, as arrows. The description of the individual quartz c-axis fabrics follows the classification of Hofmann (1974). He distinguishes seven different maxima: maximum I—in the centre of the stereonet; maximum II—between the centre and the poles of the stereonet; maximum III—between the poles and the equator along the rim of the stereonet; maximum IV—in the centre of each quadrant of the stereonet; maximum V—at both poles of the stereonet; maximum VI—on the equator of the stereonet, halfway between the centre of the stereonet and its rim; maximum VII—at the E and W end of the equator of the stereonet.

E_S value (Nadai 1963) also does not exceed 0.944. The strain is therefore surprisingly low, even in the immediate contact with the pluton. Again, as already done for the AMS ellipsoids, the orientation tensors for the three main axes of all measured ellipsoids were computed to get the average x, y and z axes for an average ellipsoid. The largest normalized eigenvalue represents the average x (y or z) axis with its eigenvector representing its orientation. All three axes are very well defined (Fig. 5). The mean x axis with its moderately steep plunge to the S (eigenvalue: 0.7415, eigenvector: 178/58) indicates oblique, subvertical material transport, supporting the kinematics observed in the field. The mean x axis together with the mean y axes (eigenvalue: 0.7653, eigenvector: 032/30) defines the mean xy plane of the strain ellipsoid, which lies, as expected, parallel to the foliation of the samples. Therefore, the mean z axis (eigenvalue: 0.9217, eigenvector: 296/15), as the pole of the xy plane, lies parallel to the poles of the foliation planes measured in the field.

Quartz c-axis measurements

Quartz c-axis measurements were carried out on ten samples from a cross-section on the west side of the JBP. The profile extended about 3 km away from the pluton–wall rock contact. In each sample 300 measurements were carried out on a four-axis Zeiss U-stage on one oriented thin section. The data obtained were plotted in a

Description and interpretation of the quartz c-axis fabrics in the aureole rocks of the JBP

Quartz c-axis measurements confirm the field observations, but also yield a much more detailed insight into the deformation processes in and near the contact aureole. The following different features can be observed with increasing distance from the pluton (Fig. 6). In a c. 500 m wide zone directly at the contact to the JBP obliquely oriented yz girdles (composed of maxima I or II and V) and type II cross girdles (consisting of maxima I and III) are the dominant quartz c-axis fabrics. According to Paschier & Trouw (1996) maxima V are due to basal slip in $\langle a \rangle$ and maxima I can be referred to prism slip in $\langle a \rangle$. The combined activation of basal and prism slip systems is typical for medium to high temperature conditions, i.e. 350–650 °C (Paschier & Trouw 1996). Tullis et al. (1973) emphasize the direct relation between the opening angle of type II cross girdles and their formation temperatures: the higher the temperature, the higher the opening angle. The opening angles of the measured type II cross girdle lie between 75° and 90°, indicating formation temperatures of at least 550 °C. This temperature is too high for the regional greenschist-facies metamorphism present in the White Inyo Mts, but matches perfectly the contact metamorphism related to the emplacement of the JBP. All observed quartz c-axis fabrics in the innermost aureole can therefore be related to this event. Several authors (Carter et al. 1964; Green et al. 1970; Tullis et al. 1973; Schmid & Casey

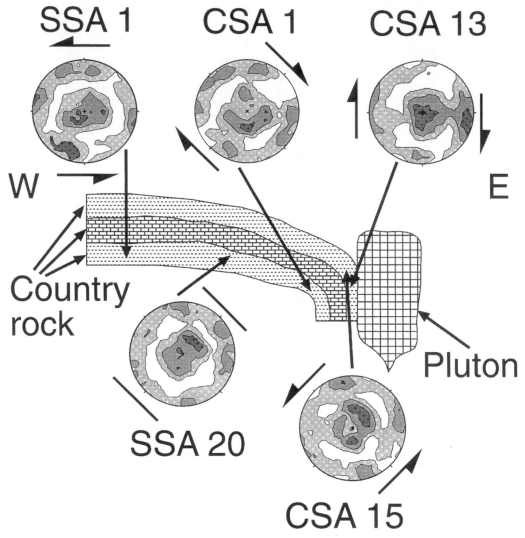

Fig. 6. Typical quartz *c*-axis fabrics from the aureole at the western rim of the JBP. Arrows represent the shear sense deduced from the quartz *c*-axis fabrics and, as arrows or lines, the orientation of the lineations on the individual samples. Sample CSA 13 was taken 250 m away from the pluton, maximum density is 5.00; CSA 15: 350 m/4.33; CSA 1: 950 m/5.00; SSA 20: 2300 m/5.00; SSA 1: 3000 m/4.67.

1986) relate type II cross girdles to coaxial, constrictional strain or interpret them at least as linear fabrics. Other linear features as the strain ellipsoids, plotting within the field of apparent constriction, or the numerous moderate to steep mineral stretching lineations are confirmed by the quartz *c*-axis measurements. The oblique orientation of the crystallographic quartz fabrics to the tectonic *x*, *y* and *z* coordinates and their dependence on steep mineral lineations indicates the presence of a component of rotational deformation. Unfortunately this rotational

deformation shows contradictory shear senses even in samples from locations close to each other, so that, for example, sample CSA 13 (250 m away from the JBP) shows a dextral shear sense (pluton-side down relative to the aureole) and sample CSA 15 (350 m away from the JBP) indicates a sinistral shear sense (pluton-side up relative to the aureole).

With increasing distance to the pluton, within a zone between 500 m and 1000 m away from the pluton, the quartz *c*-axis fabrics are still dominated by a combination of maxima I or II and V.

The difference is that part of them are now not related to moderately steep but also to horizontal mineral lineations. Those related to half steep lineations (especially CSA 1, 950 m away from the JBP) indicate transport of the aureole down relative to the rest of the host rock. Those related to horizontal lineations show dextral shear sense indicating material transport parallel to the pluton contact.

About 1 km away from the pluton the character of the quartz c-axis fabrics changes. Instead of type II cross girdles and yz girdles, xy girdles consisting of maxima I and obliquely oriented maxima VIII become the dominant fabrics. Only one sample (SSA12: c. 1.7 km away from the contact) shows a weak type II cross girdle additional to the maxima VIII and another sample (SSA20: c. 2 km away from the pluton) has a maximum I and a maximum II as the dominant fabric. Sample SSA 1 (c. 3 km away from the pluton) is dominated by oblique maxima VIII. The weak type II cross girdle in SSA 12 and the Maxima I/II in SSA 20 can still be related to combined basal and prism slip due to the intrusion of the JBP. The necessary temperatures for the construction of those fabrics (350–650 °C) were attained even c. 2 km from the contact as is proved by the occurrence of cordierite even 1.8 km from the pluton (cordierite growth needs c. 500 °C according to Seifert (1970)). It is more difficult to explain the maxima VIII, which can be found inside (SSA 12) as well as outside (SSA 1) the thermal aureole. According to Paschier & Trouw (1996) obliquely oriented maxima VIII should be interpreted as shear fabrics, originated at temperatures higher than 650 °C. Because of the distance of those samples from the pluton this assumption does not seem to be very probable; at least the fabrics should not indicate higher temperatures than those taken closer to the pluton. A better solution might be to interpret the maximum VIII dominated fabrics as the results of external rotation. Fairbairn (1949) and Griggs et al. (1960) proposed this process to produce lower grade quartz fabrics. It remains unclear, if the maxima VIII is older or younger than the other observed fabrics. The fact that maxima VIII are found also outside the aureole and assuming that the stress exerted from the JBP on the country rocks at a distance > 1 km is not high enough to obliterate an older quartz c-axis fabrics, the maxima VIII can be related to the pre-intrusive regional deformation, which is a low-temperature (green-schist-facies) event and would fit to the 'external rotation' interpretation.

Discussion

Field observations as evidence for diapirism or dyking

I will try to follow the field criteria of Clemens et al. (1997), although they seem to be too simple and fit modelled hot-Stokes diapirs better than composite, viscoelastic diapirs. Therefore, the field observations do not match fully the criteria of Clemens et al. (1997). Clemens et al. (1997) demand narrow, high-temperature shear zones directly at the pluton–wall rock interface or, in the case of deeper seated plutons, in the aureole together with steep lineations as hints for large scale downward movement of aureole material relative to the ascending diapir. No narrow high-temperature shear zones were found directly at the contact between the wall rocks and the pluton, but in the inner aureole mylonite zones with S–C fabrics consisting of fibrolitic sillimanite were observed. These mylonites show a pluton-side up kinematic and prove downward movement of aureole material relative to the pluton at the peak of the contact metamorphism and therefore during the emplacement of the JBP. Steep mineral lineations in the inner aureole support this interpretation. The observed funnel structure is not the rim syncline that Clemens et al. (1997) postulate, but is an argument as good as a rim syncline for diapirism, because in the 'funnel' country rock material is transferred downward, what makes space for the suggested JBP diapir. There is no way to produce such a funnel by dyking. Moreover both criteria, which should, according to Clemens et al. (1997), be observed inside the pluton to call it a diapir, are fulfilled for the JBP: steep magmatic and magnetic lineations were observed and a magmatic/magnetic foliation parallel to the pluton–wall rock contact was found. On the other hand none of the general criteria for dyking are fulfilled by the JBP. The observed dykes radiate from the already constructed magma chamber into the host rock and do not construct the entire chamber. No major sheeted complexes were found, but two big batches of magma, the Joshua Flat Monzonite and the Beer Creek Granodiorite make the pluton. No extensional fault zones of Jurassic age or older occur at either side of the JBP. Therefore, the JBP can be regarded as a diapir, or better as a batch of nested diapirs, because three distinct igneous phases could be established.

Field relationships and AMS: arguments for an intrusion of nested diapirs and magma chamber expansion

One of the intrusions, the Marble Canyon Diorite, was found as xenoliths only in the other two igneous phases and is therefore regarded as the oldest intrusive. Idiz (1981) observed the same rock type in the Marble Canyon Pluton, about 10 km south of the JBP (Fig. 1). There it makes up about 30% of the exposed pluton and lies as a ring-like structure in a matrix of Joshua Flat Monzonite and at the rim of this pluton. Idiz (1981) interpreted the Marble Canyon Diorite as a precursor of the Joshua Flat Monzonite as well. He interpreted the Marble Canyon Diorite as a sill which was later intruded and brecciated by the Joshua Flat Monzonite. Because the Marble Canyon Diorite was found only as disconnected blocks in the JBP, no interpretation of the intrusion mode of the Marble Canyon Diorite is possible. As described above the relations between the Joshua Flat Monzonite and the Beer Creek Granodiorite are much more complicated, but the presence of monzonitic xenoliths in the granodiorite and the lack of granodioritic xenoliths in the monzonite lead to the conclusion that the Joshua Flat Monzonite represents the second and the Beer Creek Granodiorite the third and last intrusive phase. What makes the relation between the Joshua Flat Monzonite and the Beer Creek Granodiorite more complicated is the existence of mingling and mixing structures between both (Fig. 3d). This means that granodiorite intruded the monzonite in a partly molten state. It might be that during the intrusion of the Beer Creek Granodiorite into the still melt containing Joshua Flat Monzonite, magma chamber expansion took place. Magma chamber expansion may be an explanation for the strong magmatic foliation in the monzonite relative to the weak magmatic foliation in the granodiorite as well as for the high anisotropy of the AMS ellipsoids measured in the monzonite relative to those of the granodiorite (Fig. 4). Moreover, it can be responsible for the concentric foliation pattern at the pluton's rim, for oblate fabrics in the inner aureole (namely the chocolate-tablet boudins) and for the generally oblate character of the AMS ellipsoids. Although all structures described can be explained by late-stage magma chamber expansion, the last three arguments are not necessarily connected with it. The concentric pattern made up by the magmatic foliation at the rim of the JBP can also be due to boundary-layer effects such as the wall, Magnus and Bagnold

effects (Barriere 1981), which all can align crystals parallel to the pluton–wall rock interface or either to a cooling front propagating from the contact into the pluton. Oblate fabrics, such as the chocolate–tablet boudins in the inner aureole and the AMS ellipsoids, indicate a flattening event during the emplacement and deformation history of the JBP, but do not necessarily require late-stage magma chamber expansion. Nevertheless late-stage magma chamber expansion seems to be the only way to produce the observed intensity differences of the magmatic and magnetic fabrics in the monzonite and granodiorite, respectively.

The meaning of fabrics and foliations: implications for vertical material transfer

Emplacement-related structures indicate that multiple processes played a role during the emplacement of the JBP. Most of these structures and features have one thing in common: they are steeply or even vertically oriented. This is so in the aureole for all vertical lineations and the steepening of stratigraphic units towards the pluton and, in the pluton, for steep magmatic and magnetic lineations and for stoped blocks, which indicate vertical, downward directed material transport as all other features referred to before. Moreover, all structures in the inner part of the aureole are ductile high-temperature structures and very often associated with the typical contact metamorphic minerals cordierite and sillimanite, and are related to the emplacement of the JBP and not part of the much colder and older regional metamorphic events. Because the cordierite lineations appear on the S1 planes, which are bent down around the pluton to form the funnel structure, it is suggested that this structure developed during the emplacement of the JBP. A further argument for the coeval character of the funnel is the occurrence of sillimanite in mylonite zones, which lie parallel to steeply dipping S1 planes and can be related to the construction of the funnel. A possible explanation for the funnel is found in the steep mineral stretching lineations (cordierite), which indicate ductile vertical material transport in the contact aureole. To originate the observed funnel, country rock material has to flow ductilely from the roof region of the ascending pluton towards its source region, a process recently discovered by Stein (pers. comm.) and named by him as 'ductile downward flow'. This vertical material transfer process could also solve the space problem during the ascent and emplacement of the pluton. Ductile downward flow seems to be a space making process, which is active during the entire ascent and emplacement history of a pluton or even a batholith. As soon as enough magma has accumulated in a pluton's source region in the lower crust or the upper mantle, it rises buoyantly and leaves virtually a cavity behind that has to be filled immediately due to the high lithostatic pressure in that depth. The easiest way to fill this 'hole' is to transfer material vertically, i.e. from the sides and the roof regions of the ascending pluton into the area the magma batch has just left. Material transfer takes place by ductile flow. Host rock material is 'sucked' into a funnel-like megastructure around the pluton, which is constructed as the pluton rises. Other observed structures in the aureole fit in this scenario. The heat necessary for the ductile behaviour of the wall rocks is supplied by the pluton, heat transfer occurred by conduction as well as by convection with fluids derived from the pluton and from the aureole due to dehydration reactions during the prograde contact metamorphism. As demonstrated by the melt-producing reaction silli manite + biotite + quartz → cordierite + alkali feldspar + melt + H_2O temperatures in the immediate contact area reached c. 730 °C (Le Breton & Thompson 1988). Therefore, partial melting also played a role for the ductile downward flow, although only in a 250 m wide zone around the pluton. According to Norton & Knight (1977) fluids originating during contact metamorphic dehydration reactions are transferred towards the pluton, where they are heated up and transported to the pluton's roof region. Obviously the roof region could be weakened by these fluid flow processes and become more sensitive for the ductile downward flow as long as fluid transport takes place through the bulk volume of country rocks and not on discrete shear zones or similar anisotropies. Fluids can also be helpful in enhancing the effectiveness of the shear movements (McNulty 1995) necessary to transport country rock along the sides of the rising pluton downward.

Some observed structures do not fit into the scenario of ductile downward flow. Especially moderately steep or even horizontal lineations, as well as the moderately steep axes of the strain ellipsoids, do not fit very well to this clearly vertically oriented material-transfer process. But also isoclinal, cylindrical folds with steep axes parallel to mineral lineations are hard to explain by a process where stretching, but not pure constriction, is an important ingredient. Furthermore, chocolate-tablet boudins require flattening. All these structures could have been produced during a transpressional deformation phase, which could succeed the phase of pure

ductile downward flow. Stein & Paterson (1996) already proposed this conclusion for similar synintrusional structures at the southern rim of the JBP. Due to the changing geometry of the rising pluton over time the downward moving aureole material at a certain level (e.g. the level of recent exposure) is exposed to a changing strain field, where not only rotation, stretching and thinning of strata, but oblique and horizontal, contact-parallel strike-slip transport accompanied by flattening, i.e. transpression, plays an important role. On the other hand, flattening-type structures, such as the chocolate-tablet boudins, could also be produced by late-stage magma chamber expansion during the intrusion of later magma batches, i.e. the Beer Creek Granodiorite. Another remaining problem could be solved assuming that several structures were layed out late during ductile downward flow: most kinematic indicators in the inner aureole, including part of the quartz c-axis fabrics and even the S–C fabrics by sillimanite show the pluton side to move up and the rest of the aureole to move down. This makes a coupling of the pluton with its host rock necessary, which seems to be very unlikely considering the high viscosity contrast between the pluton and the country rock. In case all these structures are late during ductile downward flow it can be assumed that the rim portion of the JBP was already near or even below the solidus. That means that the viscosity contrast between the pluton and its aureole was not very high and a coupling between both would be possible. The steep magmatic and magnetic lineations in the JBP could be incorporated in the ductile downward flow and interpreted as a magmatic flow fabric that indicates the vertical buoyancy-driven magma transport, which caused the downward movement of the aureole material.

The observation of discordant contacts, numerous country rock xenoliths in the JBP and dykes radiating from the pluton into its host rocks can be explained only by stoping. Moreover, the relative low E_S values computed even for those strain measurements close to the pluton (e.g. 0.899 at 500 m away from the contact) may point to the fact that the innermost, strongest deformed aureole was removed by stoping. At the interface between the magma and its host rocks the thermal gradient is very steep. This forces the relatively cool country rock to expand very rapidly and to break open so that dykes from the pluton can propagate into the surrounding material and get out blocks of the country rock. The blocks fall into the magma chamber, where they either become assimilated or, depending on the magma viscosity, sink to the bottom of the magma chamber and make some space for the rising pluton (Paterson & Miller 1998). One observation in the stoped blocks is of particular importance for the role stoping played during the emplacement of the JBP. Many of the xenoliths contain isoclinal folds (Fig. 3e) and boudin structures related to ductile deformation connected with late stages during ductile downward flow. The fact that these ductile structures were found in stoped blocks means that the pluton intruded its own ductile aureole and that stoping was active late during the emplacement of the JBP. If stoping still works even late during the magmatic history, when the thermal gradient between pluton and wall rock is already shallow and the viscosity of the magma relatively high, it must be even more productive during the early stages of emplacement.

Summary: time sequence of emplacement processes

It was shown that numerous emplacement processes played a role during the intrusion of the JBP: dyking and stoping, ductile downward flow accompanied by fluid release from the host rock and the pluton and partial melting of aureole material. Because the intrusion of the JBP occurred in three stages, I tried to define timing and relative importance of these processes during the stages of emplacement.

Stage 1: Intrusion of the Marble Canyon Diorite. Because the Marble Canyon Diorite was found only as xenoliths in the other two intrusive phases, no emplacement process was observed directly for this stage of the intrusion history.

Stage 2: Intrusion of the Joshua Flat Monzonite. Stoping followed by dyking is proved to be an important material transfer process for the emplacement of the Joshua Flat Monzonite by the existence of discordant contacts between the monzonite and the aureole rocks, numerous stoped blocks, found as far as 2–3 km away from the wall rock contact and numerous monzonitic dykes. Ductile downward flow played an important role during the entire emplacement history of the JBP and especially during the intrusion of the Joshua Flat Monzonite. Obviously this stage is the phase, during which the funnel around the pluton was established together with sillimanite-bearing mylonite zones and steep cordierite lineations. The steep magnetic and magmatic lineations are interpreted as representing vertical magma transport during the rise of the diapir. Evidence for partial melting that accompanied and enhanced the ductile downward flow was observed in the innermost aureole (up to 250 m away from the pluton–wall rock contact).

Stage 3: Intrusion of the Beer Creek Granodiorite. Stoping accompanied by dyking appears to be still important as is evidenced by numerous granodioritic to granitic dykes and stoped blocks. Country rock xenoliths with a ductile structural inventory prove that the JBP intruded its own ductilely deformed (i.e. by ductile downward flow) aureole. During the intrusion of the Beer Creek Granodiorite the ductile downward flow probably moved from the aureole into the Joshua Flat Monzonite. As the granodiorite ascended into the already constructed magma chamber the monzonite was forced downward, but also displaced towards the sides and the top of the magma chamber, what led to magma chamber expansion. The successive intrusion of nested diapirs supplied additional heat to the aureole and supported vertical material transfer there.

Conclusions

Field observations and laboratory data (AMS, quartz c-axes, strain) show that the JBP consists of nested diapirs, which intruded in the sequence Marble Canyon Diorite–Joshua Flat Monzonite–Beer Creek Granodiorite. The emplacement of the pluton was possible only because multiple material-transfer processes, such as stoping and dyking, ductile downward flow, partial melting and magma chamber expansion were working together at different times. Two of these space-making processes contributed to the rise of diapirs into the magma chamber: ductile downward flow and stoping acting during the entire emplacement history of the pluton. Ductile downward flow was assisted and enhanced by partial melting in the innermost aureole and increased ductility due to the convective and conductive heat flow induced by the ascending pluton. Magma chamber expansion played a role only during the last stage of the JBP emplacement. The JBP shows that diapirism is a model for pluton emplacement, because it involves many space-making material transfer processes, which can have a positive feedback on each other. Ductile downward flow and stoping, especially, seem to be very powerful material-transfer processes, which might be responsible for major crustal recycling during ascent and emplacement of the Sierra Nevada batholith.

The research was supported by grants of the DAAD, the Landesgraduiertenförderung Baden-Württemberg and the DFG. Special thanks go to T. Fink, R. B. Miller and S. R. Paterson, who put their field data at my disposal, to R. O. Greiling, E. Stein and S. R. Paterson for discussion and comments, to C. Fernandez and A. Castro for their helpful reviews and to S. Hetzler for his personal support.

References

BARRIERE, M. 1981. On curved laminae, graded layers, convection currents and dynamic crystal sorting in the Ploumanac'h (Brittany) subalkaline granite. *Contributions to Mineralogy and Petrology*, **77**, 214–224.

BATEMAN, P. C. 1992. *Plutonism in the central part of the Sierra Nevada batholith, California*. United States Geological Survey Professional Paper, **1483**.

BUDDINGTON, A. F. 1959. Granite emplacement with special reference to North America. *Geological Society of America Bulletin*, **70**, 671–747.

CARTER, N. L., CHRISTIE, J. M. & GRIGGS, D. T. 1964. Experimental deformation and recrystallisation of quartz. *Journal of Geology*, **72**, 687–733.

CLEMENS, J. D., PETFORD, N. & MAWER, C. K. 1997. Ascent mechanisms of granitic magmas: causes and consequences. *In*: HOLNESS, M. B. (ed.) *Deformation-enhanced Fluid Transport in the Earth's Crust and Mantle*. Chapman & Hall, London, 144–171.

DUNNE, G. C., GULLIVER, R. M. & SYLVESTER, A. G. 1978. Mesozoic evolution of rocks of the White, Inyo, Argus and Slate ranges, eastern California. *In*: HOWELL, D. G. & McDOUGALL, K. (eds) *Mesozoic paleogeography of the western United States*. Society of Economic Paleontologists and Mineralogists, Los Angeles, 89–207.

ERNST, W. G. 1996. Petrochemical study of regional/contact metamorphism in metaclastic strata of the central White-Inyo Range, eastern California. *Geological Society of America Bulletin*, **108**, 1528–1548.

FAIRBAIRN, H. B. 1949. *Structural petrology of deformed rocks*. Addison-Wesley Press Inc., Cambridge, Mass.

GILLESPIE, J. G. 1979. U-Pb and Pb-Pb ages of primary and detrital zircons from the White Mountains, eastern California. *Geological Society of America Abstracts with Programs*, **11**, 79.

GREEN, H. W., GRIGGS, D. T. & CHRISTIE, J. M. 1970. Syntectonic and annealing recrystallisation of fine-grained quartz-aggregates. *In*: PAULITSCH, P. (ed.) *Experimental and natural rock deformation*. Springer, Berlin, 272–335.

GRIGGS, D. T., TURNER, F. J. & HEARD, H. C. 1960. Deformation of rocks at 500°–800°C. *In*: GRIGGS, D. T. & HANDIN, J. W. (eds) *Rock Deformation*. Geological Society of America Memoirs, **79**, 39–104.

HANSON, R. B., SALEEBY, J. B. & FATES, D. G. 1987. Age and tectonic setting of Mesozoic metavolcanic and metasedimentary rocks, northern White Mountains, California. *Geology*, **15**, 1074–1078.

HOFMANN, J. 1974. *Das Quarzteilgefüge von Metamorphiten und Anatexiten, dargestellt am Beispiel des Osterzgebirges*. Freiberger Forschungshefte, **C297**.

HROUDA, F. 1982. Magnetic anisotropy of rocks and its application in geology and geophysics. *Geophysical Survey*, **5**, 37–82.

—— & LANZA, R. 1989. Magnetocrystalline anisotropy of rocks and massiv ores: a mathematical model study and its fabric implications. *Physics of Earth and Planetary Interiors*, **56**, 337–348.

IDIZ, E. F. 1981. *Geology of the Marble Canyon Plutonic Complex*. PhD thesis, University of California, Los Angeles, USA.

LE Breton, N. & THOMPSON, A. B. 1988. Fluid absent (dehydration) melting of biotite in metapelites in the early stages of crustal anatexis. *Contributions to Mineralogy and Petrology*, **99**, 226–237.

LISLE, R. J. 1985. *Geological strain analysis: A manual for the R_f/Φ method*. Pergamon Press, Oxford.

McKEE, E. H. & CONRAD, J. E. 1996. A tale of 10 plutons—revisited: Age of granitic rocks in the White Mountains, California and Nevada. *Geological Society of America Bulletin*, **108**, 1515–1527.

McNULTY, B. A. 1995. Shear zone development during magmatic arc construction: The Bench Canyon shear zone, central Sierra Nevada, California. *Geological Society of America Bulletin*, **107**, 1094–1107.

NADAI, A. 1963. *Theory of flow and fracture of solids*. McGraw-Hill, New York.

NELSON, C. A. 1962. Lower Cambrian-Precambrian succession, White-Inyo Mountains, California. *Geological Society of America Bulletin*, **73**, 139–144.

—— 1966. *Geologic map of the Blanco Mountain quadrangle, Inyo and Mono Counties, California*. United States Geological Survey Geologic Quadrangle Map, **GQ-529**, Scale 1:62 500.

——, HALL, C. A. JR. & ERNST, W. G. 1991. Geologic history of the White-Inyo Range. *In*: HALL, C. A. JR. (ed.) *Natural History of the White-Inyo Range, California*. University of California Natural History Guides, **55**, 42–47.

NORTON, D. & KNIGHT, J. 1977. Transport phenomena in hydrothermal systems: cooling of plutons. *American Journal of Science*, **277**, 937–981.

PASCHIER, C. W. & TROUW, R. A. J. 1996. *Microtectonics*. Springer, Berlin.

PATTERSON, S. R. & FOWLER, T. K. JR. 1993. Re-examining pluton emplacement processes. *Journal of Structural Geology*, **15**, 781–784.

—— & MILLER, R. B. 1998. Stoped blocks in plutons: paleo-plumb bobs, viscometers, or chronometers. *Journal of Structural Geology*, **20**, 1261–1272.

RAMSAY, J. G. 1967. *Folding and fracturing of rocks*. McGraw-Hill, New York.

SCHEIDEGGER, A. E. 1965. On the statistics of the orientation of bedding planes, grain axes, and similar sedimentological data. *United States Geological Survey Professional Paper*, **525c**, 164–167.

SCHMID, S. M. & CASEY, M. 1986. Complete fabric analysis of some commonly observed quartz c-axis patterns. *In*: *Mineral and Rock Deformation: Laboratory Studies*. AGU Geophysical Monographs, **36**, 263–286.

SEIFERT, F. 1970. Low-temperature compatibility relations of cordierite in haplopelites of the system K_2O-MgO-Al_2O_3-SiO_2-H_2O. *Journal of Petrology*, **11**, 73–99.

STEIN, E. & PATERSON, S. R. 1996. Country rock displacement during emplacement of the Joshua Flat Pluton, White-Inyo mountains, California. *In*: ONCKEN, O. & JANSSEN, C. (eds) *Basement Tectonics*. Kluwer Academic Publishers, Dordrecht, **11**, 35–49.

STERN, T. W., BATEMAN, P. C., MORGAN, B. A., NEWELL, M. F. & PECK, D. L. 1981. *Isotopic U-Pb ages of zircon from the granitoids of the central Sierra Nevada, California*. United States Geological Survey Professional Paper, **1185**.

SYLVESTER, A. G., MILLER, C. F. & NELSON, C. A. 1978. Monzonites of the White-Inyo Range, California, and their relation to the calc-alkalic Sierra Nevada batholith. *Geological Society of America Bulletin*, **89**, 1677–1687.

TARLING, D. H. & HROUDA, F. 1993. *The magnetic anisotropy of rocks*. Chapman & Hall, London.

TULLIS, J., CHRISTIE, J. M. & GRIGGS, D. T. 1973. Microstructures and preferred orientations of experimentally deformed quartzites. *Geological Society of America Bulletin*, **84**, 297–314.

Petrology, magnetic fabric and emplacement in a strike-slip regime of a zoned peraluminous granite: the Campanario–La Haba pluton, Spain

A. ALONSO OLAZABAL, M. CARRACEDO & A. ARANGUREN*

*Departamento de Mineralogía y Petrología, Universidad del País Vasco, PO 644,
E-48080 Bilbao, Spain (e-mail: npbalola@lg.ehu.es)*

*Departamento de Geodinámica, Universidad del País Vasco, PO 644, E-48080 Bilbao, Spain

Abstract: This paper reports field, petrological and structural data of the peraluminous cordierite–andalusite-bearing Campanario–La Haba granite. Crystallization age is constrained by Rb–Sr whole-rock dating at 309 ± 6 Ma (with $(^{87}Sr/^{86}Sr)_i = 0.70739 \pm 0.00038$) and took place during late Hercynian tectonic events. The pluton shows a petrographic zonation, although there are no marked differences in chemical compositions between the margin and the centre of the intrusion. Petrography, mineralogical data and geochemical modelling indicate melt generation by partial melting of a metasedimentary protolith, probably with some mantelic contribution as shown by its low Sr_i value, followed by emplacement at $P < 3$ kbar. The internal structure of the pluton resulted from the lateral spreading in the stretching direction given by an N120–130E dextral strike-slip zone and the external geometry seems to be strongly conditioned by faults (Riedel R type fractures) formed in the host rocks. This emplacement model agrees with that defined for the adjacent Extremadura granitic plutons and for the Los Pedroches batholith suggesting the existence of a dextral regional shear-zone in the South branch of the Central Iberian Zone.

The study of different plutons and batholiths in contrasted geodynamical settings and the application of new ideas and techniques contribute to the better understanding of the processes involved in the generation, ascent and emplacement of granitoids. Techniques as the anisotropy of magnetic susceptibility (AMS) measurements (Bouchez *et al.* 1997), gravity surveys (Vigneresse 1988; Améglio *et al.* 1997), numerical image analysis techniques (Allard & Benn 1989; Launeau *et al.* 1990; Panozzo-Heilbronner 1992) and physical modelling of intrusions (Román-Berdiel *et al.* 1995, 1997; Benn *et al.* 1998), together with the traditional methods of structural geology, help us to solve the internal structure of plutons and to establish the relationships between the three dimensional geometry and the tectonic model active at the time of emplacement. Structural controls on granite emplacement have been extensively discussed and documented by numerous works (Castro 1986; Hutton 1988; Pitcher 1993; Vigneresse 1995).

Traditionally, the study of granitoids comprises either petrological or structural research, but interdisciplinary work is needed to establish coherent petrogenetic and emplacement models. Recently, new interdisciplinary papers are published in which a combined geochemical and structural investigation was performed (Hecht *et al.* 1997; Vigneresse & Bouchez 1997). These works allow comparison of the structural patterns related to the lithotypes distribution to the position of feeder zones. The aim of these studies is to define more realistic emplacement models of the successive magma pulses.

This paper reports a petrological and structural study of the Campanario–La Haba pluton which contributes to a better understanding of the petrogenesis and emplacement mechanism of this massif. The first part of this work comprises field, petrography and geochemical characterization of this peraluminous granite. The second part outlines the internal structure of the pluton based on the anisotropy of magnetic susceptibility (AMS). Finally, the structural data are compared with the analogue modelling of magma intrusion during strike-slip deformation to propose a coherent emplacement model.

Geological setting

The Campanario–La Haba pluton (CHP) is located in the most meridional sector of the Central Iberian Zone (CIZ) of the Hercynian

From: CASTRO, A., FERNÁNDEZ, C. & VIGNERESSE, J. L. (eds) *Understanding Granites:*
Integrating New and Classical Techniques. Geological Society, London, Special Publications,
168, 177–190. 1-86239-058-4/99/$15.00 © The Geological Society of London 1999.

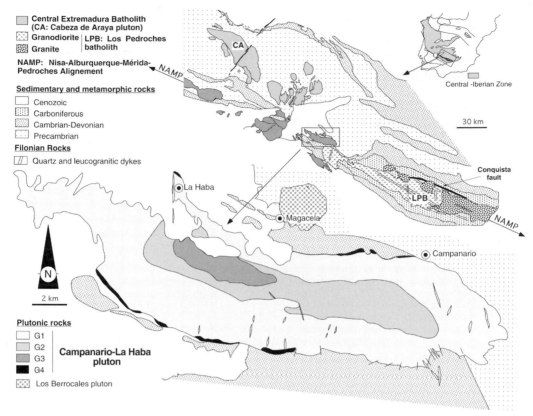

Fig. 1. Geological map of the Campanario–La Haba pluton showing the distribution of the main lithotypes. On the upper corner, a schematic map of the southern end of the Central Iberian Zone with the CHP location.

Iberian massif (Julivert *et al.* 1972). This sector of the CIZ is an anchi- to epizonal domain characterized by detritic materials of Precambrian to Carboniferous age and by the existence of numerous plutons of intermediate to acidic composition (Fig. 1). Two consecutive Hercynian deformation phases resulted in the overall structure of the area. The first phase (D1), of Namurian age, gave rise to WNW–ESE-oriented upright folds with several kilometres wavelength and an associated vertical axial plane schistosity (S1). The second phase (D2) had a local development and was characterized by vertical axis folds with an associated axial plane schistosity (S2) and subvertical shear zones in granitoids (Castro 1986; Quesada *et al.* 1987; Díez Balda *et al.* 1990). The Hercynian regional metamorphism was low or very low in most of the sector reaching the biotite zone only in lowermost portions of the Precambrian formations (Quesada *et al.* 1987).

During the Hercynian tectono-metamorphic events widespread plutonic activity took place. Two magmatic complexes are distinctive in this

sector: the Central Extremadura batholith (CEB) (Castro 1986) and the Nisa–Alburquerque–Mérida–Pedroches magmatic alignment (NAMPA) (Aparicio *et al.* 1977) (Fig. 1).

The CEB is constituted by three granitoid groups: granitoids of quartz-diorite affinities, alkali-feldspar granites and calcalkali-feldspar granitoids. This last type is the most abundant and is typified by the Cabeza de Araya pluton (CA). The CA is composed by K-feldspar megacrysts and cordierite bearing porphyritic granites, coarse-grained non-porphyritic granites and fine-grained two mica leucogranites (Corretgé *et al.* 1985). Pluton emplacement in the CEB is related to mega-tension gashes (parallel to the main foliation in the host rocks) associated to a dextral regional shear zone developed during D2 (Castro 1986; Amice 1990; Vigneresse & Bouchez 1997; Castro & Fernández 1998).

The NAMPA comprises several batholitic-scale granite and granodiorite plutons in which the Los Pedroches composite batholith (LPB) is included. A large biotite + amphibole granodiorite pluton and several, smaller, variably porphyritic, biotite + cordierite monzogranite intrusions compose the LPB (Carracedo 1991; Larrea 1998). The batholith emplacement was controlled by a dextral N130E transtensional shear zone of crustal scale (Aranguren *et al.* 1997b). The Campanario–La Haba pluton (CHP), located immediately at the NW of the Los Pedroches batholith, is one of the peraluminous granites that has been traditionally included within the granitic unit of this batholith (Fig. 1).

Field relationships and petrographic descriptions

The CHP (Fig. 1) exhibits an elongated shape (about 30 × 6 km) parallel to the main WNW–ESE Hercynian trend. It became emplaced into Precambrian to Devonian age metasedimentary materials during the last events of the Hercynian orogeny (*c* 300 Ma), after the main deformation event (D1). It does not display any D2-related structures. The low-grade regional metamorphism has been overprinted by contact metamorphism to the hornblende-hornfels conditions in a thermal aureole of 300–1800 m width. The host rock assemblages (cordierite ± andalusite) indicate shallow conditions of emplacement ($P < 3$ kbar and $T = 550–650\,°C$). Neogene materials cover discordantly the NW part of the pluton.

Four petrographic units may be distinguished based on the texture and the different modal mineralogy (Fig. 1). The first three units are disposed in concentric zones parallel to the elongated shape of the pluton. The outermost facies is composed by a coarse-grained porphyritic biotite granite (G1) characterized by the presence of K-feldspar megacrysts (2–15 cm), cordierite phenocrysts (1–3 cm) and bipyramidal quartz crystals (1–2 cm). Inward, a fine–medium-grained porphyritic granite (G2) crops out. Mineralogically, this lithotype is similar to G1 but it differs by the smaller number and size of all phenocrysts and by the greater abundance of the fine-grained groundmass. The contacts between G1 and G2 are intrusive and they are marked by a transition zone (from a few metres to 100 m in width) in which there are numerous enclaves of G1 in G2, although enclaves of G2 in G1 are not uncommon. Locally, these contacts are gradational defined by a slow and progressive decrease of the number and size of phenocrysts and by a greater abundance of the matrix towards G2. The innermost non-porphyritic biotitic granite (G3) crops out in the centre of the pluton, enclosed by

G2. It is a fine-grained equigranular granite, with occasional cordierite and K-feldspar phenocrysts and secondary muscovite. The contacts between G2 and G3 are transitional as evidenced by the progressive decrease in the amount of K-feldspar and cordierite phenocrysts. No solid state deformation has been observed at any contact. All the lithotypes contain pegmatitic differentiates with turmaline and miarolitic cavities. Leucogranites (G4) intruded at the boundary between the country rocks and G1 or in several small dykes throughout the whole pluton. Field relationships suggest a nearly coeval intrusion of the G1, G2 and G3 magmas.

The mineral composition and modal contents of the different facies are very similar. K-feldspar ($Or_{72–97}$) can occur as large mega-phenocryst ($<15 × 8$ cm) and as anhedral grains in the groundmass, commonly Carlsbad-twinned and perthitic. Plagioclase ($An_{45–2}$) generally occurs as well formed phenocryst (<2 cm) or small crystals in the groundmass, with continuous to oscillatory zonings. Albitic rims and chessboard textures are occasionally observed in K-feldspar megacrysts, showing the activity of albitization processes. Quartz phenocrysts occur as idiomorphic (1–3 cm) crystals and as aggregates of anhedral crystals in the matrix. The aluminous biotite (<3 mm) is relatively rich in Fe (Mg no. $\approx 0.28–0.46$) and normally exhibits alteration to chlorite and muscovite. It also shows the typical compositions of igneous biotite crystallized from peraluminous magmas (Nachit *et al.* 1985; Abdel Raman 1994) (Fig. 2a). Cordierite appears both as euhedral phenocryst (<3 cm) and as anhedral grains. It displays a variable alteration with the development of green-brown biotite and muscovite aggregates and pinnite. Cordierites show Mg no. of 0.36–0.48 composition and Na_2O content of 1.62–1.27 wt%. Cordierite is considered to be a magmatic phase according to its textural relationships. Other criteria are its homogeneous distribution throughout the intrusion and the observation that its grain size varies in direct relation to the grain-size variations in the other minerals. Cordierite could result from cotectic magmatic reactions under fluid undersaturated or saturated conditions (type 2c or 2d; Clarke 1995). Accessory minerals are apatite, zircon, ilmenite, monazite, andalusite and tourmaline and commonly occur as inclusions in the principal minerals of the different lithotypes.

The magmatic fabric (magmatic foliation) within G1 and G2 is easily recognizable by the shape preferred orientation of the K-feldspar megacrysts especially and, also by biotite flakes and cordierite phenocrysts. This magmatic foliation strikes parallel to the long axis of the pluton and to the main Hercynian foliation trend. Biotite schlieren and compositional banding marked by the accumulation of K-feldspar megacrysts often occur in G1. These structures were acquired during flow in the magmatic stage and are parallel to the magmatic foliation.

Two types of enclaves may be distinguished: (1) country-rocks xenoliths metamorphosed up to hornblende-hornfels conditions; (2) biotitic microgranular enclaves (10 cm–1.5 m) of tonalitic compositions.

Geochemistry: major and trace elements and isotopic data

The CHP has a restricted SiO_2 range (*c.* 70.68–73.72 wt%) and is characterized by high K_2O (3.92–5.22 wt%) content and by a peraluminous character, as shown by its high A/CNK (0.90–1.23) (Table 1). The CHP chemical characterization is summarized in Fig. 2. Compositionally, the different petrographic units show similar geochemistry although slight differences may be observed in certain major and trace elements. The main G1 and the G3 granites

Table 1. *Chemical compositions of the different petrographic units of CHP*

	G1 (*n* = 12)				G2 (*n* = 6)				G3 (*n* = 3)			
	m	M	μ	σ	m	M	μ	σ	m	M	μ	σ
SiO_2	70.84	73.72	72.43	0.99	70.68	72.72	71.81	0.76	72.00	73.60	72.83	0.80
TiO_2	0.27	0.35	0.30	0.03	0.35	0.48	0.40	0.06	0.28	0.30	0.29	0.01
Al_2O_3	14.22	15.38	14.56	0.46	14.55	15.29	15.00	0.27	14.50	15.05	14.84	0.30
$Fe_2O_3^t$	1.60	2.56	2.02	0.35	2.10	2.60	2.33	0.19	1.50	2.00	1.75	0.25
MgO	0.46	0.69	0.55	0.10	0.59	0.87	0.70	0.11	0.40	0.67	0.53	0.14
CaO	1.02	1.30	1.15	0.09	1.35	1.84	1.55	0.21	0.97	0.98	0.97	0.01
MnO	0.04	0.05	0.05	0.00	0.04	0.05	0.04	0.00	0.04	0.05	0.04	0.01
Na_2O	3.23	5.20	3.89	0.77	3.19	3.72	3.47	0.22	3.26	3.32	3.29	0.03
K_2O	3.94	5.22	4.55	0.52	3.92	4.43	4.11	0.19	4.69	5.02	4.85	0.17
P_2O_5	0.21	0.25	0.23	0.01	0.20	0.23	0.22	0.01	0.24	0.24	0.24	0.00
LOI	0.43	1.03	0.65	0.26	0.44	1.20	0.65	0.28	0.63	1.25	0.89	0.32
Total	99.00	101.68	100.38	0.89	99.09	101.38	100.28	0.87	99.95	100.84	100.54	0.51
FeO	1.44	2.30	1.82	0.31	1.89	2.34	2.10	0.17	1.35	1.80	1.57	0.23
Rb	150	315	235	64	130	287	191	53	195	316	250	61
Sr	57	97	74	14	91	196	143	35	50	93	68	22
Ba	168	325	248	66	347	567	455	76	260	470	387	112
Sc	3	5	4	1	4	5	4	0	3	4	3	1
V	14	21	18	3	22	33	28	5	15	22	19	3
Cr	0	10	4	4	0	11	6	4	0	11	6	5
Co	71	91	84	7	72	112	92	14	98	100	99	1
Ni	0	3	1	2	0	3	1	1	0	5	2	3
Y	6	18	11	5	7	14	10	3	5	13	9	4
Nb	9	16	12	3	8	12	9	2	9	12	10	2
Zr	104	150	125	18	126	154	139	11	106	142	124	18
La	11.1	31.0	19.0	7.7	17.2	30.4	24.2	4.9	18.5	29.0	22.3	5.8
Ce	25.4	67.0	41.4	17.7	41.4	68.1	53.9	10.8	34.6	69.0	50.4	17.4
Nd	12.0	34.0	19.2	9.6	16.2	32.4	25.0	6.7	11.9	34.0	23.1	11.0
Eu	0.3	0.7	0.5	0.2	0.4	0.7	0.6	0.1	0.2	0.6	0.4	0.2
Dy	1.3	3.3	2.1	0.8	1.6	2.5	2.1	0.4	1.2	2.4	1.8	0.6
Yb	0.7	1.5	1.0	0.3	0.8	1.2	0.9	0.2	0.7	1.0	0.8	0.2
A/CNK	0.90	1.22	1.10	0.12	1.10	1.23	1.16	0.05	1.17	1.22	1.19	0.03
ΣREE	62.5	139.2	90.4	32.0	94.3	155.5	120.1	24.4	75.6	137.2	107.1	30.9
I.D.	85.84	92.02	88.33	2.31	82.87	85.56	84.30	1.15	87.34	89.17	88.32	0.92

G1, coarse-grained granite; G2, fine-grained granite; G3, equigranular biotite.
n = no. of samples.
m, minimum; M, maximum; μ, average; σ, standard deviation.

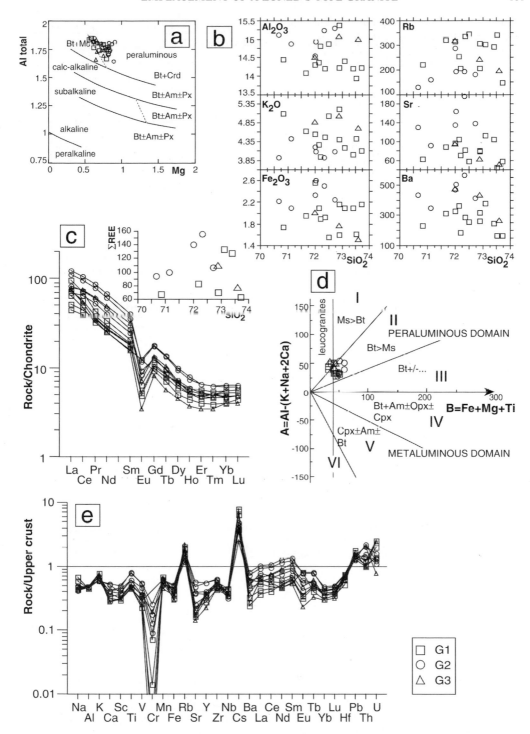

Fig. 2. Mineral and whole-rock geochemical diagrams. (**a**) Typology of the biotites of CHP (Nachit *et al.* 1985). (**b**) Some representative major- and trace-element variation diagrams. (**c**) A–B diagram (Debon & Le Fort 1983) showing the aluminous association of the CHP. (**d**) Chondrite-normalized REE diagram (Evensen 1978). (**e**) Average upper crust-normalized diagram (Taylor & McLennan 1981).

Table 2. *Whole rock isotopic data for the CHP lithotypes*

Sample	Rb (ppm)	Sr (ppm)	$^{87}Rb/^{86}Sr$	$^{87}Sr/^{86}Sr$	2s
S-25	322.5	95.2	9.8411	0.750578	9
S-28	207.6	78.4	7.6872	0.741297	10
CH-95-9	194.5	196.2	2.8714	0.720141	10
CH-40	181.1	137.8	3.8069	0.723847	8
CH-32	307.0	81.0	11.0163	0.755916	12

are chemically similar and gently more acidic than G2. The G2 facies exhibits higher Fe_2O_3, TiO_2, Al_2O_3, CaO and lower K_2O and P_2O_5 contents compared with G1 and G3. As far as the trace elements are concerned, the G2 also display higher contents in Sr, Ba, Sc and V (Table 1, Fig. 2b).

The REE content varies between 62 and 156 ppm. The G1 and G3 granites have similar REE concentrations and lower than G2, but all of them display a striking uniformity in the shape of their REE patterns. The REE content variation does not correlate with SiO_2 variations (Fig. 2c). REE-normalized patterns are distinctly more enriched and fractionated for LREE $[(La/Sm)_N = 2.48$–$4.92]$ than for HREE $[(Gd/Yb)_N = 1.89$–$2.95]$ and commonly exhibit negative Eu anomalies (Eu/Eu* = 0.3–0.53).

The A–B diagram (Debon & Le Fort 1983) reveals the aluminous character of the association, which does not exhibit a clear evolutive trend (Fig. 2d). This type of association is commonly related to the partial melting of metasedimentary crustal material (Debon & Le Fort 1983). The upper crust-normalized multi-element diagram (Taylor & McLennan 1981) (Fig. 2e) also illustrates the similarity of all the studied granitic types. Patterns in this diagram are similar to those of syn- or postcollisional S type granites (Thompson *et al.* 1984). The Ba, Sr, Nb, LREE, Sc and V anomalies could result of the partial melting process and/or be inherited from the source rocks.

The CHP has been dated by whole-rock K–Ar method at 305 ± 10 Ma (Penha & Arribas 1974). Rb–Sr data for five samples of G1, G2 and G3 yield an age of 308.7 ± 5.7 Ma (MSWD = 1.3) with an initial ratio of 0.70739 ± 0.00038 (Table 2, Fig. 3). The studied samples show a similar isotopic relation. Therefore, they could be comagmatic or, alternatively, they could come from a similar source material. The partial melting of a metasedimentary crustal protholith is the most likely mechanism to produce the peraluminous granites of the CHP. However, the Sr_i ratio (0.707) of this pluton is less radiogenic than those typical S granites of metasedimentary crustal origin, indicating that these granites may

not have been exclusively derived from crustal sources. The intermediate Sr_i datum between mantelic values at 300 Ma (0.704) and crustal values (> 0.710) (Defalque *et al.* 1992) suggests a hybrid crust–mantle origin for this granite (Barbarin 1990; Castro *et al.* 1991, 1994). The widespread occurrence of biotitic microgranular enclaves within the lithotypes and the presence of hybrid zones at the contacts support the mixing/mingling processes between them.

Discussion of some petrological aspects

The homogeneity in mineral and whole-rock chemistry for the different facies is outlined by the chemical overlapping of the different granitic and mineral compositions. The lack of linear inter-element relationships in many plots as well as the lack of changes in mineral compositions (especially biotite, Fig. 2a) suggests that the differentiation mechanisms commonly invoked to explain the chemical variations in felsic plutons, as the fractional crystallization, restite unmixing or magma mixing, can not be considered to explain the pluton zonation pattern from the outermost facies (G1) towards the centre of the pluton (G3). Therefore, it is proposed that the overall zonation pattern resulted from cooling and crystallization of at

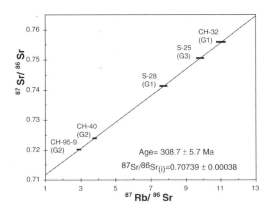

Fig. 3. Rb–Sr isochron diagram for the different lithotypes of CHP.

least two magmatic injections (forming G1 and G2), of similar composition, coming from the partial melting of a similar protolith.

The pluton shows a petrographic zonation, with each lithotype (G1, G2 and G3) exhibiting its own textural characteristics. The petrographic variations of the pluton might be explained by different undercooling degrees and water content of the melts. Groundmass and phenocryst size variations between G1 and G2 can be explained as a result of the undercooling caused by temperature differences between the two magmas. However, both magmas are almost contemporary. Therefore, other processes such as *in situ* decompression and water-content variations have been considered. The *in situ* decompression verified during the growth of phenocrysts in G2, could have been accompanied by boiling, sudden reduction in water pressure and, therefore, by displacement of the liquidus to the highest temperatures. As a consequence, nucleation rate was increased giving place to an abundant groundmass with fine- to medium-sized crystals (Candela 1997). By contrast, the possible low water content in G1, volumetrically the most important facies of the pluton, could have favoured the growth of K-feldspar megacrysts (Swanson 1977; Candela 1997). Experimental work on samples of different typology, such as the Manaslu and Gangotri granites (Scaillet *et al.* 1995), or the Strathbogie granite (Clemens & Wall 1988), shows that biotite is often the liquidus phase for H_2O-rich conditions, while for lower water contents the K-feldspar crystallizes before biotite. This suggests largest water contents for the magma corresponding to G3 compared to G2, and therefore, lower liquidus temperatures for G3. Accordingly, the undercooling produced by fluid release probably took place after the crystallization of megacrysts in G2 and during the crystallization of biotite in G3. This could explain the porphyritic texture in G2 and the equigranular texture in G3. Besides, the liquidus crystallization of biotite in G3 prevented cordierite crystallization.

Structure: anisotropy of magnetic susceptibility

Material and method

The magmatic fabrics of plutonic rocks show, in general, low anisotropies and are difficult to measure by traditional structural methods (Bouchez 1997), especially the magmatic lineation, the most important kinematic indicator. Nevertheless, these weak fabrics are 'visualized' quickly and easily by low-field magnetic fabric measurements. Actually, the anisotropy of magnetic susceptibility (AMS) is a powerful tool in tectonic studies (Hrouda 1982; Rochette *et al.* 1992; Borradaile & Henry 1997) and today it is the only technique able to tackle with guaranties the structural analysis of granitoid domains (Bouchez *et al.* 1997). The structural mapping of many plutons and batholiths during the last decade demonstrates: (a) the validity of the AMS technique to study granitoid rocks (Bouchez *et al.* 1997), (b) the remarkable fabric homogeneity over large distances of these rocks (Aranguren *et al.* 1997b; Olivier *et al.* 1997) and (c) the parallelism between the magmatic (mineral fabric) and magnetic fabrics (Guillet *et al.* 1983; Darrozes *et al.* 1994). AMS measurements provide both magnitudes and orientations of the axes $K_1 \geqslant K_2 \geqslant K_3$ of the AMS ellipsoid. A magnetic fabric is characterized by magnetic lineation (K_1) and magnetic foliation (perpendicular to K_3).

The CHP was structured in magmatic state, i.e. there are no appreciable signs of solid state deformation (Paterson *et al.* 1989). The AMS technique is used to unravel the internal magmatic structure of the CHP and to propose its emplacement model. This structural study comprises 106 drilling stations regularly distributed throughout the pluton. The result given for each drilling location represents the average of four samples ($d = 2.5$ cm and $h = 2.2$ cm) cut from two drilling cores. The magnetic measurements were carried out using a Kappabridge KLY-2 susceptometer, working at low alternating field (4×10^{-4} T; 920 Hz) with a detection limit of about 5×10^{-8} SI.

Magnetic susceptibility and anisotropy

According to its mineralogy (lack of magnetite) and the composition of the petrographic units, the CHP exhibits a typical paramagnetic character (Rochette 1987; Rochette *et al.* 1992): the magnetic susceptibility ($K = 1/3(K_1 + K_2 + K_3)$) is low, ranging from 1.8 to 13.7×10^{-5} SI (Fig. 4). The mean values and the variation range of K show a great homogeneity and similarity for all lithotypes, in agreement with their Fe contents (Fig. 2b) (Gleizes *et al.* 1993). Only the leucogranites with low values of K appear separated from the other lithotypes. The slightly higher susceptibility mean value of G2 can be explained by its higher iron-content.

The anisotropy degree ($P = K_1/K_3$) reflects the intensity of the shape preferred orientation of paramagnetic minerals (biotite + cordierite). This parameter, expressed in percentage and

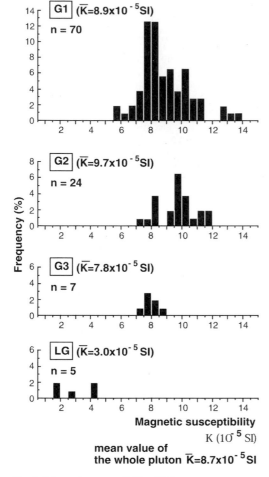

Fig. 4. Magnetic susceptibility (K) frequency histograms of the main lithotypes, showing mean K values. $n =$ number of sites.

corrected by elimination of the diamagnetic contributions of quartz and feldspar (assumed to be constant and isotropic; $D = -1.4 \times 10^{-5}$ SI, Hrouda 1986; Rochette *et al.* 1992), is calculated as follows: $P_{para}\% = 100 \; [((K_1 + D)/(K_3 + D)) - 1]$. The $P_{para}\%$ values are very low and range from 0.8 to 4.2 (except for two samples with $P_{para}\% = 4.9$ & 7.1), varying the 70% of sampling stations between 1.5 and 3%. These values ($<5\%$) are characteristic of fabrics acquired during magmatic state.

The P mapping does not display any additional information because there are not any preferential distribution of the highest P values ($>3\%$), for example with respect to the sigmoidal termination of the pluton or to the pluton borders. However, the 81% of the sites with $P_{para}\% > 3\%$ are located preferentially

within the G1 lithotype. In any case, the highest P values ($P_{para}\% = 4.9$ & 7.1) are located in the SE termination and related to G1. According to numerous AMS studies in granitic plutons, the highest P values are related with solid-state deformation zones, which record twice or higher P values than those in areas structured in magmatic state. The CHP exhibits low P values according to the lack of the solid state deformation and, therefore, the P values show small variations throughout all the pluton.

The magnetic susceptibility and the total anisotropy of CHP are compared with the values of Cabeza de Araya pluton (Amice 1990) and the granitic plutons of Los Pedroches batholith (Aranguren *et al.* 1997*b*) (Table 3). All of them show very similar values. They are typically paramagnetic and are structured in magmatic state ($P_{para}\% < 5$). However, K and $P_{para}\%$ value-ranges of Cabeza de Araya are slightly wider.

Directional data of magnetic fabric

In paramagnetic granites, the AMS is induced by the iron-bearing silicates, therefore, mainly biotite $\pm$ cordierite in the CHP. The cordierite also contributes to the magnetic fabric (in G1 and G2) and it could be problematic due to its inverse magnetic fabric (K_3 parallel to the C elongation axis of the cordierite prism). Cordierite appears as megacrysts easy to be avoided during drilling and, besides, it is rare to find it fresh or unaltered. Cordierite is usually replaced by micaceous products (muscovites and green biotites) obviously changing its magnetic behaviour and reducing (to zero in totally altered crystals) its contribution to the magnetic fabric. Inversion between K_1 and K_3 axes and scattering in the fabric diagrams was only found in some of the samples with fresh cordierite. Besides, the parallelism between the magnetic foliation and the G1–G2 contact demonstrates that cordierite does not significantly influence the G1 and G2 magnetic fabric. The CHP magnetic fabric is therefore controlled and dominated by the shape preferred orientation of biotite.

The *magnetic foliations* are generally gently to steeply dipping and distributed in zone axis around the lineation (Fig. 5a, stereonet for K_3). The foliation trajectories exhibit a concentric pattern parallel to the lithotype contacts and outward dipping is observed on the pluton scale as well as on the facies scale. G1 shows steeply dipping foliations, indicating that the contact with the country rocks is quite vertical. However, G2 and G3 display gently plunging foliations

Table 3. *Comparative study of scalar and directional AMS data*

	Campanario–La Haba	Cabeza de Araya (Amice 1990; Vigneresse & Bouchez 1997)	Pedroches granites (Aranguren *et al.* 1997*b*)
K $(10^{-5}$ SI)	Mean value = 8.7 Range from 1.8 to 13.7 84% between 7 and 11.5	Mean value = 7.4 Range from 2.3 to 27.5 69% between 3 and 10	Mean value = 8.6 Range from 1.6 to 17.4 70% between 6 and 12
P_{para}%	Mean value = 2.38 Range from 0.8 to 4.9 69% between 1.5 and 3.0	Mean value = 3.46 Range from 1.4 to 8.6 78% between 2.0 and 5.0	Mean value = 2.24 Range from 0.9 to 5.7 83% between 1.0 and 3.0
Magnetic lineation	Subhorizontal to subvertical Mean value (306/11)	Subhorizontal N137E to subvertical	Subhorizontal Mean value (225/01)
Magnetic foliation	Subvertical to subhorizontal	Subvertical	Subhorizontal

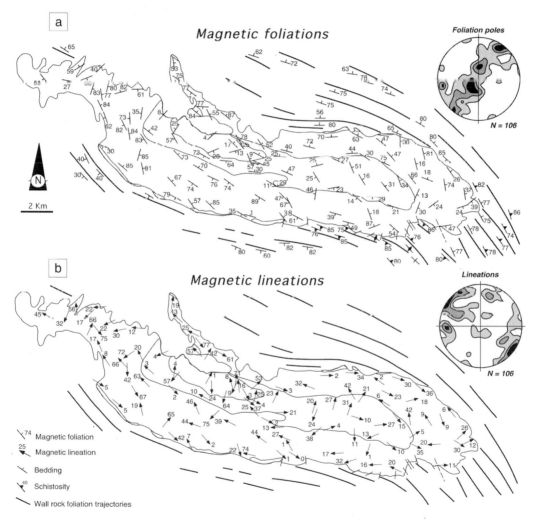

Fig. 5. Structural maps of magnetic foliations (**a**) and lineations (**b**) and the corresponding orientation diagrams (contour intervals in multiples of a uniform distribution) based on 106 stations (Schmidt, lower-hemisphere). Each map also shows the foliation trajectories in country rocks.

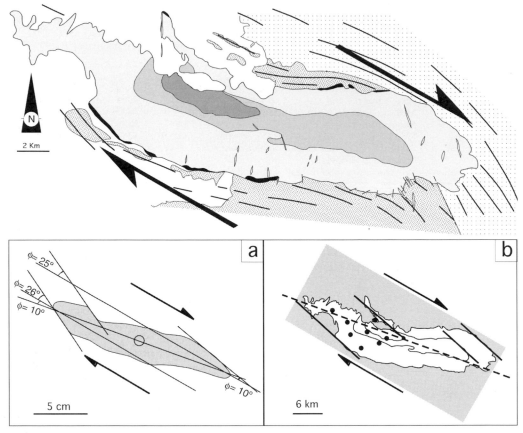

Fig. 6. Kinematic of the CHP emplacement showing the sigmoidal pattern of the pluton and country rock foliation according to the proposed shear sense. (**a**) Experimental model (horizontal section) of pluton emplacement in strike-slip zones (Roman-Berdiel *et al.* 1997). (**b**) The Riedel-type fractures associated with dextral strike-slip zone control the external and internal petrographic geometry of the pluton. The filled points represent subvertical lineations ($>60°$).

defining elongated domes, one centred in G3 and the other in the eastern half of G2.

The *magnetic lineation* is parallel to the elongation of the pluton (mean value = 306/11) (Fig. 5b). This tendency is clear along the borders of the pluton. The magnetic lineations are mostly gently plunging (Fig. 5b, stereonet for K_1). However, steeply plunging lineations ($>60°$; 11% of the sites) are also observed and are located in the western half of the pluton around G3 facies. These subvertical lineations are distributed within the three lithotypes (Figs 5b and 6b), suggesting the existence of a root zone or magma feeding zone in this area. On the other hand, gently plunging lineations oblique to the pluton elongation are also found. These magnetic lineations are situated in G2 and G3 lithotypes associated to concentrically

distributed subhorizontal foliations. At the presently exposed level, these areas could have been the roof zones of small intrusions or magma batches of the G2 and G3 lithotypes inside the G1.

These magnetic data are compared with the available magnetic data of the adjacent Cabeza de Araya pluton (Amice 1990) and the Los Pedroches batholith (Aranguren *et al.* 1997*b*) (Table 3). The compared scalar data (K and P_{para}%) are very similar. The directional data show obvious similarities between the CHP and Cabeza de Araya, according to the mean orientation of the magnetic lineation (*c.* N130E) and to the general distribution of the magnetic foliation. However, the magnetic foliations of the granitic plutons of LPB are subhorizontal and the magnetic lineations are scattered showing a mean value of 225/1 (Aranguren *et al.* 1997*b*). This orientation is practically perpendicular to the magnetic lineation orientation in the CHP and Cabeza de Araya. On the other hand, the Pedroches granodiorite (Aranguren *et al.* 1997*b*) shows the predominance of flat-dipping magnetic foliations. However, the magnetic lineations are subhorizontal and exhibit a mean east-west orientation (274/2). This lineation differs from the lineation orientation in the other granitic plutons (CHP, Cabeza de Araya and the granitic unit of the LPB).

Emplacement model

The deformation controls the magma emplacement in the crust and several recent studies show the clear relationships existing between pluton geometry and the regional tectonic environment at the time of intrusion (Hutton 1988; Vigneresse 1995). Plutons emplaced in transcurrent regimes are typically ellipsoidal in shape, crop out as elongated bodies parallel to the main structures in the country rocks, and show subhorizontal lineations (parallel to the pluton long axis) and moderate to steep dipping foliations (Vigneresse 1995).

The external geometry of the CHP and the country rock foliation trajectories draw a sigmoidal pattern, which is also shown on the cartographic geometry of the granitic lithotypes (Fig. 6). The sigmoidal geometry of this pluton is similar to those ones obtained in experimental models of granite intrusions in the brittle upper crust along strike-slip zones (Fig. 6a) (Román-Berdiel *et al.* 1997). These experiments show that: (a) the intrusions are always elongated in the extension field defined by the strike-slip regime and the emplacement was controlled by this direction since the first stages of intrusion; (b) the fractures formed in the cover conditioned the lateral expansion of the intrusions; this produced local parallelism between the faults and the contacts of the intrusions.

The CHP is emplaced in the upper brittle crust and the emplacement of the lithotypes occurs in the same magmatic and kinematic context. According to the distribution of the facies, G1 was the first magma emplaced and it spread parallel to the principal stretching direction associated with the strike slip regime. This direction controlled the emplacement of all the facies. G2 was emplaced as a second magma batch when G1 was partially crystallized. The character of the contact between these two facies depends on the rate of crystallization of G1, exhibiting intrusive or transitional contacts. The G2 lithotype was not crystallized at the time of the G3 emplacement as revealed by the transitional contacts between them. It should be stressed that the transitional and intrusive nature of contacts suggest that the entire CHP granite sequence could have been emplaced within a relatively short range of time with overlap between consecutive intrusive magma batches.

The lack of ductile deformation in the host rocks suggests that the opening of the necessary space for the intrusion has been made by the displacement of rigid blocks along brittle faults. The pluton external geometry seems to be strongly conditioned by faults (Riedel R type fractures) formed in the country rock (Fig. 6b), as shown by the experimental models (Román-Berdiel *et al.* 1997; Benn *et al.* 1998; Dauteuil & Mart 1998). On a small scale, the cartographic shape of the granitic types also reflects the control that these kind of faults exert on the emplacement and the geometry of the late-lithotypes (Fig. 6b). These faults might have locally constrained the lateral spread of the magma and controlled the final geometry of the pluton. In experimental models, the injection of the fluid (the magma) is located to the centre of the intrusion (Fig. 6a). Taking into account that the western end of the pluton is covered by the Cenozoic deposits, and according to the lithotype distribution and subvertical lineations (>60°) (Fig. 6b), the root zone could be approximately centred in G3.

The final geometry of the CHP shows the kinematics of the strike-slip regime: it is a sigmoidal and elongated (N100E) pluton emplaced in an N120–130E dextral shear-zone (Fig. 6). This pluton was built by the emplacement of successive magma batches which were structured by lateral expansion parallel to the regional stretching direction during brittle shear deformation. The proposed emplacement model agrees with the one defined for the Central Extremadura batholith (Castro 1986; Amice 1990; Vigneresse & Bouchez 1997; Castro & Fernández 1998) and for the granodioritic unit of the Los Pedroches batholith (Aranguren *et al.* 1997*b*) (Fig. 1). These models suggest the existence of a dextral regional strike-slip zone parallel to the regional foliation in the south branch of the CIZ. The granitic unit emplacement of the Los Pedroches batholith was a late occurrence relative to this main transcurrent regime, because these granitic plutons were emplaced later than the granodioritic body of the batholith and cut the Conquista fault (Fig. 1). This fault is a dextral transtensional ductile shear zone with a normal component related to a late extensional evolution of the CIZ (Aranguren *et al.* 1997*a*).

Conclusions

The Campanario–La Haba pluton comprises relatively homogeneous cordierite and andalusite peraluminous granites emplaced at $P < 3$ kbar in a metasedimentary host rock. The emplacement occurred during late-Hercynian deformation events $(309 \pm 6 \text{ Ma})$. The chemical homogeneity of the facies indicates that the lithotypes correspond to at least two different magmatic pulses derived from partial melting of a similar metasedimentary protolith, although low Sr_i value may indicate some mantellic contribution. The lithotypes textures reflect changes in the undercooling conditions due to *in situ* decompression and to the different water-undersatured conditions.

The structural control is important on the emplacement of the Campanario–La Haba pluton. The CHP is a synkinematic pluton related to a dextral strike-slip regime associated to the D2 phase of Hercynian deformation. This strike-slip regime allowed the development of extensional fractures which acted as ascending zones for magma, until its emplacement in a high structural level. There, the space appeared as a result of block displacements along brittle fractures. This emplacement is related to the existence of a NW–SE dextral megashear-zone parallel to the regional foliation in the South of the Central Iberian Zone.

This work is part of A. Alonso Olazabal's PhD thesis at the Basque Country University which has been mainly supported by a study grant from the Basque Country Government. We thank also L. A. Ortega and M. Menéndez for sample preparation and mass spectrometer measurements. We are greatly indebted to A. Castro and an anonymous referee for reviewing and improving this paper. This study was financially supported by the research projects UPV130.310-EB207/96, PB96-1452-C03-03 and UPV001.310-EB003/95.

References

ABDEL RAMAN, A. M. 1994. Nature of biotites from alkaline, calc-alkaline and peraluminous magmas. *Journal of Petrology*, **35**, 525–541.

ALLARD, B. & BENN, K. 1989. Shape preferred orientation analysis using digitized images on a microcomputer. *Computers and Geosciences*, **15**, 441–448.

AMÉGLIO, L., VIGNERESSE, J. L. & BOUCHEZ, J. L. 1997. Granite pluton geometry and emplacement mode inferred from combined fabric and gravity data. *In*: BOUCHEZ, J. L., HUTTON, D. H. W. &

STEPHENS, W. E. (eds) *Granite: From Segregation of Melt to Emplacement Fabrics*. Kluwer Academic Publishers, Amsterdam, 199–214.

AMICE, M. 1990. *Le complexe granitique de Cabeza de Araya (Estremadure, Espagne). Zonation, structures magmatiques et magnétiques, géométrie. Discussion du mode de mise en place.* PhD Thesis, University of Toulouse.

APARICIO, A., BARRERA, J. L., CASQUET, C., PEINADO, M. & TINAO, J. M. 1977. El plutonismo hercínico postmetamórfico en el SO del Macizo Hespérico (España). *Boletín Geológico y Minero*, **88-6**, 497–500.

ARANGUREN, A., CUEVAS, J., TUBÍA, J. M., CARRACEDO, M. & LARREA, F. J. 1997*a*. La faille de Conquista: caractérisation structurale et cinématique d'un cisaillement ductile dextre à composante normale dans le domaine sud de l'Arc ibéro-armorican. *Comptes Rendus de l'Académie des Sciences de Paris*, **325**, 601–606.

——, LARREA, F. J., CARRACEDO, M., CUEVAS, J. & TUBÍA, J. M. 1997*b*. The Los Pedroches batholith (Southern Spain): poliphase interplay between shear zones in transtension and setting of granites. *In*: BOUCHEZ, J. L., HUTTON, D. H. W. & STEPHENS, W. E. (eds) *Granite: From Segregation of Melt to Emplacement Fabrics*. Kluwer Academic Publishers, Amsterdam, 215–229.

BARBARIN, B. 1990. Granitoids: main petrogenetic classifications in relation to origin and tectonic setting. *Geological Journal*, **25**, 227–238.

BENN, K., ODONNE, F. & DE SAINT BLANQUAT, M. 1998. Pluton emplacement during transpression in brittle crust: new views from analogue experiments. *Geology*, **26**, 1079–1082.

BORRADAILE, G. J. & HENRY, B. 1997. Tectonic applications of magnetic susceptibility and its anisotropy. *Earth Sciences Reviews*, **42**, 49–93.

BOUCHEZ, J. L. 1997. Granite is never isotropic: an introduction to AMS studies of granitic rocks. *In*: BOUCHEZ, J. L., HUTTON, D. H. W. & STEPHENS, W. E. (eds) *Granite: from Segregation of Melt to Emplacement Fabrics*. Kluwer Academic Publishers, Netherlands, 95–112.

——, HUTTON, D. H. W. & STEPHENS, W. E. (eds) 1997. *Granite: from Segregation of Melt to Emplacement Fabrics*. Kluwer Academic Publishers, Netherlands, 358 pp.

CANDELA, P. A. 1997. A review of shallow, ore-related granites: textures, volatiles, and ore metals. *Journal of Petrology*, **38**, 1619–1633.

CARRACEDO, M. 1991. *Contribución al estudio del batolito de Los Pedroches (Córdoba)*. PhD Thesis, University of Basque Country.

CASTRO, A. 1986. Structural pattern and ascent model in the central Extremadura batholith, Hercynian belt, Spain. *Journal of Structural Geology*, **8**, 633–645.

—— & FERNÁNDEZ, C. 1998. Granite intrusion by externally induced growth and deformation of magma reservoir. The example of the Plasenzuela pluton, Spain. *Journal of Structural Geology*, **20**, 1219–1228.

——, MORENO-VENTAS, I. & DE LA ROSA, J. D. 1991. H-type (hybrid) granitoids: a proposed revision of the granite-type classification and nomenclature. *Earth-Science Reviews*, **31**, 237–253.

——, —— & —— 1994. Rocas plutónicas híbridas y mecanismos de hibridación en el Macizo Ibérico Hercínico. *Boletín Geológico y Minero*, **105-3**, 285–305.

CLARKE, D. B. 1995. Cordierite in felsic igneous rocks: a synthesis. *Mineralogical Magazine*, **59**, 311–325.

CLEMENS, J. D. & WALL, V. J. 1988. Controls on the mineralogy of S-type volcanic and plutonic rocks. *Lithos*, **21**, 53–66.

CORRETGÉ, L. G., BEA, F. & SUAREZ, O. 1985. Las características geoquímicas del batolito de Cabeza de Araya (Cáceres, España): implicaciones petrogenéticas. *Trabajos de Geología, Universidad de Oviedo*, **15**, 219–238.

DARROZES, J., MOISY, M., OLIVIER, P., AMÉGLIO, L. & BOUCHEZ, J. L. 1994. Structure magmatique du granite du Sidobre (Tarn, France): de l'échelle du massif à celle de échantillon. *Comptes Rendus de l'Académie des Sciences, Paris*, **318**, 243–250.

DAUTEUIL, O. & MART, Y. 1998. Analogue modelling of faulting pattern, ductile deformation, and vertical motion in strike-slip fault zones. *Tectonics*, **17**, 303–310.

DEBON, F. & LE FORT, P. 1983. A chemical-mineralogical classification of common plutonic rocks and associations. *Transactions of the Royal Society of Edinburgh*, **73**, 135–149.

DEFALQUE, G., DEMAIFFE, D., DUMONT, P. & LALIEUX, PH. 1992. Le batholite de Los Pedroches (Sierra Morena). Etudes cartographique, pétrographique, géochimique, géochronologique et métallogénétique. *Annales de la Société géologique de Belgique*, **115**, 77–89.

DÍEZ BALDA, M. A., VEGAS, R. & GOZÁLEZ LODEIRO, F. 1990. Structure. Central Iberian Zone (Part IV). *In*: DALLMEYER, D. & MARTÍNEZ GARCÍA, E. (eds) *Pre-Mesozoic geology of Iberia*. Springer-Verlag, Berlin, 172–188.

EVENSEN, M. M., HAMILTON, P. J. & O'NIONS, R. K. 1978. Rare earth abundances in chondritic meteorites. *Geochimica et Cosmochimica Acta*, **42**, 1199–1212.

GLEIZES, G., NÉDÉLEC, A., BOUCHEZ, J. L., AUTRAN, A. & ROCHETTE, P. 1993. Magnetic susceptibility of the Mont-Louis Andorra ilmenite-type granite (Pyrennes): a new tool for the petrographic characterization and regional mapping of zoned granite plutons. *Journal of Geophysical Research*, **98**, B3, 4317–4331.

GUILLET, P., BOUCHEZ, J. L. & WAGNER, J. J. 1983. Anisotropy of magnetic susceptibility and magnetic structures in the Guérande granite massif (France). *Tectonics*, **2**, 419–429.

HECHT, L., VIGNERESSE, J. L. & MORTEANI, G. 1997. Constraints on the origin of zonation of the granite complexes in the Fichtelgebirge (Germany and Czech Republic): evidence from a gravity and geochemical study. *Geologische Rundschau*, **86**, S93–S109.

HROUDA, F. 1982. Magnetic anisotropy of rocks and its application in geology and geophysics. *Geophysical Surveys*, **5**, 37–82.

—— 1986. The effect of quartz on the magnetic anisotropy of quartzite. *Studia Geophysica et Geodetica*, **30**, 39–45.

HUTTON, D. H. W. 1988. Granite emplacement mechanism and tectonic controls: inferences from deformation studies. *Transactions of the Royal Society of Edinburgh*, **79**, 245–255.

JULIVERT, M., FONTBOTE, J. M., RIBEIRO, A. & NABAIS CONDE, L. E. 1972. *Mapa tectónico de la península Ibérica y Baleares*, E 1:1.000.000 IGME, Madrid.

LARREA, F. J. 1998. *Caracterización petrológica y geoquímica del sector oriental del batolito de Los Pedroches*. PhD Thesis, University of Basque Country.

LAUNEAU, P., BOUCHEZ, J. L. & BENN, K. 1990. Shape preferred orientation of object populations: automatic analysis of digitized images. *Tectonophysics*, **180**, 201–211.

NACHIT, H., RAZAFIMAHEFA, N., STUSSI, J. M. & CARRON, J. P. 1985. Composition chimique des biotites et typologie magmatique des granitoides. *Comptes Rendus de l'Académie des Sciences, Paris*, **301**, 813–818.

OLIVIER, PH., DE SAINT-BLANQUAT, M., GLEIZES, G. & LEBLANC, D. 1997. Homogeneity of granite fabrics at the metre and dekametre scales. *In*: BOUCHEZ, J. L., HUTTON, D. H. W. & STEPHENS, W. E. (eds) *Granite: from Segregation of Melt to Emplacement Fabrics*. Kluwer Academic Publishers, Amsterdam, 113–128.

PANOZZO-HEILBRONNER, R. 1992. The autocorrelation function: an image processing tool for fabric analysis. *Tectonophysics*, **212**, 351–370.

PATERSON, S. R., VERNON, R. H. & TOBISCH, O. 1989. A review of criteria for the identification of magmatic and tectonic foliations in granitoids. *Journal of Structural Geology*, **11**, 349–363.

PENHA, M. & ARRIBAS, A. 1974. Datación geocronológica de lagunos granitos uraníferos españoles. *Boletín Geológico y Minero*, **85**, 271–273.

PITCHER, W. S. 1993. *The nature and origin of granite*. Chapman & Hall, London.

QUESADA, C., FLORIDO, P., GUMIEL, P., OSBORNE, J., LARREA, F. J., BAEZA, L., ORTEGA, M. C., TORNOS, F., SIGÜENZA, J. M., QUEREDA, J. M., BAÑÓN, L. & DE LA CRUZ, E. 1987. *Mapa geológico-minero de Extremadura*. Consejería de Industria y Energía de la Junta de Extremadura.

ROCHETTE, P. 1987. Magnetic susceptibility of the rock matrix related to magnetic fabric studies. *Journal of Structural Geology*, **9**, 1015–1020.

ROCHETTE, P., JACKSON, M. & AUBOURG, C. 1992. Rock magnetism and the interpretation of anisotropy of magnetic susceptibility. *Review of Geophysics*, **30**, 209–226.

ROMÁN-BERDIEL, T., GAPAIS, D. & BRUN, J. P. 1995. Analogue models of laccolith formation. *Journal of Structural Geology*, **17**, 1337–1346.

——, —— & —— 1997. Granite intrusion along strike-slip zones in experiment and nature. *American Journal of Science*, **297**, 651–678.

SCAILLET, B., PICHAVANT, M. & ROUX, J. 1995. Experimental crystallization of leucogranites magmas. *Journal of Petrology*, **36**, 664–706.

SWANSON, S. E. 1977. Relation of nucleation and crystal-growth rate to the development of granitic textures. *American Mineralogist*, **62**, 966–978.

TAYLOR, S. R. & MCLENNAN, S. M. 1981. The composition and evolution of the continental crust: rare earth element evidence from sedimentary rocks. *Philosophical Transactions of the Royal Society, London*, **A301**, 381–399.

THOMPSON, R. N., MORRISON, M. A., HENDRY, G. L. & PARRY, S. J. 1984. An assessment of the relative roles of crust and mantle in magma genesis: an elemental approach. *Philosophical Transactions of the Royal Society, London*, **A310**, 549–590.

VIGNERESSE, J. L. 1988. Forme et volume des plutons granitiques. *Bulletin de la Société Gélogique de France*, **8**, 897–906.

—— 1995. Control of granite emplacement by regional deformation. *Tectonophysics*, **249**, 173–186.

—— & BOUCHEZ, J. L. 1997. Successive granitic magma batches during pluton emplacement: the case of Cabeza de Araya (Spain). *Journal of Petrology*, **38**, 1767–1776.

Brittle behaviour of granitic magma: the example of Puente del Congosto, Iberian Massif, Spain

CARLOS FERNÁNDEZ & ANTONIO CASTRO

Departamento de Geología, Universidad de Huelva, E-21819 Palos de la Frontera, Huelva, Spain (e-mail: fcarlos@uhu.es)

Abstract: The magmatic structures appearing in the Puente del Congosto granitic outcrop (central Iberian Massif) are described in this work. They are interpreted as the result of a complex interplay between viscous (Newtonian) and brittle behaviour of granitic magma, which allowed the newer magma pulses to intrude and deform older magma batches. A model of Newtonian magma intruding a linear viscoelastic host rock may be extended with some confidence to the case of magma into magma emplacement. The resulting structures combine the characteristics of dykes and diapirs. The formation of large batholiths might be initiated or entirely accomplished by this process. In order to investigate the influence of the general stress conditions characteristic of a given tectonic regime on the strength of granitic magma, an over-simplified macroscopic model considering mixed Newtonian and brittle behaviour has been developed in this work. The brittle response is simulated by the Modified Griffith criterion, so that only rough estimates of the critical differential stresses for brittle magma behaviour can be gained. The results of this model suggest that the brittle response of viscous granitic magmas is possible for any type of tectonic regime (specially under contractional tectonics). A comprehensive, physically sound model for this viscous–brittle behaviour of granitic magma is not yet available. Integrated theoretical, experimental and field-based studies are the best way to arrive at such a complete model.

Research in granites and granitic magmas is nowadays a multidisciplinar task, and increasing attention is paid to the rheological aspects of granites (e.g. McBirney & Murase 1984; Fernandez & Barbarin 1991; Dingwell *et al.* 1993; Vigneresse *et al.* 1996). Rheology is the science dealing with deformation and flow of materials (Ranalli 1995). Granitic magmas as well as their country rocks are able to deform and flow, as shown by structures observable at all scales. Therefore, a clear understanding of rheology is the prerequisite for studying the mechanics of segregation, ascent and emplacement of granitic magmas into the continental crust. Various kinds of ideal material response are considered to operate in natural magmas over the appropriate range of geological conditions (lithostatic pressure, temperature, deformation and deformation rate, tectonic stress, water content, melt percentage and magma composition). The basic models involve Newtonian and non-Newtonian viscosity, viscoelasticity and viscoplasticity (e.g. Fernandez & Barbarin 1991). Below some specific volume fraction of crystals (first rheological threshold of Fernandez & Barbarin 1991) magmas behave as Newtonian (or power-law)

fluids. The larger the crystal content, the larger the viscosity until the magma develops a yield strength and behaves as a Bingham viscoplastic material (or power-law fluid with yield strength). A second rheological threshold appears for larger crystal contents that limits viscoplastic behaviour from that of a rigid or elastic solid. Brittle response of granitic magmas is supposed to be possible only during these later stages of magma crystallization, when the high percentage of crystals allows the magma to support high levels of differential stresses (Bouchez *et al.* 1992; Vigneresse *et al.* 1996). Instead, experimental and theoretical studies by Dingwell (1997) demonstrated that the sudden application of a stress may result in the brittle failure of the melt in high-level granitic magmas at low pressure. This is favoured by the presence of other phases as crystals or vesicles, although they need not constitute large volume fractions of the magma. The method proposed by Dingwell (1997) requires viscoelastic behaviour and explains the brittle response as a consequence of intersecting the glass transition after significant total strains of the magma and for given high strain rates.

From: CASTRO, A., FERNÁNDEZ, C. & VIGNERESSE, J. L. (eds) *Understanding Granites: Integrating New and Classical Techniques.* Geological Society, London, Special Publications, **168**, 191–206. 1-86239-058-4/99/$15.00 © The Geological Society of London 1999.

Field observations indicate that new improvements are required in the theoretical and experimental control of fracture processes in magmas (e.g. Fernández *et al.* 1997). Is the granitic magma able to suffer brittle failure as a liquid, i.e. for crystal fractions under the first rheological threshold? Is brittle behaviour of magmatic systems the result of instantaneous, energetic loads of seismic origin, or is it rather a consequence of the local concentration of high pore pressures? Alternatively, are the stresses developed at the tip of dykes of felsic or mafic magma pulses able to break magma (as was initially suggested by Walker 1969)? Recent developments of the theory of dyking (e.g. Clemens & Mawer 1992; Rubin 1993; Petford *et al.* 1993; Petford 1996; Weinberg 1996) and new works on experimental determination of breaking strengths (e.g. Webb & Dingwell 1990; Gerth & Schnapp 1996; Dingwell 1998) are the starting point for a comprehensive treatment of this problem.

A granitic outcrop located in the Central System batholith (Iberian Massif, Spain) is studied in this work. The observed structures seem indicative of brittle failure of granite magmas with low crystal percentages and subsequent intrusion of new magma batches. This process could be responsible for the generation of large granitic batholiths. A summary and evaluation of the available mechanical models and experimental work relevant to the physical understanding of this process is presented in the discussion. However, no satisfactory answer to the problem of brittle behaviour of melts has yet been advanced. This goal could only be accomplished by integrating new theoretical models with field observations and the use of experimental techniques.

Granitic batches in the Puente del Congosto complex

Excellent outcrops of granitic rocks appear in the Iberian Massif that show evidence of magma breaking and successive emplacement of new magma batches within older magma pulses. Observations in such key outcrops provide the empirical basis for the development of theoretical models and, in turn, this field evidence must be used to test the predictions of the models. One of these outcrops is the Puente del Congosto complex in the Spanish Central System in which we have carried out some detailed mapping and petrological studies.

Geological setting

The Puente del Congosto outcrop is located in the central part of the Iberian Massif (Fig. 1), within the vertical-folds domain of the Central Iberian zone (Julivert *et al.* 1972; Díez-Balda *et al.* 1990). It belongs to the Central System batholith composed of anatectic granites, hybrid facies with granodiorites and tonalites, and basic rocks (Castro *et al.* 1990, 1991; Moreno-Ventas *et al.* 1995). The detailed emplacement mechanisms responsible for this large batholith are far from being well understood. However, the study of the solid-state fabrics and structures in the batholith and its host-rocks led Doblas (1991) to propose a bulk extensional tectonic scenario with a N–S-oriented extension direction. This tectonic evolution could be responsible for the emplacement of the granitic bodies as suggested by Casquet *et al.* (1988). The emplacement of large masses of granodiorites of the Central Iberian zone in relation to extensional tectonics was first proposed by Oen (1970).

Structural and petrological description

A review of the structural and petrological characteristics of the Puente del Congosto complex may be found in Fernández *et al.* (1997). This outcrop is a segment of the northern contact of the Central System batholith. Here, the contact with the metamorphic host has an E–W direction (Fig. 1). The most abundant facies are biotite granodiorites, with subordinate tonalites, diorites, gabbros and granites. Magmatic microgranular enclaves of tonalite to quartz-diorite composition are common within the granodiorites. Mineral assemblages in country-rock xenoliths are dominated by the association Crd–Sil–Bt–Qtz–Kfs in which Grt is absent indicating a maximum pressure of *c.* 400 MPa for the emplacement of the granodiorites in this region.

Several large granitic to gabbroic bodies can be traced in the mapped area (Fig. 1). They are elongated in an E–W direction, parallel to the contact with the metamorphic rocks. D2 and D3 structures in the host-rocks are cross-cut by the contact with the plutonic rocks (this is specially notorious in the northern part of the map in Fig. 1). Kilometre-scale, polygonal fragments of low-grade metasediments appear included between and within the plutonic batches. Structures in the metasedimentary fragments are disrupted by the contacts with the plutonic rocks. Internal contacts between the distinct igneous facies dip more than 40° to the north. The magmatic foliation describes complex patterns

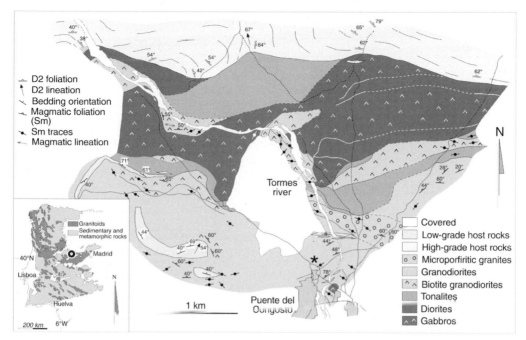

Fig. 1. Geological map of the Puente del Congosto granitic outcrop. Inset shows the location of this area (bold circle) in the central part of the Iberian Massif, to the west of Madrid. The asterisk marks the position in the map of the outcrop depicted in Fig. 4.

but it is statistically (although not in the detail) parallel to the average orientation of S2 in the host-rocks (Figs 1 and 2a).

In order to better understand the disposition and significance of the magmatic structures, detailed mapping was made on the bed of the Tormes river, from the village of Puente del Congosto towards the north (Figs 2 and 3). More than six batches of granodiorites, tonalites and gabbros crop out in this area. The traces of the magmatic foliation show a notable independence from one batch to another. The magmatic foliation is basically defined by elongate micro-granular enclaves and schlieren, with a minor contribution from the orientation of feldspar megacrysts. Magmatic lineations are poorly defined and highly variable in orientation from place to place. Contacts between batches disrupt the magmatic foliation of the older pulse, whereas the foliation in the newer pulse adapts smoothly to the contact (Figs 2 and 3). Two sets of conjugate magmatic shear zones (Fernández *et al.* 1997) developed in the proximity to the contacts (Fig. 2b). The traces of the magmatic foliation and minor aplitic bodies (Fig. 2c) as well as the long axes of xenoliths (Fig. 2d) lie in the obtuse angle between both systems of shear zones. Elegant patterns of magmatic folia-tion embracing more competent inclusions,

commonly gabbro blobs, can be observed (northern part of the maps in Figs 2 and 3). Complex mixing zones are extremely elongated in the more deformed areas around the gabbro mega-inclusions (Fig. 3). It is important to realize that this heterogeneity is limited to the interior of each body, with the plutonic contacts cross-cutting the mixing zones, schlieren and other structures.

Interpretation

Mixing zones and other structures (such as schlieren and elongated enclaves) appearing within the granitic bodies in the Puente del Congosto complex indicate that they were compositionally heterogeneous in origin, i.e. before their arrival at the final emplacement level. The nature and geometry of the contact between granitic bodies and their host rocks, as well as the shape and orientation of xenoliths at all the scales, suggest emplacement along brittle fractures. Coeval sinking of stoped xenolith or gabbro blocks cannot be excluded. The growth of this part of the batholith was accomplished by episodic intrusion of magmatic batches, whose relative age relationships reveal that the succes-sive pulses did not systematically follow the contact with the host rocks, but instead they

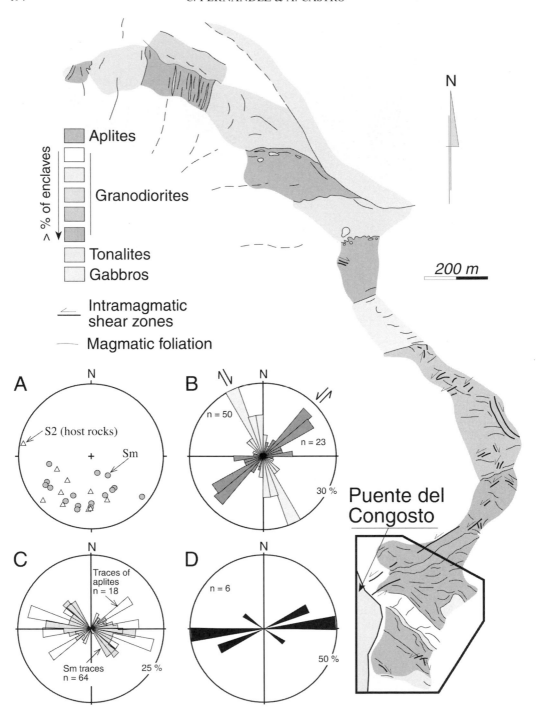

Fig. 2. Detailed geological sketch along the bed of the Tormes river, to the north of Puente del Congosto. The marked area is shown with more detail in Fig. 3. (**a**) Plot in equal-area, lower-hemisphere projection of magmatic foliation (Sm) in granitic rocks and D2 foliation in the host rocks (S2). (**b**) Rose diagram showing the orientation on a horizontal plane of conjugate intramagmatic shear zones. The shear sense is also indicated (arrows). (**c**) Rose diagram illustrating the orientation on a horizontal plane of aplitic dykes and traces of magmatic foliation (Sm). (**d**) Rose diagram with the orientation in a horizontal plane of the long axes of xenoliths.

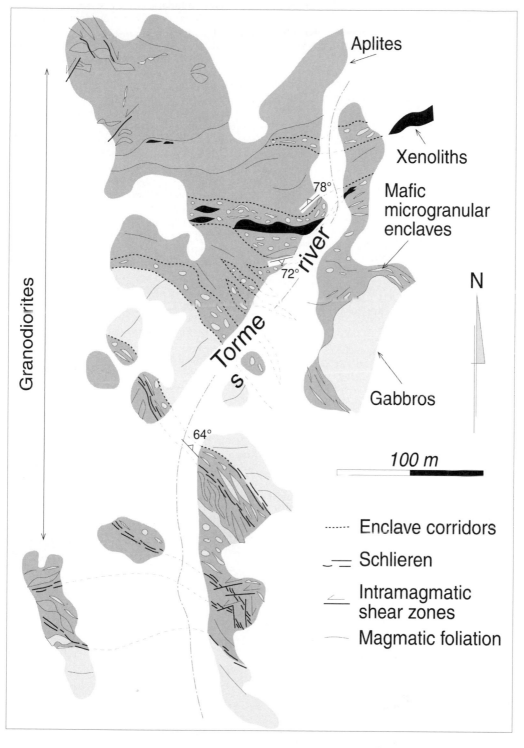

Fig. 3. Detailed sketch of magmatic structures and intrusive relations in several granitic batches located to the east-northeast of the Puente del Congosto village. For location see Fig. 2.

found their own way through the previous magma batches. This interpretation is based on the relative disposition of magmatic structures disrupted by the contacts between granitic bodies (Figs 1, 2 and 3). This fact is not commonly observed in granitic outcrops around the world and, therefore, it is generally believed that the magmatic foliation is a late structure (e.g. Fowler & Paterson 1997). This is not the case in Puente del Congosto, and a number of criteria suggests that intrusion of magma into magma via the opening of brittle fractures effectively took place here. These criteria are:

(a) the reorientation suffered by the magmatic foliation (schlieren and elongated mafic enclaves) of the older batch near a contact;
(b) the development of magmatic shear zones, specially within the older magma batch in the vicinity of a contact;
(c) the presence of granodioritic and aplitic dykes affected by magmatic folds and shear zones;
(d) feldspar megacrysts are never broken or displaced by a contact;
(e) synmagmatic mafic dykes that give place to swarms of microgranular enclaves and zones of mixing and mingling between acid and mafic magma (Fernández et al. 1997).

A granodiorite dyke intruding an older granitic batch is shown in Fig. 4 as an example of structure suggesting brittle behaviour of viscous magma. The dyke has an average thickness of about 20 cm and a length of at least 15 m, both measured in plan view. The dyke show pegmatite margins: a very important feature from a mechanical point of view, as will be discussed

later. The boundaries of the dyke cross-cut the magmatic foliation of the granitic host at a low angle. The magmatic foliation is defined by schlieren and the preferred orientation of feldspar megacrysts. Folds and shear zones affected the dyke as well as its host as evidenced by the folded schlieren, with fold wavelengths which vary as a function of the schliere thickness (right part of Fig. 4), in accordance with the classical theories of single layer buckling of a viscous layer in a matrix of lower viscosity (e.g. Ramberg 1960; Biot 1961). Shear zones with nearly constant spacing and sinistral apparent displacement limit dyke segments and displace the magmatic foliation in the host granite. A previous episode of local dyke stretching and necking may explain the regular thickening and thinning of the dyke in the deformed zones (Fig. 4, centre and right). The 'fish-hook' style of the deformed dyke suggests that the position of folds and shear zones was controlled by pre-existing boudins and pinch-and-swell structures (Price & Cosgrove 1990). The described sequence of deformational events is shown in the idealized sketches of Fig. 5: a granodiorite magma batch (magma 2) intruded an older body (magma 1) and developed a magmatic foliation (Fig. 5a), followed by the brittle emplacement of a granodiorite-pegmatite dyke (Fig. 5b), which was followed by stretching of the dyke (Fig. 5c) and, finally, by folding and shearing (Fig. 5d). According to the criteria by Paterson et al. (1989), both the dyke and granite host were deformed while in the magmatic state, as evidenced by the generalized absence of plastic intracrystalline microstructures. The feldspar megacrysts are euhedral or sub-euhedral and do

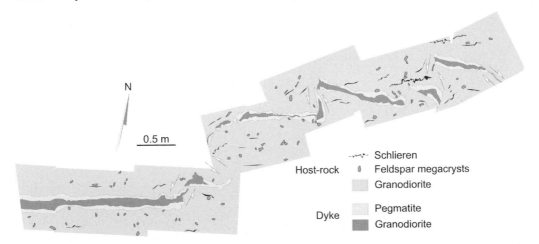

Fig. 4. Granodiorite-pegmatite dyke disrupted, folded and sheared. Drawn from a series of photographs taken in the location marked with asterisk in Fig. 1.

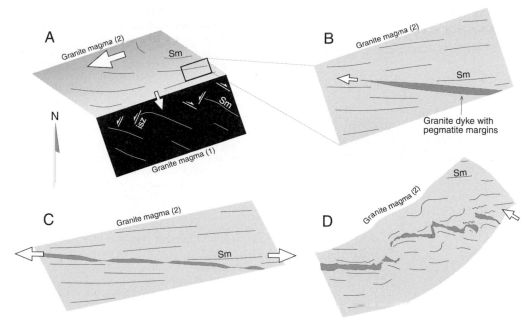

Fig. 5. (a) to (d) Interpretative evolution of the dyke shown in Fig. 4 (see text for an explanation). In (a) the large symmetric arrow indicates the supposed horizontal component of magmatic flow, and the small symmetric arrow marks the pushing effect of the new magma batch (granite magma 2) on the older one (granite magma 1). Asymmetric arrows depict the shear sense in the intramagmatic shear zones developed in the granite magma 1. (b) The emplacement of a granite-pegmatite dyke into granite magma 2. Dyke propagation is as indicated by the arrow. (c) and (d) Deformation in the magmatic state of both the dyke and the host magma. Sm: magmatic foliation; isz: intramagmatic shear zones. Not to scale.

not show *submagmatic* microfractures as those described by Bouchez *et al.* (1992). Therefore, it seems quite probable that the granite magma were deformed with a low percentage of mega-crytals and other solid inclusions (schlieren, mafic enclaves). Solid-fraction contents around 20% in volume may be deduced from a simple modal estimation of schlieren and megacrysts in the area shown in Fig. 4. The interpreted evolution of this dyke suggests that the granite magma was able to fail as a brittle material, before and after it flowed as a viscous material.

In summary, these criteria indicate that the magma was relatively free from crystals when fractured, except perhaps a small fraction of large feldspar megacrysts and other solid inclusions. These characteristics coincide with those expected for magmas under the first rheological threshold (Fernandez & Barbarin 1991; Fernandez & Gasquet 1994). It is also suggested that these magma bodies were able to suffer an important deformation by magmatic flow after each fracturing episode. Accordingly, the granitic magma batches were probably in the Newtonian field when fractured.

All these observations and interpretations, together with the elongated shape of the individual magma batches (Fig. 1), and the coincident E–W orientation of the contacts between batches, the aplitic dykes, the external contact with host rocks, and the long axes of xenoliths, may be explained in a model of episodic magma emplacement in fractures related to a N–S directed, extensional tectonic regime (Fig. 6).

Discussion

There is a considerable geological interest in the evaluation of the feasibility of granitic magmas with a low crystal fraction to fail brittley, as shown by the example of the Puente del Congosto area. The influence of parameters like the lithostatic pressure or the strain rate must be accurately determined. As a first approximation, magmas with a low crystal fraction may be considered as Newtonian viscous materials (e.g. Dingwell *et al.* 1993; Fernandez & Gasquet 1994). On the other hand, brittle failure requires that the stress intensity reaches some critical value at a particular crack tip (e.g. Engelder

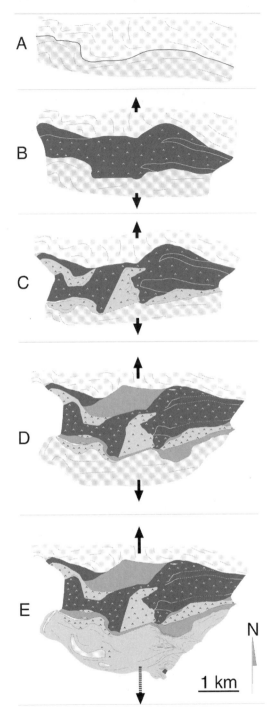

Fig. 6. Schematic diagram showing the idealized evolution of granitic (and gabbroic) batches in the Puente del Congosto outcrop shown in Fig. 1, as deduced from the study of their intrusive relations. The arrows show the deduced position of the horizontal extensional axis.

1993; Atkinson & Meredith 1987). Combining both material responses a sort of *viscous–brittle material* results. Conventional silicon putty used in analogue experiments exhibits such a behaviour when subjected to the appropriate stress, strain and strain rate (Price & Cosgrove 1990). The feasibility of combining both mechanical responses was discussed for magmatic systems by Corretgé & Martínez (1978) and by Shaw (1980). It requires the presence of imperfections, crystals, bubbles or minute cracks which act as the necessary stress concentrator (e.g. Ingraffea 1987). A viscous–brittle material may flow as a Newtonian viscous fluid without restriction, unless the stress state reaches the critical stress-intensity factor characteristic of the material. Under these circumstances, the mechanical conditions are satisfied for the nucleation and propagation of a fracture in an otherwise viscous magma. New magma batches may exploit the fracture and ascend quickly as dykes through older magmatic pulses, resulting in auto-intrusion structures that may finally give rise to large granitic plutons. The stress conditions in the broken magma, at both sides of the inflating dyke, allows the broken magma to deform again as a Newtonian fluid. This may result in deformation of fracture walls due to the dragging or pushing-aside effects of the intruding new batches of magma. This qualitative interpretation must be tested against a mechanical model of such a mixed behaviour in order to determine their governing parameters and to characterize the tectonic regimes in which it might occur. Although the unified model of a viscous–brittle material is not yet available, an approximate understanding of the problem could benefit from some previous theoretical developments.

Viscous magma intruding a viscoelastic magmatic host

Rubin (1993) presented a solution to the problem of a pressurized dyke intruding a linear viscoelastic medium. His results show that dykes and diapirs are just end-members for intruding viscous magma, with a complex interplay between elastic and viscous behaviours in the host rocks. Following Rubin (1993), two dimensionless ratios are important in governing dyke evolution: first, the elastic response of the host rock (p_o/G), and second, the viscous response of the host rock $(\eta_m/\eta_r)^{1/3}$, where p_o is the excess magma pressure, η_m and η_r are the magma and host-rock viscosity, respectively, and G is a measure of the elastic stiffness, equivalent to the elastic shear modulus divided by one minus the Poisson ratio.

Experimental investigations have shown that silicic melts exhibit both liquid-like and solid-like behaviours that may be adequately described by viscoelastic models (e.g. Dingwell et al.. 1993; Richet & Bottinga 1995, Webb & Dingwell 1995). Therefore, it seems possible to use the model by Rubin (1993) to place physical constraints to the process of brittle failure of a granitic magma followed by the intrusion of a new magma batch.

First, it is necessary to examine the most plausible values for the excess pressure p_o and the other relevant parameters in the Rubin (1993) model. The pressure at the upper part of a magmatic intrusion may be estimated from classical equations (e.g. Philpotts 1990) and depends on the depth to the magmatic reservoir, among other parameters. As stated above, the Puente del Congosto granitoids were emplaced at 400 MPa of lithostatic pressure. In this case, an excess pressure of 5–25 MPa is predicted for magmatic chambers at 5–10 km below the emplacement level. An estimation of G for viscoelastic granitic magma gives a range of 29–53 GPa, as deduced from the experimentally determined values of the unrelaxed short-timescale shear modulus of granitic melts (Bagdassarov et al. 1993). These estimations yield limiting values for p_o/G between 10^{-3} and 10^{-4}. Viscosity values in the range from 10^2–10^3 to 10^9–10^{10} Pa s are commonly cited as characteristics of viscous behaviour of granitic magmas (e.g. Fernandez & Barbarin 1991; Vigneresse et al. 1996; Dingwell 1998). Recent experimental determinations of Newtonian viscosity of silicic melts indicate a non-linear dependence of viscosity on temperature and water content, with marked differences between peralkaline, peraluminous and metaluminous granitic melts (e.g. Hess & Dingwell 1996; Dingwell et al. 1998; Holtz et al. 1999). Therefore, it seems reasonable to establish a wide range of viscosity ratios of about 10^{-5} to 10^{-7}, in order to simulate the emplacement of a hot, low-viscosity magma into a crystallizing, old magma batch. Moreover, high-viscosity ratios are favoured by compositional differences (as observed in Puente del Congosto, Fig. 1) or expected contrasts in water contents among the intruding magma batches.

Second, a comparison between the elastic and viscous dimensionless ratios relevant to the case of Puente del Congosto and the predictions of the theoretical model by Rubin (1993) (Fig. 7) indicates that a low-viscosity, new granitic magma batch ($\eta_{nm} \approx 10^3$ Pa s) is physically able

to intrude as a dyke into a viscous host granitic magma ($\eta_{om} \approx 10^8$ to 10^{10} Pa s). The predicted aspect ratios for the resulting dykes are of about 10^{-1} (Fig. 7), a value which coincides with the horizontal aspect ratios measurable in the map of the Puente del Congosto area (Fig. 1). Interestingly, the geometry of most of the granitic bodies in Fig. 1 shows the concave outward shape characteristics of dykes intruding a viscoelastic medium (Rubin, 1993). This contrasts with the more elliptical sections characteristics of dykes intruded in purely elastic rocks. The host-magma deformation would be predominantly viscous for more than 90% of the dyke walls (Fig. 7), and this accounts for most of the magmatic structures described in the granitic batches of Puente del Congosto. For instance, the magmatic conjugate shear zones are one of the consequences of the deformation in the dyke walls (Fig. 2). The arrangement of these shear zones (Fig. 2b) is kinematically incompatible with the deduced

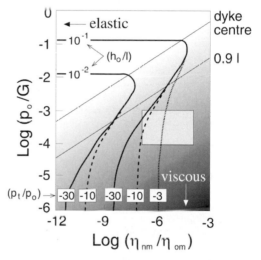

Fig. 7. Rubin (1993) plot showing in a logarithmic scale the elastic dimensionless number (p_o/G) against the viscous dimensionless number (η_{nm}/η_{om}). p_o is the excess magma pressure; G is the elastic stiffness; η_{nm} and η_{om} are the viscosities of the new and old magma batches, respectively. The old, host magma behaves as an elastic body in upper left and as a viscous material in lower right. The curves represent a contour plot of the new magma dyke aspect ratio (h_0/l), according to the equations by Rubin (1993), where h_0 is the thickness at dyke centre and l is the dyke length. These curves are plotted for several (p_t/p_o) values, where p_t is the pressure at the tip of the dyke. Diagonal straight lines mark the limit values where viscous displacements exceeds elastic displacements at dyke centre and over 90% of dyke (0.9l). The rectangular area with a dotted pattern shows the range of estimated parameters for magma batches in the Puente del Congosto region.

N–S extension direction. The pushing-aside exerted by the intruding batches may be considered as a better explanation for these structures. Viscous flow will continue after each transient fracturing episode. This explains why the magmatic foliation is truncated and, at the same time, it is deformed near the contact with the newer pulses. Structures resulting from viscous flow and brittle failure can become repeatedly superimposed and complex magma-into-magma emplacement structures may result. An intermediate term between dykes and diapirs would be adequate to describe the origin and evolution of these features (Rubin 1993). The observation of granitic dykes with pegmatite margins (Fig. 4) is relevant as it confirms the assertion by Rubin (1993) that water-rich, low viscosity phases could be important in propagating the dyke rapidly in a viscous medium.

The analysis by Rubin (1993) neglects fracture resistance at the tip of the dyke, where the behaviour of the host is essentially elastic. In fact, stress concentrations at the tip of fractures are known to favour the emplacement of dykes in an elastic host rock (Pollard & Segall 1987; Clemens & Mawer 1992). Experimental determinations of the critical stress-intensity factor (K_c) of silicic magma under different temperature and pressure conditions are not available (Dingwell 1998), although Gerth & Schnapp (1996) gave values in the range of 1.2 to 1.8 MPa $m^{1/2}$ for a crystal-poor obsidian glass at room temperature. For a definition of K_c see e.g. Atkinson & Meredith (1987) or Ingraffea (1987). Taking this critical value as a first-order approximation it is possible to estimate the minimum dyke length (l_c) necessary to neglect fracture resistance, as $l_c^{1/2} \gg K_c / p_0$ (Lister & Kerr 1991). This yields a critical l_c of 2–130 mm for the parameters used in the analysis of the Puente del Congosto area. These values are to be considered with caution, considering that the relevant property data (K_c) are not yet available. However, minor imperfections of lengths covering this order of magnitude are available in granitic magmas, including the external bondaries of feldspar megacrysts, bubble arrays or the cracks generated by shear inhomogeneities in the shear thinning regime previous to the brittle failure of the melt (Dingwell & Webb 1990).

A simple macroscopic fracture criterion approach

The model discussed above says nothing about the influence of the tectonic regime on the magma brittle failure. To investigate the influence of general stress conditions on the strength of rocks and magmas it is necessary to use macroscopic fracture criteria that are empirical or semiempirical (Scholz 1990). These criteria account only for a first-order mechanical approximation to the strength properties of geological materials and they predate the models based on the fracture mechanics approach (e.g. Ingraffea 1987). As an example, the generalized forms of the Griffith criterion are based on a correct description of the physical mechanisms of failure, although a degree of empiricism is involved, as evidenced by the determination of some relevant parameters like tensile strength or friction coefficient.

According to this, the equations for an ideal viscous and brittle material subjected to a three-dimensional flow might be expressed as follows:

$$2\eta D'_{ij} = T'_{ij} \quad \text{for} \quad F < 0 \qquad (1)$$

$$2\eta D'_{ij} = FT'_{ij} \quad \text{for} \quad F \geqslant 0 \qquad (2)$$

where η is a physical constant analogous to the viscosity coefficient of a fluid, D'_{ij} is the deviatoric rate-of-deformation tensor (also called deviatoric stretching tensor), T'_{ij} is the deviatoric stress tensor, and F is a dimensionless measure of the far-field stress level:

$$F = (1 - k), \qquad (3)$$

so that if $k > 1$ then $F < 0$ and by equation (1) the material behaves as a Newtonian fluid, but if $k \leqslant 1$ then $F \geqslant 0$ and sudden failure characterized by initiation and propagation of a fracture is allowed to proceed and to coexist with Newtonian flow (equation (2)). Therefore, k is a measure of the stress threshold for brittle failure in a melt. The fracture criterion considered here is the *Modified Griffith Theory* proposed by McClintock & Walsh (1962). It can be easily shown that under this criterion k takes the following value:

$$k = \frac{8T_0(\sigma_1 + \sigma_3)}{(\sigma_1 - \sigma_3)^2} \qquad (4)$$

under tensile conditions, and

$$k = \frac{C_0 + \sigma_3[(\mu^2 + 1)^{1/2} + \mu]^2}{\sigma_1} \qquad (5)$$

under compressive conditions, where σ_1 and σ_3 are the maximum and minimum principal stresses, respectively, T_0 is an empirical coefficient of tensile strength, C_0 is the uniaxial compressive strength and μ is the coefficient of internal friction of the material. The convention of sign more usual in continuum mechanics is adopted here, so that tensile stresses are reckoned positive. Therefore, the lithostatic pressure is always negative.

Equations (1) and (2) are formally analogous to those for viscoplastic materials (e.g. Malvern 1969). This viscous–brittle model is also similar to the elasto-plastic approximation to faulting developed by Gerbault et al. (1998) to account for some previously unexplained features of brittle failure. The constitutive equations (1) and (2) are only valid for homogeneous, isotropic and incompressible materials. Ideal magmas relatively free from bubbles and crystals may be considered as homogeneous melts (Richet & Bottinga 1995) with isotropic viscosity. A necessary prerequisite to accomplish the assumption of incompressibility is a constant volume deformation. As the model attempts to describe the rapid evolution of successive magmatic batches, isothermal and isobaric conditions must ensure density and viscosity stability within the granite magma. These formulations do not take account of any dependence of the frictional sliding on temperature or strain rate.

Evaluation of equations (2) indicates that critical T'_{11} for brittle failure are insensitive to variations in viscosity ($\leqslant 10^{12}$ Pa s) or strain rate ($\leqslant 10^0$ s^{-1}). Therefore, viscosity and strain rate may be neglected to simulate the most reasonable geological conditions prevalent in plutonic contexts. Under pure, simple or general shear (De Paor 1983), equations (2) simplify to only one equation:

$$T'_{11} = \pm 2\sqrt{T_0 T_{33}} \qquad (6)$$

under tensile conditions, and

$$T'_{11} = \frac{C_0 + (p-1)T_{33}}{(p+1)}, \qquad (7)$$

under compressive conditions, where $T_{11} = -T'_{22}$ is the only non-zero component of the deviatoric stress tensor, T_{33} is one of the principal total stresses as well as the lithostatic pressure, and

$$p = [(\mu^2 + 1)^{1/2} + \mu]^2. \qquad (8)$$

Equations (6) and (7) are formally identical to those predicting brittle failure in solid rocks. Figure 8 summarizes the results of this simplified model. The tensile strength measured by Webb & Dingwell (1990) in rhyolite melts subjected to fibre elongation experiments is of about 316 MPa

(see also Dingwell 1997). Large water contents are known to drastically reduce the tensile strength of vesicular melts down to values of about 2 MPa (Dingwell 1998). In Fig. 8a two set of curves could be traced following the equations (6) and (7), but only the curve with the minimum T'_{11} for each T_0 has been considered. These curves allow a quick estimation of the critical deviatoric stresses for lithostatic pressures from 100 to 1000 MPa. Figure 8b shows the influence of the friction coefficient on the critical deviatoric stress at 400 MPa of lithostatic pressure. It can be appreciated that for tensile strengths larger than c. 120 MPa an increase in the friction angle (ϕ) reduces the critical stress. The inverse is true for tensile strengths lower than c. 120 MPa. Accordingly, the model discussed here leads to a rough estimate of the differential stresses necessary to break-up a viscous magma. Lower differential stresses are necessary to break a magma at low lithostatic pressures. Variations in the critical parameters of the model, most of them strictly empirical, such as the friction coefficient or the tensile strength, could produce important differences in the deduced critical stresses. For instance, reducing the tensile strength would originate a smooth decrease in the critical deviatoric stresses for brittle failure (Fig. 8a). For moderate to large tensile strengths, large friction coefficients impose a substantially lower critical stress for magma failure (Fig. 8b). On the other side, a small element of viscoelastic behaviour must be present immediately before the brittle failure, a feature that has not been considered in this simplified model. Therefore, the model must be considered as a first-order phenomenological approach to the problem of magma brittle failure. Remember that this viscous–brittle semi-empirical model is presented just to evaluate the geological possibility of brittle behaviour of granitic magmas in the continental crust, given the constraints imposed by the different tectonic regimes. By no means must this model be considered as an exhaustive physical description of the micromechanics of magma brittle failure.

According to Fig. 8 the differential stress (twice the critical deviatoric stress T'_{11}) necessary to break granitic magma at this confining pressure vary from 1400 MPa if $T_0 \approx 316$ MPa (Webb & Dingwell 1990) to 110 MPa if $T_0 \approx 2$ MPa (Dingwell 1998). But, is the middle continental crust able to resist these differential stresses? According to Engelder (1993) the theoretical limit to the magnitude of differential stresses in the continental crust is on the order of 1000–1500 MPa. Therefore, at a first approximation, these large differential stresses

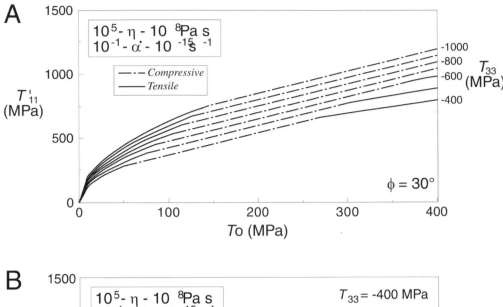

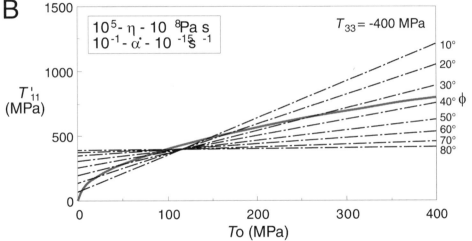

Fig. 8. Plots of the critical deviatoric stress (T'_{11}, absolute value) for brittle failure of viscous magmas against tensile strength (T_o). Tensile stress are reckoned positive. (**a**) Critical lines for different lithostatic pressures (T_{33}) according to the McClintock & Walsh (1962) criterion. It only represents the lower value of either the tensile or the compressive conditions of this criterion. (**b**) Investigates the influence of the ϕ (friction angle) value in the critical deviatoric stress, at a lithostatic pressure of 400 MPa. For a comparison, the grey line shows the critical values for the tensile conditions.

are geologically possible and the brittle behaviour is a possible type of material response for granitic magmas in the Earth's crust. However, as stated above, the bulk tectonic regime assumed to operate during the emplacement of the Central System batholith was extensional with local strike-slip shear zones interpreted as transfer structures (e.g. Doblas 1991). It is a well-known fact that frictional sliding of the crust imposes different limiting stress differences depending on the type of faulting: thrust, strike-slip and normal faults (e.g. Sibson 1974).

Equations expressing the limiting differential stress for these faulting regimes as a function of depth and friction coefficient (Ranalli & Murphy 1987; Ranalli 1995) were resolved for a typical continental crust with an average granitic composition and represented in Fig. 9. Lines of critical differential stresses for the mixed viscous–brittle behaviour are also depicted. A large incertitude is originated by the scarce number of experimental values of magma tensile strength presently available. Nevertheless, Fig. 9 shows that brittle behaviour of granitic

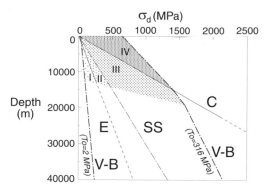

Fig. 9. Differential stress (σ_d) against depth for typical tectonic settings: C, convergence tectonics; SS, strike-slip tectonics, E, extensional tectonics. The dashed segments correspond to the crustal zones where ductile behaviour predominates according to the particular conditions established by Ranalli & Murphy (1987). This depth is not computed for the strike-slip regime. The lines of critical differential stresses deduced from the macroscopic viscous–brittle criterion are also shown (V—B). To the right the line computed for a tensile strength of 316 MPa (Webb & Dingwell 1990), to the left the line for a tensile strength of 2 MPa (Dingwell 1997). The actual critical line of a given granitic magma, with a tensile strenght intermediate between those extreme values, should be traced in the area comprised between these two lines. Sub-area I corresponds to granitic magmas with low tensile strengths, able to be broken in an extensional tectonic regime. Sub-area IV marks the position of granitic magmas unable to fail brittly in the upper 15 km of continental crust, even if subjected to a convergent tectonic regime. The other sub-areas can be interpreted in a similar way.

magmas is possible in the upper crust under extensional or strike-slip regimes only for the lowest values of tensile strength. On the other hand, stress levels in contractional tectonics at depths of about 14 km are in the field of brittle behaviour of magmas with large tensile strengths. Interestingly, the stress level at the brittle–ductile transition in the crust provides the best conditions for the brittle failure of granitic magmas in any type of tectonic regime (Fig. 9). We suggest that this could explain the location of some large batholiths, including the Central System batholith in Spain.

Final comments and conclusions

Several granite batches with an elongate shape appear in the Puente del Congosto area (Iberian Massif). The magmatic structures cropping out in this region suggest the coeval activity of

viscous flow and brittle failure of granitic magma. The model by Rubin (1993) of dykes intruding viscoelastic rocks may satisfactorily explain the main structural features of the studied area. The resulting mechanism of magma-into-magma emplacement is intermediate between dyke and diapir models. It is hoped that similar observations could be made in other natural examples. Specifically, field criteria that can be used to argue the possible applicability of this process are the observation within granitic plutons of both, systematic cross-cut relations in magmatic structures and pervasive deformation, or even transposition, of brittle features and older magmatic fabrics near an internal contact. A model including Newtonian behaviour and a brittle fracture criterion (Modified Griffith Theory) has been put forward to explain the evolution of granitic magma in tectonic contexts. For most of the geological conditions prevailing in such environments, the rheological response is independent of strain rate and magma viscosity. Independence of strain rate is a consequence of the over-simplified nature of the model, which does not account for the elastic component in the host magma behaviour. According to the results of the macroscopic fracture criterion, the differential stresses necessary to fracture a viscous magma are in the range of possible differential stresses in the middle crust. Continental convergent zones provide the best place to break a granitic magma solely under the influence of tectonic stresses, but this process is also possible in extensional or strike-slip zones, specifically at the brittle–ductile transition in the crust.

The episodic formation of a batholith according to the magma-into-magma emplacement mechanism is shown in the idealized sketch of Fig. 10. This figure explains the growth of granitic batholiths by the episodic supply of magma from a feeder. Newer magma batches find their room by first breaking and then pushing-apart the emplaced magma bodies, that still retain a Newtonian rheology. The field example described in this work is informative about the importance of such a complex interplay between viscous flow and brittle behaviour in granitic magmas. The macroscopic model of the previous section shows that this process is at least possible in the crust. However, it should be clear from the discussion that we are far from a unified physical model of the ability of silicic melt to fracture under the conditions of the upper continental crust. The model by Rubin (1993) shoud be considered as a valuable start point to this goal: a pressurized dyke intruding a viscoelastic magmatic medium.

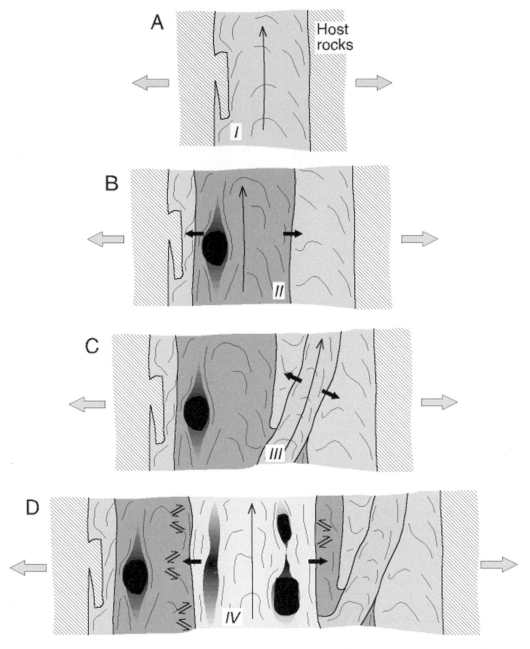

Fig. 10. Idealized sketch showing the sequential evolution of the magma-into-magma emplacement mechanism. (a) The first magma pulse (*I*) breaks the host rocks apart and intruded into the middle continental crust. (b), (c) and (d) New magma batches (*II*, *III* and *IV*) with a contrasted composition, perhaps coming from different depths in the source layer, exploit magma fractures and ascend through the previously emplaced pulses, deforming them near the intrusive contacts where intramagmatic shear zones develop.

Future advances are only possible as the result of an interdisciplinary work. New theoretical models must incorporate a better appraisal of the fracture resistance of the magma, including other complexities such as subcritical crack growth (Atkinson & Meredith 1987) or inelastic response in the fracture process zone (Ingraffea 1987), with a careful control of the range of strain rate values over which brittle failure is possible. At the same time, accurate experimental investigations are needed to place limits to the values of the most important rheological parameters of granitic magmas, like viscosity and fracture toughness, and these parameters must be measured under a complete range of geologically relevant conditions. The importance of non-linear elastic and viscous rheologies must be carefully analysed. Lastly, geophysical and field-based research must be used as a test for theoretical models and experimental data, as well as a guideline for future developments.

We are indebted to L. G. Corretgé for his comments and suggestions about the rheology of granites. Careful reviews by D. B. Dingwell and an anonymous referee considerably improved a first version of the manuscript, and are gratefully acknowledged. Financial support from the D.G.I.C.Y.T. (Project PB94-1085), the Junta de Andalucía (RNM-0120), and the University of Huelva is gratefully acknowledged.

References

ATKINSON, B. K. & MEREDITH, P. G. 1987. The theory of subcritical crack growth with applications to minerals and rocks. In: ATKINSON, B. K. (ed) Fracture mechanics of rock. Academic Press, London, 111–166.

BAGDASSAROV, N. S., DINGWELL, D. B. & WEBB, S. L. 1993. Effect of boron phosphorus and fluorine on shear stress relaxation in haplogranite melts. European Journal of Mineralogy, 5, 409–425.

BIOT, M. A. 1961. Theory of folding of stratified viscoelastic media and its implications in tectonics and orogenesis. Geological Society of America Bulletin, 72, 1595–1620.

BOUCHEZ, J. L., DELAS, C., GLEIZES, G., NÉDÉLEC, A. & CUNEY, M. 1992. Submagmatic microfractures in granites. Geology, 20, 35–38.

CASQUET, C., FÚSTER, J. M., GONZÁLEZ CASADO, J. M., PEINADO, M. & VILLASECA, C. 1988. Extensional tectonics and granite emplacement in the Spanish Central System. A Discussion. Fifth Workshop on the European Geotraverse Project, Proceedings, 65–76.

CASTRO, A., MORENO-VENTAS, I. & DE LA ROSA, J. D. 1990. Microgranular enclaves as indicators of hybridization processes in granitoid rocks, Hercynian belt, Spain. Geological Journal, 25, 391–404.

——, —— 1991. H-type (hybrid) granitoids: a proposed revision of the granite-type classification and nomenclature. Earth Science Reviews, 31, 237–253.

CLEMENS, J. D. & MAWER, C. K. 1992. Granitic magma transport by fracture propagation. Tectonophysics, 204, 339–360.

CORRETGÉ, L. G. & MARTÍNEZ, F. J. 1978. Problemas sobre estructura y emplazamiento de los granitoides: aplicación a los batolitos hercínicos del Centro-Oeste de la Meseta Ibérica. Cuadernos del Seminario de Estudios Cerámicos de Sargadelos, 27, 113–134.

DE PAOR, D. G. 1983. Orthographic analysis of geological structures. I: Deformation theory. Journal of Structural Geology, 5, 255–277.

DÍEZ BALDA, M. A., VEGAS, R. & GONZÁLEZ-LODEIRO, F. 1990. Central-Iberian zone: autochtonous sequences, structures. In: DALLMEYER, R. D. & MARTÍNEZ GARCÍA, E. (eds) Pre-mesozoic geology of Iberia. Springer-Verlag, Berlin, 172–188.

DINGWELL, D. B. 1997. The brittle-ductile transition in high-level granitic magmas: material constraints. Journal of Petrology, 38, 1635–1644.

—— 1998. Recent experimental progress in the physical description of silicic magma relevant to explosive volcanism. In: GILBERT, J. S. & SPARKS, R. S. J. (eds) The physics of explosive volcanic eruptions. Geological Society, London, Special Publications, 145, 9–26.

—— & WEBB, S. L. 1990. Relaxation in silicate melts. European Journal of Mineralogy, 2, 427–449.

——, BAGDASSAROV, G. Y., BUSSOD, G. Y. & WEBB, S. L. 1993. Magma rheology. In: LUTH, R. W. (ed.) Handbook on experiments at high pressure and applications to the Earth's mantle. Mineralogical Association of Canada, Short Courses, 21, 131–196.

——, HESS, K. U. & ROMANO, C. 1998. Extremely fluid behavior of hydrous peralkaline rhyolites. Earth and Planetary Science Letters, 158, 31–38.

DOBLAS, M. 1991. Tardi-Hercynian extensional and transcurrent tectonics in Central Iberia. Tectonophysics, 191, 325–334.

ENGELDER, T. 1993. Stress regimes in the lithosphere. Princeton University Press, New Jersey.

FERNANDEZ, A. & BARBARIN, B. 1991. Relative rheology of coeval mafic and felsic magmas: Nature of resulting interaction processes and shape and mineral fabrics of mafic microgranular enclaves. In: DIDIER, J. BARBARIN, B. (eds) Enclaves and Granite Petrology. Elsevier, Amsterdam, 263–275.

—— & GASQUET, D. R. 1994. Relative rheological evolution of chemically contrasted coeval magmas: example of the Tichka plutonic complex (Morocco). Contributions to Mineralogy and Petrology, 116, 316–326.

FERNÁNDEZ, C., CASTRO, A., DE LA ROSA, J. D. & MORENO-VENTAS, I. 1997. Rheological aspects of magma transport inferred from rock structures. In: BOUCHEZ, J. L., HUTTON, D. H. W. & STEPHENS, W. E. (eds) Granite: from segregation of melt to emplacement fabrics. Kluwer, Amsterdam, 75–91.

FOWLER, T. K., JR. & PATERSON, S. R. 1997. Timing and nature of magmatic fabrics from structural relations around stoped blocks. Journal of Structural Geology, 19, 209–224.

GERBAULT, M., POLIAKOV, A. N. B. & DAIGNIERES, M. 1998. Prediction of faulting from the theories of elasticity and plasticity: what are the limits? Journal of Structural Geology, 20, 301–320.

GERTH, U. & SCHNAPP, J. D. 1996. Investigation of the mechanical properties of natural glasses using indentation methods. *Chemie der Erde*, **56**, 398–403.

HESS, K. U. & DINGWELL, D. B. 1996. Viscosities of hydrous leucogranitic melts: a non-Arrhenian model. *American Mineralogist*, **81**, 1297–1300.

HOLTZ, F., ROUX, J., OHLHORST, S., BEHRENS, H. & SCHULZE, F. 1999. The effects of silica and water on the viscosity of hydrous quartzofeldspathic melts. *American Mineralogist*, **84**, 27–36.

INGRAFFEA, A. R. 1987. Theory of crack initiation and propagation in rock. *In*: ATKINSON, B. K. (ed.) *Fracture mechanics of rock*. Academic Press, London, 71–110.

JULIVERT, M., FONTBOTÉ, J. M., RIBEIRO, A. & CONDE, A. 1972. *Mapa Tectónico de la Península Ibérica y Baleares, 1:1 000 000*. Instituto Geológico y Minero de España, Madrid.

LISTER, J. R. & KERR, R. C. 1991. Fluid-mechanical models of crack propagation and their application to magma transport in dykes. *Journal of Geophysical Research*, **96**, 10 049–10 077.

MALVERN, L. E. 1969. *Introduction to the mechanics of a continuous medium*. Prentice–Hall, Englewood Cliffs, New Jersey.

MCBIRNEY, A. R. & MURASE, T. 1984. Rheological properties of magmas. *Annual Review of Earth and Planetary Sciences*, **12**, 337–357.

MCCLINTOCK, F. A. & WALSH, J. B. 1962. Friction on Griffith cracks under pressure. *Fourth United States National Congress of Applied Mechanics, Proceedings*, 1015–1021.

MORENO-VENTAS, I., ROGERS, G. & CASTRO, A. 1995. The role of hybridizaton in the genesis of Hercynian granitoids in the Gredos massif. Inferences from Sm/Nd isotopes. *Contributions to Mineralogy and Petrology*, **120**, 137–149.

OEN, I. S. 1970. Granite intrusion, folding and metamorphism in Central Portugal. *Boletín Geológico y Minero*, **81**, 271–298.

PATERSON, S. R., VERNON, R. H. & TOBISCH, O. T. 1989. A review of criteria for the identification of magmatic and tectonic foliations in granitoids. *Journal of Structural Geology*, **11**, 349–363.

PETFORD, N. 1996. Dykes or diapirs. *Transactions of the Royal Society of Edinburgh: Earth Sciences*, **87**, 105–114.

——, KERR, R. C. & LISTER, J. R. 1993. Dike transport of granitoid magmas. *Geology*, **21**, 845–848.

PHILPOTTS, A. R. 1990. *Principles of igneous and metamorphic petrology*. Prentice–Hall, Englewood Cliffs, New Jersey.

POLLARD, D. D. & SEGALL, P. 1987. Theoretical displacements and stresses near fractures in rock: with applications to faults, joints, veins, dikes, and solution surfaces. *In*: ATKINSON, B. K. (ed.) *Fracture mechanics of rock*. Academic Press, London, 277–349.

PRICE, N. J. & COSGROVE, J. W. 1990. *Analysis of geological structures*. Cambridge University Press, Cambridge.

RAMBERG, H. 1960. Relationships between lengths of arc and thickness of ptygmatically folded veins. *American Journal of Science*, **258**, 36–46.

RANALLI, G. 1995. *Rheology of the earth*. Chapman & Hall, London.

—— & MURPHY, D. C. 1987. Rheological stratification of the lithosphere. *Tectonophysics*, **132**, 281–295.

RICHET, P. & BOTTINGA, Y. 1995. Rheology and configurational entropy of silicate melts. *In*: STEBBINS, J. F., MCMILLAN, P. F. & DINGWELL, D. B. (eds) *Structure, dynamics and properties of silicate melts*. Mineralogical Society of America, Reviews in Mineralogy, **32**, 67–93.

RUBIN, A. M. 1993. Dikes vs diapirs in viscoelastic rocks. *Earth and Planetary Science Letters*, **117**, 653–670.

SCHOLZ, C. H. 1990. *The mechanics of earthquakes and faulting*. Cambridge University Press, Cambridge.

SHAW, H. R. 1980. The fracture mechanisms of magma transport from the mantle to the surface. *In*: HARGRAVES, R. B. (ed.) *Physics of magmatic processes*. Princeton University Press, New Jersey, 201–264.

SIBSON, R. H. 1974. Frictional constraints on thrust, wrench, and normal faults. *Nature*, **249**, 542–544.

VIGNERESSE, J. L., BARBEY, P. & CUNEY, M. 1996. Rheological transitions during partial melting and crystallization with application to felsic magma segregation and transfer. *Journal of Petrology*, **37**, 1579–1600.

WALKER, G. P. L. 1969. The breaking of magma. *Geological Magazine*, **106**, 166–173.

WEBB, S. L. & DINGWELL, D. B. 1990. NonNewtonian rheology of igneous melts at high stresses and strain rates: experimental results for rhyolite, andesite, basalt and nephelinite. *Journal of Geophysical Research*, **95**, 15 695–15 701.

—— & —— 1995. Viscoelasticity. *In*: STEBBINS, J. F., MCMILLAN, P. F. & DINGWELL, D. B. (eds) *Structure, dynamics and properties of silicate melts*. Mineralogical Society of America, Reviews in Mineralogy, **32**, 95–119.

WEINBERG, R. F. 1996. Ascent mechanisms of felsic magmas: news and views. *Transactions of the Royal Society of Edinburgh: Earth Sciences*, **87**, 95–103.

Origin of megacrysts in granitoids by textural coarsening: a crystal size distribution (CSD) study of microcline in the Cathedral Peak Granodiorite, Sierra Nevada, California

MICHAEL D. HIGGINS

Sciences de la Terre, Université du Québec à Chicoutimi, Chicoutimi, G7H 2B1, Canada
(e-mail: mhiggins@uqac.uquebec.ca)

Abstract: Microcline megacrysts in the Cathedral Peak Granodiorite and other parts of the Tuolumne Intrusive Suite were formed by textural coarsening (Ostwald Ripening) of earlier formed crystals. The early-formed crystals nucleated and grew in an environment of increasing undercooling, probably during the ascent of the magma. Emplacement of the magma into warm host rocks promoted textural coarsening. Crystals smaller than a certain size (the critical size) dissolved in the interstitial melt whilst larger crystals grew. Microcline was most sensitive to this effect as the magma temperature was buffered close to its liquidus for a long period by the release of latent heat of crystallization. Positive feedback between textural coarsening and magma permeability channelled the flow of interstitial melt to produce a heterogeneous distribution of megacrysts. Megacryst growth was halted when cooling resumed at the end of the intrusive cycle. K-feldspar nucleation was then renewed and K-feldspar crystals grew to form part of the groundmass. It was the particular thermal history of this pluton that promoted textural coarsening—chemically similar plutons that lack megacrysts probably did not have the pause during cooling that was necessary for the development of this texture.

Euhedral K-feldspar megacrysts up to 20 cm long are a striking and relatively common component of many granitoid plutons, but their origin is still controversial. Vernon (1986) reviewed the subject and found the consensus to be that megacrysts are igneous in origin, that is, they grew from a granitic melt and are phenocrysts. However, he noted that there was a minority view that megacrysts are post-magmatic and grew from a circulating water-dominated fluid. An extreme version of this hypothesis proposes that such granites are entirely metasomatic (granitization).

A magmatic origin is supported by many features, such as crystal shape, crystal fabric, chemical composition, zonation and inclusions (Vernon 1986). The low population density of the megacrysts has been ascribed by Vernon (1986) and others to 'nucleation difficulties', which delay nucleation until the temperature is close to the solidus. It is necessary to discuss briefly experimental work relevant to this problem.

The experimental studies of Swanson (1977) were launched by the commonly held belief that the textures of granites are controlled by patterns of nucleation and growth of phases, an idea that will be shown later to be partly incorrect.

Swanson (1977) and Fenn (1977) measured the nucleation and growth rates of plagioclase, K-feldspar and quartz crystallizing from glassy charges during geologically rapid cooling. Fenn (1977) found that there was a significant time delay between the cooling of the charge below the liquidus and the appearance of the crystals of K-feldspar that he termed the 'incubation period'. This is the effect referred to by the term 'nucleation difficulties'. This effect may be important for volcanic rocks, but cannot be important for slowly-cooled plutonic rocks. Textural evidence from rocks also negates this idea. Plagioclase commonly crystallizes before K-feldspar (Naney 1983) and K-feldspar can overgrow plagioclase (anti-rapakivi texture), hence nucleation of K-feldspar cannot be a problem. Therefore, the low population density of K-feldspar megacrysts should not be ascribed to nucleation difficulties. It will be shown later that this problem can be resolved if the nucleation density is distinguished from the final population density of the megacrysts. That is, that there has been resorbtion of many K-feldspar crystals.

A metasomatic origin for the megacrysts has been suggested by some authors because of their

From: CASTRO, A., FERNÁNDEZ, C. & VIGNERESSE, J. L. (eds) *Understanding Granites: Integrating New and Classical Techniques.* Geological Society, London, Special Publications, **168**, 207–219. 1-86239-058-4/99/$15.00 © The Geological Society of London 1999.

occurrence, albeit rarely, in more mafic enclaves, country rocks and cutting across the borders of aplite veins (see review by Vernon 1986). Dickson (1996) proposed that metasomatism may be driven by deformation. Although many occurrences can be explained by elaborations of the igneous theory, others appear to require metasomatism. The argument then runs that if some crystals are metasomatic, then all may be (Collins 1988).

Here, I propose that the dominant process forming the megacrysts is textural coarsening of earlier formed small K-feldspar crystals during which most original crystals were sacrificed to feed the growth of a select few. This process can occur in crystal-poor or crystal-rich mushes, where it is mediated by residual silicate liquids (Higgins 1998), as well as in metamorphic rocks where it is mediated by water-rich fluids (Cashman & Ferry 1988), hence accounting for the presence of megacrysts in both granites and country rocks and reconciling the opposing theories.

Megacrysts are fundamentally a textural phenomenon, hence the approach here has been to quantify aspects of the texture that can reveal the origin of the megacrysts. The crystal-size distribution method (CSD) has been applied to some volcanic and plutonic rocks (e.g. Cashman 1990), but it has not yet been applied in many published granite studies.

The Cathedral Peak Granodiorite

The Cathedral Peak Granodiorite (CPG) is the most voluminous part of the concentrically zoned Tuolumne Intrusive Suite, in the Sierra Nevada of California (Figs 1, 2; Bateman & Chappell 1979). The intrusion of this suite commenced with the emplacement of quartz-dioritic magma, which is still preserved in the periphery of the suite. This magma started to crystallize on the walls, but before solidification was complete new, more felsic, magma was introduced into the centre of the pluton. It eroded parts of the solidified diorite and expanded the chamber to the northwest. This new unit again started to crystallize from the walls inwards, first with an equigranular texture, and later as a porphyritic rock, with K-feldspar megacrysts. Parts of this unit are preserved as the Half Dome Granodiorite. Again, complete solidification was interrupted by the emplacement of new magma, the Cathedral Peak Granodiorite. As before this process led to the erosion of parts of the earlier units and expansion of the chamber to the north. The final unit, the Johnson Granite Porphyry, must have been

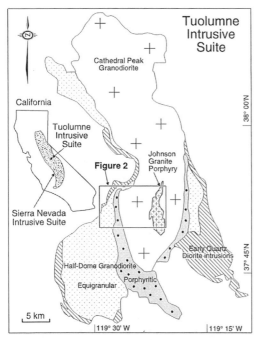

Fig. 1. Geological map of the Tuolumne Intrusive Suite, Sierra Nevada, California (after Bateman & Chappell 1979). The quartz-diorite intrusions that form the outer parts of the intrusion were emplaced first. The Half-Dome granodiorite was subsequently intruded into the partly solidified interior of the pluton. The Cathedral Peak granodiorite followed in the same way. The final intrusion was the Johnson Creek Porphyry. There is an overall progression from mafic to felsic compositions.

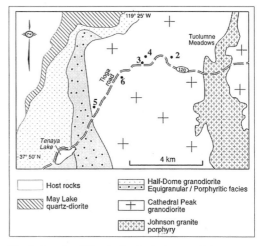

Fig. 2. Detail of Fig. 1. Geological map of the Tuolumne Intrusive Suite near Tuolumne Meadows along route 120 (after Bateman & Chappell 1979). Numbers refer to sample sites.

emplaced when the CPG was almost completely solid as the contacts are sharp. Bateman & Chappell (1979) considered that this unit was sub-volcanic on the basis of the porphyritic texture, the fine-grained mesostasis and the broken nature of many crystals. The overall progression of the suite is from more mafic to more felsic magmas. All units of this suite are fresh and almost completely undeformed in the solid-state.

An important component of the Cathedral Peak Granodiorite, adjacent parts of the Half-Dome Granodiorite and the Johnson Granite Porphyry is K-feldspar megacrysts. They are most abundant in the Cathedral Peak Granodiorite where they vary from an overall mode of 10% near the outer edge to 2% near the inner border. Bateman & Chappell (1979) considered that the modal variation in the megacrysts was balanced by changes in the K-feldspar mode of the matrix, and hence that the overall quantity of K-feldspar in the rock remains constant at about 20%.

The megacrysts are distributed heterogeneously with high concentrations in 'nests'

and linear zones up to a meter wide that meander for tens of metres across the outcrop (Fig. 3a, b, c). The distribution of these megacryst-rich areas is similar in both vertical and horizontal sections. A brief reflection will confirm that all three-dimensional sheets have linear intersections with surfaces, whereas string-like forms have point intersections. Therefore, the observed distributions of the megacrysts on the outcrop surfaces suggests that the volumes rich in megacrysts have sheet and string-like forms in three dimensions.

The megacrysts are composed of microcline. The mean of 200 analyses was Or_{88} but individual analyses varied from Or_{97} to Or_{76} (Kerrick 1969). The crystals contain subhedral plagioclase as included grains and exsolved patches (mean composition An_{33}; Kerrick 1969). The wide range in composition of the megacrysts and the included plagioclase preclude the use of a two-feldspar geothermometer. Subhedral hornblende, minor biotite and rare quartz are also present. Included minerals are commonly concentrated into euhedral-shaped concentric zones (Fig. 3d). Many of the megacrysts have Carlsbad twinning.

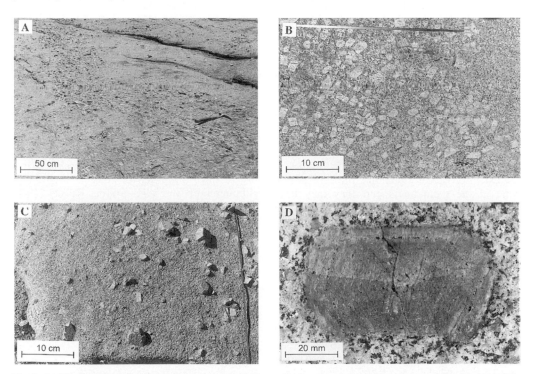

Fig. 3. (**a**) Weathered outcrop of the Cathedral Peak Granodiorite, showing the heterogeneous distribution of the microcline megacrysts. Megacryst-rich areas tend to be linear or in patches ('nests'), with similar patterns in both horizontal and vertical sections. (**b**) 'Nest' of megacrysts on glacially polished surface. (**c**) Diffuse megacrysts on an eroded surface. (**d**) Single megacryst exposed on a glacially polished surface. Inclusions of amphibole in the crystal are concentrated in euhedral zones. Plagioclase is not visible in this photograph. The matrix is slightly enriched in mafic minerals near the crystal.

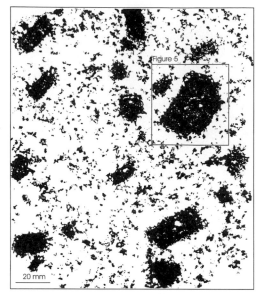

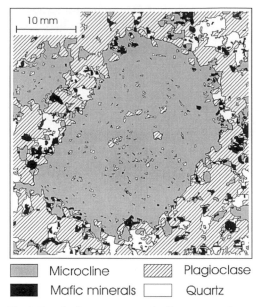

Microcline ▨ Plagioclase
Mafic minerals ☐ Quartz

Fig. 4. Distribution of microcline in a slab from near sample site 6. A sawn surface was stained with sodium cobaltinitrate (Hutchison 1974), then scanned at 10 lines per millimetre and the orange colour of the microcline digitally extracted. Other minerals have been omitted for clarity.

Fig. 5. Detail of a megacryst and adjacent groundmass from Fig. 4. All mineral species were separated on the basis of their colour values in the stained slab. Intergranular areas could not always be successfully identified and have been included with quartz.

The shape of the megacrysts varies with their size. Large megacrysts are euhedral (Fig. 3b, c, d) with an aspect ratio of 1:1.6:2.6, as measured from the dimensions of weathered-out crystals. The surfaces of the megacrysts are rough and slabs stained with sodium cobaltinitrate (Hutchison 1974) reveal that there are sinuous extensions of microcline from the megacrysts into the groundmass (Figs 4, 5). The shapes of the extensions greatly resemble those of microcline in the groundmass and they may have grown at the same time. Smaller crystals are more rounded, but with the same extensions as the larger crystals. The smallest megacrysts are just clots of microcline in the groundmass. The groundmass is richer in mafic minerals and poorer in K-feldspar adjacent to the edges of some of the megacrysts. The megacrysts in the CPG resemble K-feldspar megacrysts in other granitoids, hence a general explanation for their origin must be found (Vernon 1986).

The groundmass is dominated by rounded crystals of quartz and plagioclase up to 10 mm long (cores typically An_{35}; Kerrick 1969). Microcline (typically Or_{88}) occurs as small anhedral crystals interstitial to the quartz and plagioclase (Figs 4, 5). Biotite and hornblende are also interstitial and are much finer than the other phases. Here, biotite is more abundant than hornblende.

Although both the megacrysts and the groundmass microcline have similar K/Na ratios their Ba contents are very different. Kerrick (1969) found celsian contents of 1–2% in the megacrysts, but only 0.25% or less in the groundmass microcline. He also found a Ba-poor rim on one megacryst, but it is not clear if it is one of the extensions into the groundmass or a euhedral zone. Similarly, the obliquity of the microcline in the groundmass was two to three times that of adjacent megacrysts, although both were very variable. Kerrick (1969) concluded from the obliquity that the megacrysts crystallized at higher temperatures than the groundmass microcline.

Quantitative textural measurements

Observations of textures are a fundamental part of igneous petrology, but the quantification of these textures necessary to model petrogenetic processes is still rarely applied (see review in Cashman 1990). The goal of this study is to understand the genesis of the megacrysts, but the groundmass microcline cannot be neglected. Differences in scale between the two populations of crystals necessitates two different quantification techniques: the megacrysts were studied in the field and the groundmass crystals in stained slabs up to 15 cm long.

Megacrysts were studied quantitatively in the Cathedral Peak Granodiorite component of the suite because they are such an important component, but it should not be forgotten that megacrysts also occur in the other components. Five outcrops located across the width of the Cathedral Peak Granodiorite were measured (Fig. 2). Some outcrops still had their glacial polish, but in others weathering has produced a knobby surface defined by the megacrysts (Fig. 3). Several of the analysed areas had 'nests' of megacrysts, whereas others were more uniform with a lower population density. Megacrysts were recognized by their euhedral shape. All visible megacrysts were measured using a ruler, but the smaller megacrysts were not so euhedral as the larger ones and hence more difficult to distinguish from the groundmass. This measurement technique gives the dimensions of the euhedral part of the megacrysts and ignores the winding crystal extensions (Fig. 5). All these effects mean that this technique is more reliable for the larger crystals. The smallest crystal that could be measured was estimated to be 10 mm long. Areas measured ranged from 0.25 to 0.8 m² (Table 1).

Three slabs were sampled from loose material in road-cuts near the studied outcrops. The slabs were sawn, stained with sodium cobaltinitrate (Hutchison 1974) and optically scanned to form a digital image. Microcline was distinguished from other minerals in the images using colour values and the dimensions of the crystals were determined using PCImage (a version of NIH Image, a program developed at the US National Institutes of Health). It is not possible to separate touching crystals with this technique, hence there is a bias towards larger sizes than are actually present. This effect is compounded for large crystals that necessarily have faces closer to each other than smaller crystals. Therefore, this technique is more reliable for smaller crystals. Minimum measurable crystal size was 1 mm. Larger, more euhedral crystals are clearly megacrysts, hence crystals longer than 10 mm were excluded from the analysis so that the data could be compared with that gathered from the outcrops (Figs 4, 5).

The data obtained from the study of the outcrops and the slabs pertain only to the length of the intersections in two dimensions. These data must be converted to true three dimensional CSD before they can be interpreted. This conversion is not simple and is treated in the branch of mathematics called stereology (Royet 1991). Two aspects of the problem are important here: the intersection probability effect and the cut section effect. The first is simply that smaller crystals are less likely to be intersected by a plane than larger crystals. This can be compensated if the number of crystals per unit area in a size interval is divided by the mean size of the interval. The second effect is that a plane is unlikely to pass exactly through the centre of a crystal in the direction of the largest section. The nature of the fabric and its quality and orientation, as well as the section orientation with respect to the fabric and crystal shape, must be known to make these conversions. However, approximate values are adequate as the result is not very sensitive to these effects.

The interactive program CSDCorrections was used to do the stereological conversions (Higgins in prep.). The observed fabric of the megacrysts and matrix is not sufficient to make a significant difference to the CSDs, hence the rock was

Table 1. *Numbers of microcline crystals and areas measured on outcrops and slabs*

Upper limit of size category (mm)	Outcrops: megacrysts					Slabs: groundmass		
	O-2 Diffuse with nests	O-3 Nests	O-4 Diffuse	O-5 Diffuse	O-6 Nests	S-3	S-4	S-6
1–1.8						85	76	156
3.2						43	88	131
5.6						31	41	71
10.0						8	4	32
17.8	24	16	20	33	29			5
31.6	54	72	53	68	91			10
56.2	32	28	19	25	39			
100	10	3	2	7	9			
Largest grain (mm)	97	70	62	80	75			
Area measured (mm²)	500 000	250 000	500 000	800 000	500 000	4800	4800	27 000
Crystal content (%)	15	24	8	9	19	28	12	14

The crystal content of the slabs includes both the groundmass and the megacrysts.

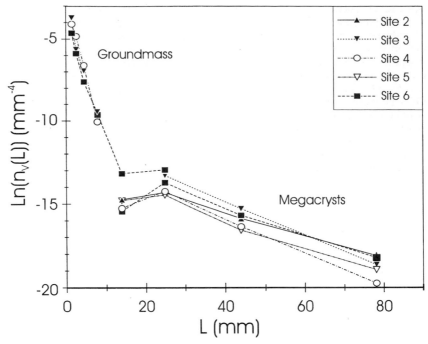

Fig. 6. Crystal size distributions of microcline crystals. The vertical axis is the natural logarithm of the population density for crystal size L, $n_V(L)$. It is the number of crystals per unit volume divided by the size interval. Hence, it has the dimensions of mm^{-4}. Crystal size, L, is the mean projected height of crystals, which is close to the intermediate dimension of parallelepipeds with similar aspect ratios to the crystals. Data were reduced using the program CSDCorrections (Higgins in prep.). Megacrysts from five outcrops were measured in the field with a ruler. Dimensions of microcline crystals in the slab seen in Fig. 4 and other slabs were determined by automatic image analysis. Open symbols are for areas with diffuse megacrysts, whereas solid symbols are for sites with 'nests' of megacrysts.

assumed to be massive. The shape of the megacrysts is easily established from separated crystals. However, the microcline in the ground-mass is commonly interstitial to quartz and plagioclase and hence is more difficult to deter-mine. For the sake of simplicity it is assumed to be the same as that of the megacrysts.

The data were plotted on a diagram of Ln(population density) versus crystal size, here-after called a crystal size distribution (CSD) diagram (Fig. 6; Marsh 1988). The population density is the number of crystals per unit volume within a certain size range divided by the width of the size range. This division is necessary so that the population density does not change with the width of the size interval. The units of population density are mm^{-4}. Logarithmic size intervals were used, with each bin $10^{0.25}$ larger than the previous bin. There were no empty bins within the CSDs, that is intervals with no crystals. There are several sources of error in this diagram, but the most important are probably that due to counting statistics for larger

crystals and for the tailing corrections in the case of the smaller crystals. However, the logarithmic vertical scale of this diagram, and the wide size bins, ensures that the errors are not significant.

The CSD data from the Cathedral Peak Granodiorite are presented in Fig. 6. Somewhat surprisingly the CSDs of both the 'nests' of megacrysts and the diffuse areas have similar shapes, but the nests have higher overall popu-lation densities for all crystal sizes that the diffuse areas. All CSDs have a peak at 25 mm. The smallest megacryst that could be reliably measured was 10 mm, hence the turn-down of the population density to the left of the peak is not an artifact. If the CSDs were straight down to 10 mm then the population density at 10 mm would have been 10 to 50 times higher, which is readily distinguishable from the actual data. The limb to the right of the peak is generally straight.

The CSDs of the groundmass microcline are quite different from those of the megacrysts (Fig. 6). All three CSDs are quite straight, right down to the smallest crystal that could be measured (1 mm) and much steeper than those of the megacrysts. One slab was sufficiently large that the abundance of smaller megacrysts could be determined, although the small numbers of large crystals did not give a high precision. The consistency of the two methods is confirmed by the overlap of the CSDs.

The CSD data of different samples are difficult to distinguish on a conventional CSD diagram (Fig. 6), hence the data have been transformed. The right side of the CSD was regressed to give the slope and intercept. The slope was then transformed into the characteristic length by the equation (Marsh 1998):

$$\text{characteristic length} = -1/\text{slope}.$$

The characteristic length has the units of length (mm in this case), but does not have a physical meaning for most systems.

Figure 7a shows a plot of characteristic length versus intercept for microcline crystals. The fields of the megacrysts and groundmass microcline are quite different. Both groups of data have negative correlations between characteristic length and intercept, but only that of the megacrysts can be interpreted as significant ($r^2 = 0.61$ for 3 degrees of freedom) as there are not enough groundmass samples. The megacryst samples from the nests lie to the right of the diffuse area samples,

suggesting that the overall megacryst mode in the sample may be important.

The CSDs can easily be corrected for modal variations. The population density, $n_V(L)$, is normally calculated from

$$n_V(L) = \frac{n}{v(L_2 - L_1)}$$

where n = number of crystals in size interval, v = volume and $L_2 - L_1$ = size interval. However, in this case we are interested in the phase population density, $n'_V(L)$. This can be calculated from

$$n'_V(L) = \frac{n}{v'(L_2 - L_1)}$$

where v' = the volume of the phase in question. The proportion of the phase, χ, is defined as

$$\chi = \frac{v'}{v}$$

hence

$$n'_V(L) = \frac{n_V(L)}{\chi}.$$

This equation applies equally to all points, so the slope will be unchanged, but the intercept will be increased by a factor of $1/\chi$.

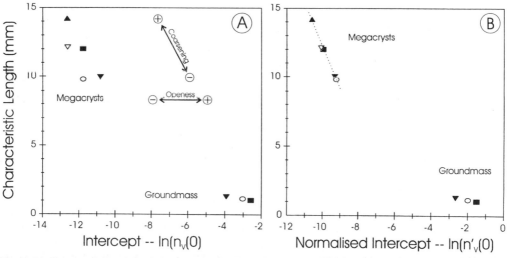

Fig. 7. (a) Characteristics of the right part of the CSDs of Fig. 6, expressed as Characteristic Length (-1/Slope) versus Intercept (see text). The symbols are as Fig. 6. The megacrysts and groundmass microcline crystals fall in totally separate parts of the diagram. The distribution of the megacryst data define a negative correction. The vector for *Coarsening* indicates the displacements of the samples that are more ($+$) or less ($-$) coarsened. *Openness* is the amount of material added during coarsening. More open areas ($+$) have received more material than less open ($-$) areas. **(b)** The quality of the correlation can be improved dramatically if the intercept values are normalized to 100% microcline (see text).

The correlation of the Characteristic Length and the Corrected Intercept is very strong for the megacrysts ($r^2 = 0.98$ for 3 degrees of freedom; Fig. 7b) is very strong and will be discussed below.

Discussion

The study of crystal size distributions in rocks has been somewhat neglected in igneous petrology but similar studies in materials science are well developed. Marsh (1988) applied the work of Randolf & Larson (1971) to igneous systems and suggested that continuous flux, steady-state crystallization in an open system of linearly increasing undercooling produced straight lines on a graph of Ln (population density) versus length (Fig. 8a). Marsh (1998) elaborated these theories for batch (closed system) crystallization. He showed that at high degrees of crystallinity the population density of small crystals is less than predicted by the simple model because of a decrease in the amount of liquid remaining, in which nucleation occurs (Fig. 8a, e). Fractional crystallization can also alter the shape of CSDs—accumulation of crystals under the influence of gravity will skew the right side of the CSD upwards (Fig. 8b, e).

The CSDs of the groundmass microcline are straight, hence could have formed by batch crystallization under conditions of linearly increasing undercooling (Fig. 6). In this case, if there was a deficiency of small crystals, then it may be concealed below the detection limit of 1 mm. However, none of the models discussed so far are capable of producing the hump-shaped CSDs of the megacrysts with a strong correlation between characteristic length and intercept (Fig. 6).

Hump-shaped CSDs are common in metamorphic rocks (Cashman & Ferry 1988), but can be found also in some plutonic and volcanic rocks (Boudreau 1987; Cashman & Marsh 1988; Resmini 1993; Higgins 1998). Such CSDs can be formed by textural coarsening (also known as Ostwald ripening, annealing, textural maturation) of crystal populations (Voorhees 1992). Such coarsening occurs because small grains have a higher surface energy per unit volume than larger grains. Therefore, to minimize energy in the system crystals smaller than a certain size, termed the 'critical size', will dissolve and 'feed' the growth of larger crystals. This can only occur when a crystal is held at a temperature close to its liquidus for a long period of time. Under these conditions the nucleation rate is zero, but growth rate is high for crystals larger than the critical size. It should be noted that this does not necessarily mean that the temperature was held constant, just that the undercooling remained small. Material is transferred from one crystal to another by diffusion, hence observable coarsening can only occur if a fluid phase is present, such as residual melt. Coarsening can also occur at lower temperatures, under metamorphic conditions, where the nucleation rate is also zero and the fluid is aqueous.

Textural coarsening can occur in closed or open systems. If the system is closed then the total quantity of a phase remains constant while its mean grain-size increases. If the system is open, for example by circulation of melt through a crystal mush, then the total quantity of a phase can increase independently of the mean grain size.

Textural coarsening is an industrially important process, but the equations governing this process are still poorly defined, especially for geologically relevant conditions. Most formulations of textural coarsening are based on the Lifshitz–Slyozov–Wagner theory (Lifshitz & Slyozov 1961). This theory assumes that all crystals communicate with a uniform medium, that is diffusion is not a limiting factor. The results of this process on a material with an initially straight CSD were examined by Higgins (1998), who showed that small grains are almost completely removed, but the slope of the CSD for large grains is almost unaffected as growth is highest for crystals just larger than the critical size (Fig. 8c). The more recent communicating neighbours (CN) theory (Dehoff 1991) is based on chemical exchange between neighbouring crystals. In this case, growth or solution of crystals is limited by diffusion. Small crystals are removed much less efficiently by this process, but the growth of crystals much larger than the critical size is favoured (Higgins 1998). The slope of right part of the CSD decreases with this process (Fig. 8d, e).

The Cathedral Peak Granodiorite CSDs show the negative correlation between slope and intercept that is predicted by the CN theory (Figs 6, 7, 8d, e; Dehoff 1991; Higgins 1998). Unfortunately, the left part of the CSD is swamped by the groundmass population of crystals, hence it is not clear if it descends steeply following the LSW theory or has a shallow slope in accordance with the CN model. The improvement in the correlation between characteristic length and intercept following normalization to 100% microcline indicates that textural coarsening here is an open process (Figs 7, 8e). That is, some areas have been strongly coarsened, but with little addition of new material (e.g. outcrops 2 and 5). Hence, the 'nests' of crystals are areas

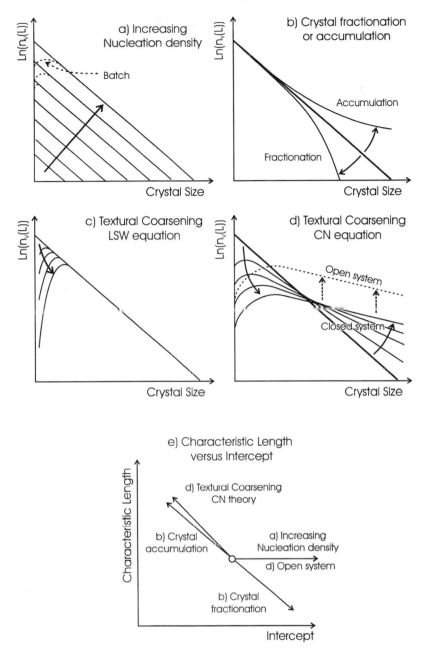

Fig. 8. Schematic crystal size distributions. (**a**) A straight-line distribution can be produced by continuous-flow steady-state (Marsh 1988). Batch crystallization during linearly increasing undercooling can produce a similar CSD, except that there will be a turn-down for small crystals (Marsh 1998). (**b**) Accumulation of crystals under the influence of gravity will favour larger grains, hence the right end of the line will be deflected upwards (Marsh 1988). (**c**) Textural coarsening following the LSW theory (Lifshitz & Slyozov 1961) will augment the number of large crystals, like accumulation, but will eliminate small crystals, giving a characteristic hump-shaped CSD (Higgins 1998). (**d**) Textural coarsening can also follow the CN theory (Dehoff 1991). In this case small crystals will not be so deficient, but the right part of the CSD will rotate, giving a negative correlation between slope and intercept (Higgins 1998). (**e**) Summary of CSD models in terms of Characteristic Length versus Intercept. The LSW textural coarsening model does not change the slope of the right part of the CSD and hence does not make a vector in this diagram.

where material has been added, but are not necessarily the most coarsened parts of the rock (e.g. outcrop 3).

Solidification of the Cathedral Peak Granodiorite

The first stage in the solidification of the Cathedral Peak Granodiorite must have been nucleation and growth of the major minerals in an environment of increasing undercooling (Fig. 9a). Cooling of the magma during emplacement cannot be the cause of the undercooling as the host rock for much of the Cathedral Peak Granodiorite is the more primitive, and hence hotter, Half-Dome Granodiorite, which was emplaced a short while before (Bateman & Chappell 1979). If cooling and nucleation followed emplacement of all components of the Tuolumne Intrusive Suite then there should not be any difference in texture between the different components. Hence, the undercooling must be

related to decreasing pressure or degassing during ascent of the magma and initial nucleation and growth of crystals must have occurred at depth. This is testified by the steeply dipping magmatic foliations in many of the Tuolumne Intrusive Suite rocks (Bateman & Chappell 1979). The initial CSD of the K-feldspar is not accessible, but it may have resembled the CSD of the later groundmass microcline.

The second phase was initiated when the Cathedral Peak Granodiorite was emplaced into the still warm Half-Dome Granodiorite. At this time heat was unable to escape and undercooling was stalled or even decreased, both in the Cathedral Peak Granodiorite and adjacent parts of the Half-Dome Granodiorite (Fig. 9b). The temperature would have been buffered by the release of latent heat of crystallization of K-feldspar, the latest, and most abundance phase. The small undercooling of the K-feldspar would have suppressed nucleation of new crystals and promoted grain-size coarsening. Crystals smaller than the critical size dissolved in the

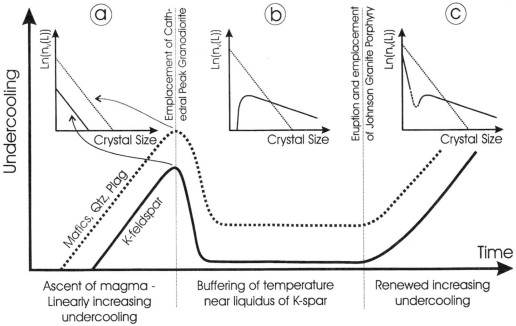

Fig. 9. Model for the development of megacrysts in the Tuolumne Intrusive Suite. (**a**) The initial phase was of nucleation and growth of all major phases. K-feldspar is the last major mineral to nucleate from granodioritic magmas, except at low pressures and water contents (Naney 1983). All other phases have been lumped together for clarity, but really have separate CSDs. Linearly increasing undercooling during this period produced linear CSDs. (**b**) Emplacement of the CPG into earlier parts of the intrusive suite stalled cooling and the temperature was buffered near the liquidus temperature of K-feldspar. Conditions were now appropriate for textural coarsening of K-feldspar and small crystals were removed. The other phases were strongly undercooled at this time and hence did not coarsen, although crystals of all sizes continued to grow. (**c**) Eruption of the Johnson Creek porphyry enabled renewed undercooling and K-feldspar again nucleated and grew to form the groundmass crystals.

interstitial magma and consequently fed larger crystals that grew to become the megacrysts. The peak of the megacryst CSDs at 25 mm indicates that the final critical size was close to this value. Earlier critical sizes were smaller.

When the CPG was emplaced quartz, plagioclase and mafic mineral crystals were already present. They nucleated before the K-feldspar and hence probably were larger than the K-feldspar at this time as crystal growth rates do not appear to vary by more than a factor of 2 or 3 between different mineral species (Cashman 1990). Buffering of the temperature at the K-feldspar liquidus kept the remaining phases much more undercooled. Maximum growth rates may have been more rapid than that of K-feldspar, but the strong undercooling for earlier-formed mineral species was not suitable for textural coarsening. New crystals of these phases crystallized instead of dissolving, hence limiting the size of these crystals. K-feldspar was not limited in this way and overtook the early phases in size, to produce the megacrysts.

Many grains of other minerals were incorporated into the growing crystals, especially those with a similar structure, such as plagioclase, for which the energy of the interface between the two minerals is low. Dissimilar grains with higher interface energy (e.g. amphiboles) were commonly swept aside to accumulate beside the megacrysts. At this point the magma would have contained large euhedral megacrysts that were growing and smaller, irregularly shaped crystals that were dissolving. A small amount of melt would have been present. The overall population density of the microcline would have been reduced by the coarsening process.

Any preferential orientation of the initial, small K-feldspar crystals produced by early flow of the magma would be maintained in the final megacrysts, even if magmatic deformation had ceased. However, grain rotations during coarsening may reduce the quality (intensity), but not the direction, of the mineral fabric (Higgins 1998).

Migration and crystallization of the interstitial melt over decimetric distances produced nests of crystals because of the positive feedback between coarsening and fluid flow. Coarsening may enhance the permeability of crystal mushes by removing small crystals that may block channels. Increased permeability will lead to focusing of fluid flow and further growth of megacrysts along the channels. The channels formed by these interactions will appear as linear regions and nests of crystals in the two-dimensional surface of the outcrop. Hence, the megacryst-rich zones were open systems. Euhedral zones of inclusions in the megacrysts may represent times when

growth rate was higher due to greater fluid flow and hence crystal transport of crystal nutrients.

Chadocrysts in mafic rocks preserve early textures (Mathison 1987; Higgins 1991). If the Cathedral Peak Granodiorite megacrysts are indeed phenocrysts then the included grains of plagioclase, mafic minerals and quartz must be chadocrysts. The important size difference between the chadocrysts and the groundmass minerals indicates that all the minerals in the groundmass, except microcline, must also have grown. The small quantity of crystals in the core of the megacrysts shows that coarsening must have started at low crystal contents, of the order of 5–10% (some early mafic minerals were swept aside by the megacryst and not incorporated).

The groundmass microcline crystals form a population distinct from the megacrysts both on the basis of their shape and size distributions (Figs 4, 5, 7). The groundmass microcline crystals cannot be a residual early population of crystals as such crystals would have been dissolved during coarsening. Straight-line CSDs like those seen in the groundmass have been observed in many volcanic rocks (Cashman 1990) and are interpreted to indicate either continuous flow, steady-state crystallization or batch crystallization in an environment of linearly increasing undercooling (Marsh 1988, 1998). Although studies of plutonic rock CSDs are rare, and hence few straight-line CSDs have been recorded, the batch crystallization model could readily be applied to the groundmass microcline. Indeed the CSDs of plagioclase in plutonic rocks have broadly similar slopes and intercepts to the groundmass microcline (Higgins 1991, 1998). Such an environment of increasing undercooling occurred at the end of the intrusive cycle of the Tuolumne Suite (Fig. 9c). It may have been related to the eruption of the Johnson Creek porphyry—loss of confining pressure may have been accompanied by loss of water from the magma. The resulting increase in liquidus temperature would have increased the undercooling.

Megacrysts occur in many other granitoids, but are also lacking in many chemically similar plutons. The model proposed here necessitates a period of low undercooling to develop the megacrysts. If this was not present then an equigranular rock would have formed. Hence, it is the cooling environment, rather than the chemical composition, that controls the development of megacrysts.

Conclusions

It has been commonly stated that igneous rock textures are produced by the interplay of nucleation and growth of crystals (i.e. Swanson 1977), but in the Cathedral Peak Granodiorite, solution of crystals was equally important. Three steps were necessary to produce the textures observed in the Cathedral Peak granodiorite. (1) Initial nucleation and growth in an environment of increasing undercooling, probably associated with the ascent of the magma. (2) Reduction of undercooling following emplacement of the Cathedral Peak Granodiorite into warm host rocks. Small crystals of K-feldspar were dissolved and larger crystals grew, reducing the overall population density of K-feldspar in the Cathedral Peak Granodiorite and the adjacent parts of the Half-Dome Granodiorite. K-feldspar coarsened more than the other crystals because it was maintained close to its liquidus for a longer period. (3) Resumption of undercooling at the end of the intrusive cycle. Early orientation fabrics produced during the first phase will be conserved in the megacrysts, even if magma movement has ceased. Other important points are as follows. (1) Although the textural coarsening process proposed here is firmly igneous in origin, it has similarities with some metasomatic processes. The megacrysts grew in a normal igneous fashion from a silicate melt at high temperature and are, indeed, phenocrysts. However, the process envisaged here involves both solution of crystals and transport of material in open systems, which are both features of metasomatic systems. (2) The low population density of the megacrysts is not due to 'nucleation difficulties' but is a result of solution of most of the early formed crystals. (3) Coarsening started at low crystal contents—5–10%. (4) The nests and linear zones of megacrysts are 'fossil' fluid channels, produced by the positive feedback between permeability and coarsening. (5) Coarsening and 'openness' of magmatic systems are independent. That is, in some areas crystals can grow very large, but their volumetric abundance will not change. Elsewhere, the abundance of crystals will increase concomitantly with coarsening. (6) Megacrysts are developed in granitoids that have had a particular cooling history. Undercooling must be buffered near the liquidus of K-feldspar for a long period so that the texture can coarsen.

I would like to thank W. Hildreth for suggesting this project. E. Sawyer, M. Roberts, D. Vanko, E. Ferre and an anonymous reviewer made helpful suggestions. I completed this project during a sabbatical at Université Blaise-Pascal, Clermont-Ferrand, France. This project was funded by the Natural Science and Engineering Research Council of Canada.

References

BATEMAN, P. C. & CHAPPELL, B. W. 1979. Crystallisation, fractionation, and solidification of the Tuolumne Intrusive series, Yosemite National Park, California. *Geological Society of America Bulletin*, **90**, 465–482.

BOUDREAU, A. E. 1987. Pattern forming during crystallisation and the formation of fine-scale layering. *In:* PARSONS, I. (ed.) *Origins of igneous layering*. D. Reidel, Dordrecht, 453–471.

CASHMAN, K. V. 1990. Textural constraints on the kinetics of crystallization of igneous rocks. *In:* NICHOLLS, J. & RUSSELL, J. K. (eds) *Modern methods of igneous petrology: understanding magmatic processes*. Mineralogical Society of America, 259–314.

—— & FERRY, J. M. 1988. Crystal size distribution (CSD) in rocks and the kinetics and dynamics of crystallization III. Metamorphic crystallization. *Contributions to Mineralogy and Petrology*, **99**, 410–415.

—— & MARSH, B. D. 1988. Crystal size distribution (CSD) in rocks and the kinetics and dynamics of crystallisation II. Makaopuhi lava lake. *Contributions to Mineralogy and Petrology*, **99**, 292–305.

COLLINS, L. G. 1988. *Hydrothermal differentiation and myrmekite; a clue to many geologic puzzles*. Theophrastus Publications, Athens, Greece.

DEHOFF, R. T. 1991. A geometrically general theory of diffusion controlled coarsening. *Acta Metallurgica et Materialia*, **39**, 2349–2360.

DICKSON, F. W. 1996. Porphyroblasts of barium-zoned K-feldspar and quartz, Papoose Flat, Inyo Mountains, California, genesis and exploration implications. *In:* COYNER, A. R. & FAHEY, P. L. (eds) *Geology and Ore Deposits of the Cordillera*. Geological Society of Nevada Symposium Proceedings, Reno/Sparks, Nevada, 909–924.

FENN, P. M. 1977. The nucleation and growth of alkalifeldspars from hydrous melts. *Canadian Mineralogist*, **15**, 135–161.

HIGGINS, M. D. 1991. The origin of laminated and massive anorthosite, Sept Iles intrusion, Quebec, Canada. *Contributions to Mineralogy and Petrology*, **106**, 340–354.

—— 1998. Origin of anorthosite by textural coarsening: Quantitative measurements of a natural sequence of textural development. *Journal of Petrology*, **39**, 1307–1325.

HUTCHISON, C. S. 1974. *Laboratory handbook of petrographic techniques*. John Wiley & Sons.

KERRICK, D. M. 1969. K-feldspar megacrysts from a porphyritic quartz monzonite, central Sierra Nevada, California. *American Mineralogist*, **54**, 839–848.

LIFSHITZ, I. M. & SLYOZOV, V. V. 1961. The kinetics of precipitation from supersaturated solid solutions. *Journal of Physics and Chemistry of Solids*, **19**, 35–50.

MARSH, B. 1988. Crystal size distribution (CSD) in rocks and the kinetics and dynamics of crystallization I. Theory. *Contributions to Mineralogy and Petrology*, **99**, 277–291.

MARSH, B. D. 1998. On the interpretation of Crystal Size Distributions in magmatic systems. *Journal of Petrology*, **39**, 553–600.

MATHISON, C. I. 1987. Pyroxene oikocrysts in troctolitic cumulates—evidence for supercooled crystallisation and postcumulus modification. *Contributions to Mineralogy and Petrology*, **97**, 228–236.

NANEY, M. T. 1983. Phase equilibria of rock-forming ferromagnesian silicates in granitic systems. *American Journal of Science*, **283**, 993–1033.

RANDOLF, A. D. & LARSON, M. A. 1971. *Theory of particulate processes*. Academic Press, New York.

RESMINI, R. G. 1993. *Dynamics of magma in the crust: A study using crystal size distribution*. PhD, Johns Hopkins University.

ROYET, J.-P. 1991. Stereology: A method for analysing images. *Progress in Neurobiology*, **37**, 433–474.

SWANSON, S. E. 1977. Relation of nucleation and crystal-growth rate to the development of granitic textures. *American Mineralogist*, **62**, 966–978.

VERNON, R. H. 1986. K-feldspar megacrysts in granites—Phenocrysts not porphyroblasts. *Earth-Science Reviews*, **23**, 1–63.

VOORHEES, P. W. 1992. Ostwald ripening of two-phase mixtures. *Annual Review of Materials Science*, **22**, 197–215.

Movement of melt during synchronous regional deformation and granulite-facies anatexis, an example from the Wuluma Hills, central Australia

E. W. SAWYER[1], C. DOMBROWSKI[1] & W. J. COLLINS[2]

[1]*Module de Sciences de la Terre, Universite du Quebec a Chicoutimi, Chicoutimi, Quebec G7H 2B1, Canada (e-mail: ewsawyer@uqac.uquebec.ca)*

[2]*Department of Geology, The University of Newcastle, Newcastle, NSW 2308, Australia*

Abstract: Granulite facies anatexis ($T \approx 900\,°C$) in the Wuluma Hills region of the Arunta Inlier was synchronous with deformation. During D3 contractional deformation strain was partitioned into S3 shear zones, which alternate with lower strain domains containing F3 fold hinges. Subsequent D4 deformation was minor and in part extensional.

Leucosomes in the S3 shear zones are principally veins oriented parallel, or subparallel, to the pervasive S3 foliation. Leucosomes in the F3 hinge domains are more complex, and occur parallel to anisotropy due to lithological layering, the pre-existing S1/2 foliation, S3 and fold axial planes (F3 and F4). Some leucosomes (generally high Na_2O, low K_2O and Rb/Sr) record melt migration paths, and other sites of melt accumulation.

All the migmatites are residual and lost melt when deformation forced melt from matrix grain boundaries, through a network of small lensoid channelways to accumulation sites in fold hinges, there larger batches of magma developed. Leucosomes in accumulation sites develop a schlieric or diatexitic appearance because inflowing melt eroded the host rocks. Later increments of D3 contractional strain overpressured the accumulated granitic magma and it migrated again to other (more stable) low pressure sites through veins generally oriented parallel to S3. Magma/melt movement stopped when the solidus was reached, or the magma reached a structurally stable site (e.g. pluton).

Four sequential steps have to occur to form a granite magma during anatexis of the continental crust. (1) The metamorphic temperatures must be high enough and affect sufficient of the crust that a large volume of felsic melt is generated. (2) The melt fraction has to be segregated from its residuum. (3) The granitic magma consisting of melt plus entrained crystals must be transported to a site where (4) magma batches can collect; the magma batches may migrate further and amalgamate to form a granite pluton. Smaller volumes of melt, or magma, trapped within the anatectic region are called leucosomes. The first two stages, melting and melt segregation occur at the site of anatexis. Although the transport and accumulation stages begin in the anatectic domain, they could end anywhere between there and the surface.

Most studies of the melt transport stage have concentrated on determining what the transfer mechanism might be (Weinberg 1996; Petford 1996). Diapirism had been considered the principal mechanism (Wickham 1987), but now buoyant rise of melt along fractures is favoured. Present concern is over whether the fractures are simply dykes (Clemens & Mawer 1992; Petford *et al.* 1993) or whether the fractures occur in the context of crustal scale shear zones (D'Lemos *et al.* 1992; Brown & Solar 1998*a*, *b*). An alternative view is that magma rises along a complex combination of existing structural channelways and new fractures (Collins & Sawyer 1996). Most of the dyke models are based on theoretical analyses (e.g. Clemens & Mawer 1992; Petford *et al.* 1993), whereas the other models are primarily based on field examples (e.g. D'Lemos *et al.* 1992; Collins & Sawyer 1996; Brown and Solar 1998*a*, *b*).

There have been few studies of the path taken by melt moving within anatectic regions (Allibone & Norris 1992; Hand & Dirks 1992; Greenfield *et al.* 1996; Sawyer 1998). In large part the complex appearance of the anatectic regions has deterred their study. Typically, the leucosomes are numerous and morphologically varied, but they can be divided into specific

From: CASTRO, A., FERNÁNDEZ, C. & VIGNERESSE, J. L. (eds) *Understanding Granites: Integrating New and Classical Techniques.* Geological Society, London, Special Publications, **168**, 221–237. 1-86239-058-4/99/$15.00 © The Geological Society of London 1999.

types, or generations, based on their relationship to the local structure and deformation history. Several recent results from migmatites have improved our understanding of the processes involved in crustal anatexis. Some migmatite leucosomes have cumulate, not melt compositions (Cuney & Barbey 1982; Barr 1985; Sawyer 1987), and hence record the location of early crystallization products left behind as melt migrated. Studies linking the geochemical composition with petrographic and field relations demonstrate how palaeosome, residuum and melt-rich rocks can be identified and related (e.g. Sawyer 1998). Thus, the means are now available for recognizing the sources, transfer paths and collection sites of melt in migmatites. It remains, however, to show how regions of melting, melt loss and melt accumulation in the source region relate to the local deformation history.

The purpose of this study is to examine the causes of the morphological complexity observed in a region of granulite facies crustal anatexis from the Wuluma Hills in the Arunta Inlier of central Australia. Specifically, mapping of the structural elements (foliations, folds, etc.) and distribution of leucosomes are combined with petrographic and geochemical techniques to identify regions of melt loss, melt transfer and melt accumulation, so that the segregation and movement of melt during the deformation of anisotropic rocks can be evaluated.

Geological setting

Regional geology

The Arunta Inlier consists of early Proterozoic volcanic and sedimentary rocks that have been separated into Divisions 1, 2 and 3 by Shaw *et al.* (1979). The three divisions are believed to be part of a single depositional cycle that began with bimodal, terrestrial volcanism, then progressed to dominantly flysch-type sedimentation and ended with shallow water, platform-type deposition (Stewart *et al.* 1984). Some of the volcanic rocks represent 1900–1700 Ma additions of juvenile mantle material (Zhao & McCulloch 1995) to the continental crust, but many of the sediments (Windrim & McCulloch 1986) and granites (Zhao & McCulloch 1995) have Nd model ages >2000 Ma, which indicates the recycling of older, probably Archaean, continental crust. The major regional deformation and granulite facies metamorphism, termed the Early Strangways Orogeny, occurred between 1780 and 1770 Ma, but other locally important granulite facies metamorphism and anatexis occurred at

1860–1820 Ma (Mt Stafford Event) and 1745–1730 Ma (Late Strangways Event) in various parts of the Arunta Inlier (Collins & Shaw 1995). Younger, lower-grade metamorphism has affected many parts of the Arunta Inlier during the middle and upper Proterozoic, and significant events that retrograded the granulites have been dated at between 1680 and 1640 Ma, *c.* 1450 Ma and *c.* 1150 Ma (Black & Shaw 1995; Collins & Shaw 1995; Zhao & Bennet 1995). Major uplift and unroofing of the Arunta Inlier occurred as a result of south-vergent thrusting during the mid-Palaeozoic Alice Springs Orogeny (Collins & Teyssier 1989).

The Wuluma Hills region

The Wuluma Hills area north of Alice Springs lies within the southern part of the Strangways Metamorphic Complex portion of the Central Tectonic Province of the Arunta Inlier (Fig. 1). Granulite-facies anatexis during the Late Strangways Event produced the Ingula migmatite suite (Shaw *et al.* 1979), which hosts a number of small granitic plutons. The origin and emplacement of the largest of these granites, the 1728 ± 3 Ma Wuluma Pluton, has been the subject of two recent studies (Collins *et al.* 1989; Lafrance *et al.* 1995). The present study area is some 8 km east of the Wuluma Granite (Fig. 2), and includes the northeastern part of the area mapped by Lafrance *et al.* (1995).

General structure

Three major deformation events have been identified in the Strangways Metamorphic Complex (Shaw *et al.* 1984) and the Wuluma Hills area (Collins *et al.* 1989; Lafrance *et al.* 1995). A D1 deformation is correlated with the Early Strangways Event, but the D2 and D3 deformations belong to the Late Strangways Event. Results from this study are in broad agreement with these findings, but there may be an additional post-D1 and pre-D3 deformation in the region based on the folding of pre-D3 foliations. However, insufficient is known about the geometry and areal extent of this deformation to warrant modification of the existing consensus on the pre-D3 structural history at this time. The recognition of a D4 deformation (Figs 2 and 3) is of significance in establishing the timing of melt movement in the migmatites.

The oldest structural element is an S1 foliation. Generally, S1 is oriented parallel to the lithological layering, but locally it is axial planar to isoclinal, rootless intrafolial F1 folds. The present regional structural pattern is largely due to

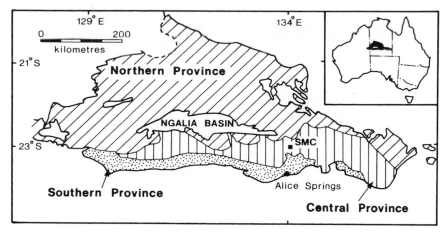

Fig. 1. Location of the Wuluma Hills study area (filled square) in the Strangways Metamorphic Complex (SMC) within the Central Province of the Arunta Inlier, central Australia.

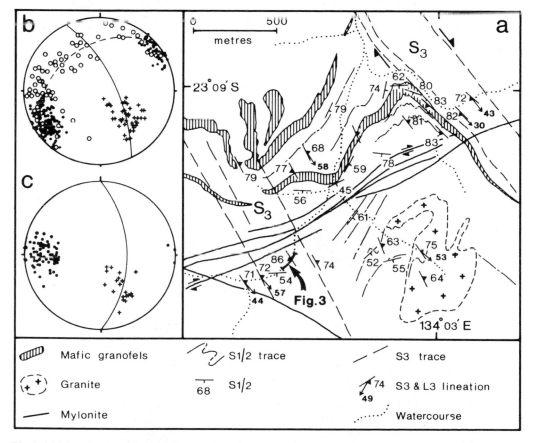

Fig. 2. (a) Map showing the F3 folding of lithological layering and S1/2 in the Ingula migmatites and the location of S3 shear zones on the fold limbs. Lambert equal area stereograms; (b) D3 structures, poles to S1/2 (open circles), poles to S3 foliation (dots) and F3 fold hinges (crosses); (c) D4 Structures, poles to S4 (dots) and F3 fold hinges (crosses).

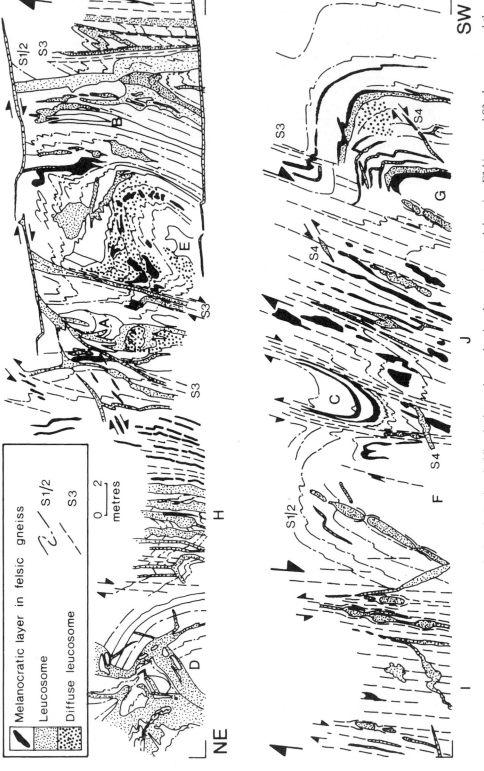

Fig. 3. Sketch map of a section (see Fig. 2 for location), viewed obliquely down-plunge, showing the two structural domains, F3 hinges and S3, shear zones and the distribution of leucosome types within them. Small leucosomes parallel to S4 are enlarged for clarity.

tight, or isoclinal, sheath-like F2 folds (Lafrance et al. 1995) that fold S1. Only in the F2 fold hinges can the S2 foliation be clearly recognized; S2 is essentially parallel to S1 on the flanks of F2 folds. Hence the main, regional foliation is a composite of the S1 and S2 foliations, and will be termed S1/2 (Fig. 2a) hereafter. A transpressive D3 deformation produced close to tight, asymmetric F3 folds with SE-trending plunges and NW-trending axial planes (Fig. 2b). In the F3 fold hinges, S3 is a weakly developed crenulation cleavage or axial planar foliation, consequently the S1/2 foliation is the stronger fabric. In contrast, S3 is penetratively developed in shear zones located on the steep limbs of F3 folds. Asymmetric F4 folds are small-scale structures with easterly-dipping axial planes and hinges with more southerly trends than the F3 folds (Fig. 2c). The F4 folds occur where the lithological layering and S1/2 have shallow dips. However, small (10–30 cm long) reverse and normal sense D4 shear zones are developed where the planar anisotropy dips steeply. The geometry of D4 structures suggests that they formed as the D3 regional transpressive deformation relaxed to a weakly extensional regime.

In the map area (Fig. 2), post-D4 deformation occurred principally along discrete NW and ENE trending shear zones. Mylonites in the shear zones contain amphibolite facies mineral assemblages, and so they post-date the regional granulite facies metamorphism.

Rock types and metamorphism

All the migmatites belong to Division 1 of Shaw et al. (1979). Five principal rock types are present.

(1) *Felsic gneiss* is the most common rock type in the area and can be divided into an Al-rich, and an Al-poor group. These two groups may correspond to pelites and felsic volcanic rocks respectively. The Al-rich gneisses all contain quartz, plagioclase, perthite, biotite (<15 modal %) and cordierite (15 modal %), plus either orthopyroxene or garnet; some samples contain trace amounts of hercynite. Al-poor gneisses are generally fine-grained and contain quartz, plagioclase, biotite and orthopyroxene; a few lack biotite. The mineral parageneses suggest that the felsic gneisses have all partially melted. The foliation in the Al-poor gneisses is a schistosity due to well oriented biotite. A gneissic banding is rarely developed, and they are notably more massive and uniform than the Al-rich gneisses. In contrast, the aluminous gneisses contain more biotite and a more prominent foliation, which typically has leucosomes parallel to it, producing a strongly banded and anisotropic rock.

(2) *Melanocratic rocks* form discontinuous layers and trains of boudins in the felsic gneisses. Many of the melanocratic rocks lack both quartz and feldspars (Fig. 4), but others contain up to 40 modal % quartz and a trace of plagioclase. Most of the melanocratic rocks contain cordierite, orthopyroxene and a small amount (<20 modal %) of biotite, but a few also contain garnet. Sillimanite is absent, but some samples contain hercynite inclusions in garnet or cordierite. Their mineralogy suggests that these rocks are the residua from the partial melting of aluminous layers in the gneisses.

(3) *Quartz-rich rocks* form massive grey-weathering layers, or more commonly boudins, in the felsic gneisses. These rocks contain 50 to 60 modal % quartz, minor amounts (<10 modal %) of biotite, and no feldspar or sillimanite, they can be divided into two groups. One group contains 30 to 40 modal % cordierite, whereas the other contains either orthopyroxene or garnet. The mineralogy suggests that they are the residua from the partial melting of quartz-rich layers in the gneiss.

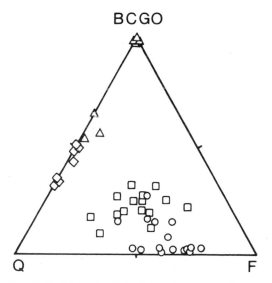

Fig. 4. Modal composition of the migmatites and leucosomes. Q, quartz; F, plagioclase + K-feldspar; BCGO, biotite + cordierite + garnet + orthopyroxene. Symbols: circles, leucosomes; squares, felsic gneisses; diamonds, quartz-rich layers; triangles, melanocratic layers.

(4) *Mafic granofels* are massive, fine to medium-grained orthopyroxene + clinopyroxene + plagioclase rocks that form persistent thin layers (Fig. 2), or trains of boudins, in the felsic gneisses. They are probably mafic sills, dykes or lava flows in Division 1. Many mafic granofels show incipient melting and contain a small proportion (< 5 vol%) of tonalitic leucosome, generally located in fractures.

(5) *Leucosomes* are believed to have crystallized from an anatectic melt or magma, and some preserve igneous textures. Textures and grain sizes vary greatly, most have hypidiomorphic–granular textures, but some are porphyritic and others pegmatite textured; grain size ranges from fine to coarse. Many contain small amounts of pink garnet which typically occurs as 2 mm to 2 cm euhedral crystals. In some cases the garnet is intergrown with quartz. In some late, vein-type leucosomes the garnets are partially replaced by either cordierite, biotite or chlorite. Other leucosomes contain skeletal cordierite crystals up to 2 cm across; orthopyroxene is less common and tends to occur in the patch leucosomes (*in situ* types) rather than in the vein leucosomes. Most leucosomes contain small amounts of biotite (< 3 modal %), but if biotite is abundant (10 modal %) it forms schlieren from which magmatic flow sense can be determined.

Lafrance *et al.* (1995) estimated a peak temperature of about 750 °C and a pressure of 5 to 6 kbar, for an assumed $a_{H_2O} = 0.5$. However, the mineral parageneses suggests that their geothermometry may have underestimated the maximum temperature, probably due to the common problem of cation exchange during the cooling of granulite terranes (Frost & Chacko 1989).

The massive, Al-poor gneisses provide the most direct evidence for high metamorphic temperatures. Within this rock type there is a progressive range from biotite + plagioclase + quartz + orthopyroxene to plagioclase + quartz + orthopyroxene in which modal biotite decreases as modal orthopyroxene increases. Moreover, small leucosomes oriented in S3 are developed in some rocks. These changes suggest that the biotite-consuming and melt-producing reaction

$$biotite + quartz + plagioclase$$
$$= orthopyroxene + melt$$

has occurred, and gone to completion in some layers. This reaction has been the subject of several experimental studies using different biotite and plagioclase compositions. Biotites from the Wuluma rocks have Mg/(Mg + Fe) > 0.6, which, when compared to the results of Vielzeuf & Montel (1994) and Patiño Douce & Harris (1998), suggests that the reaction began at temperatures above 830 °C. Vielzeuf & Montel (1994) determined the temperature interval between 'orthopyroxene in' and 'biotite out' to be between 50 and 100 °C depending upon pressure. Hence, metamorphic temperatures may have reached, or exceeded, 900 °C.

High peak temperatures can also be inferred from the residual parageneses found in quartz-rich layers and the melanocratic rocks. These have consumed all their feldspars and quartz + feldspar respectively which, assuming a pelitic protolith (a realistic assumption considering the aluminous compositions), requires temperatures in excess of 900 °C (Vielzeuf & Holloway 1988). Metamorphic pressures cannot be readily estimated. However, the preponderance of cordierite-bearing assemblages, some with garnet, most with orthopyroxene, and the sporadic presence of hercynite, suggest medium pressure (4–5 kbar) based on the results of Vielzeuf & Montel (1994).

Migmatite morphology

Figure 3 is a section mapped across the steep limb of the F3 antiform shown on Fig. 2. All the large folds on the section are meso-scale F3 structures, too small to show on Fig. 2. Two types of structural domain are evident, and coincide with the shallow-dipping and steeply-dipping limbs of F3 folds respectively. In the first type, continuous lithological layering, the S1/2 foliation and rare, small-scale isoclinal intrafolial pre-F2 folds are preserved in F3 fold closures. However, in the other type, no pre-D3 structures are recognizable; they have been transposed and overprinted by a very strongly penetrative S3 foliation which locally contains a strong SE-plunging mineral elongation lineation. Based on their field appearance, the F3 fold closures are domains of lower D3 strain, whereas the domains where S3 predominates represent shear zones and higher D3 strain.

In general the F3 synformal closures (e.g. A, B and C on Fig. 3) are much tighter than the antiformal closures (e.g. D, E, F and G). The opposed movement sense of shear zones on either flank of synform C suggests that the competent unit in the closure was expelled upwards out of the fold core as the fold tightened. In general, the presence of S3 shear zones on the limbs of the fold closures at A and C and the location of the tightest closure (B) within a narrow shear zone, suggests that the S3 shear zones nucleated on the steep limb and then migrated into the synforms as the D3 deformation progressed and tightly pinched the synforms. The resulting pattern is of well preserved F3 antiforms separated from each other by compressed synforms, or by wide, steeply dipping S3 shear zones (e.g. H, I and J, Fig. 3).

F3 hinge domains

These domains contain the widest range of leucosome types (Fig. 5a), and their distribution appears to be strongly controlled by the competence and anisotropy of their host rocks. The more competent Al-poor gneisses will be considered first.

Figure 5b shows the contact between a competent and massive Al-poor gneiss and a less competent, strongly foliated (i.e. anisotropic) Al-rich gneiss. The centre of the competent layer contains small orthopyroxene porphyroblasts with biotite-free haloes, whereas towards the contact the porphyroblasts are much larger and biotite is practically absent. The lighter colour of the competent layer near the pelite is due to a progressive increase in modal quartz as plagioclase is exhausted by the melt-producing reaction. It is noticeable that no leucosomes representing melt from the biotite-consuming reaction occur in the competent unit, but that there are abundant leucosomes parallel to S1/2 in the Al-rich gneiss. This relationship suggests that the competent, but isotropic, layer has lost melt, whereas the incompetent, but anisotropic, layer has gained it.

In another competent, but foliated (i.e. less massive) Al-poor gneiss, several bands of coarse orthopyroxene are developed parallel to S1/2. There are no leucosomes parallel to S1/2; but there are centimetre-scale leucosomes parallel to S4. However, these are too small to account for the total melt production from this unit, i.e. this competent layer has lost melt too.

The Al-rich gneisses are more anisotropic and contain many leucosomes of various forms; three broad types can be recognized based on their morphology and location. (a) Leucosomes along S-planes, (b) leucosomes in fold cores and (c) *in situ* melt patches.

Most of the strongly foliated gneisses contain many leucosomes oriented parallel to S1/2. One population consists of very thin (1–5 mm), fine-grained leucosomes that contain an S2 foliation in F2 fold closures; these belong to an earlier melting event (see Collins *et al.* 1989; Lafrance *et al.* 1995) and will not be considered further. The other leucosomes oriented parallel to S1/2

are coarser-grained and lack an internal S1/2 foliation. Tracing these leucosomes carefully across large outcrops reveals that many of them either cross S1/2 at a shallow angle (Fig. 5c), or that they step from one S1/2 plane to another. Moreover, some can be traced to root in vein-type leucosomes that occupy F3 axial surfaces, or small shear zones on the fold limbs. Leucosomes which cross S1/2 at a shallow angle and are folded coaxially with S1/2, but record F3 folds with larger interlimb angles than those in their hosts, indicate that melt injection was during F3 fold growth. Thick (> 10 cm) leucosomes that intruded parallel, or subparallel, to S1/2 generally contain an internal foliation marked by biotite flakes and schlieren, resulting from the incorporation and partial disaggregation of wall rock material as the injecting magma eroded its host; these leucosomes resemble diatexite migmatites. Some schlieren in veins that pre-date F4 contain small F3 folds, whereas in other veins, schlieren preserve F4 folds. Hence, melt injection occurred during D3 and continued to D4.

Leucosomes are also in, or else slightly oblique to, many F3 (Fig. 5d) and F4 fold axial surfaces. Commonly, these leucosomes are small, lensoid bodies in which plagioclase predominates over K-feldspar. However, some are much wider and laterally continuous, vein-like leucosomes with offshoots that are parallel, or sub parallel, to S1/2 and folded by F3. These larger veins show a range of compositions; some contain subequal amounts of plagioclase and K-feldspar, and others are very rich in K-feldspar. Veins with abundant schlieren, fragments of host rock and flow fabrics resemble diatexite migmatites (Fig. 6a), and are generally K-feldspar-rich.

The F3 fold hinges are sites of melt accumulation. Irregularly-shaped, diffuse leucosome patches (Fig. 3) with both sharp and diffuse contacts, are developed in F3 fold cores within thick Al-rich gneiss. The original fold structure can be discerned in the leucosome by the ghost structure preserved as trains of country rock fragments and schlieren detached as the melt filled the hinge, some patches resemble diatexite migmatites (Fig. 6b). Discordant contacts and variations in texture, mineralogy and inclusion/schlieren abundance within these leucosomes provide evidence of repeated influx of magma into the hinge. Successively younger melt batches are more leucocratic and contain fewer inclusions, or schlieren, and tend to be K-feldspar-rich. Melt flow directions can be determined locally from the orientation and tiling of tabular feldspars and from enclave trains. These suggest that in fold closure E (Fig. 3) melt flowed in along shear zones parallel

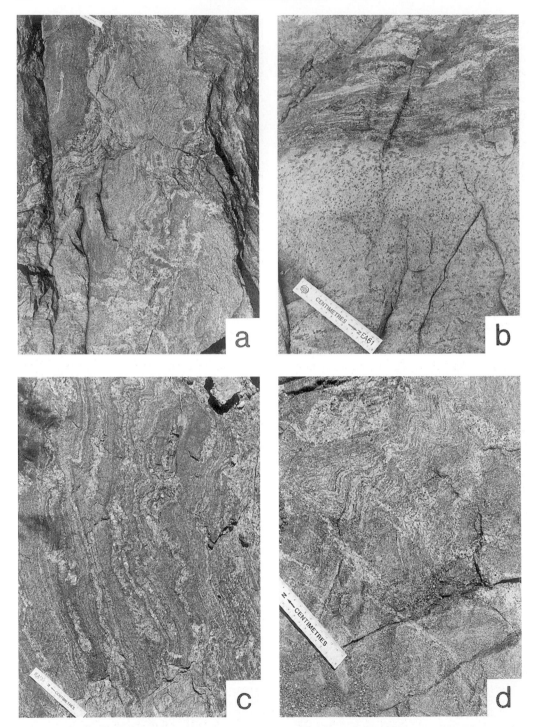

Fig. 5. Morphology of the migmatites from F3 hinge domains. (**a**) Synformal F3 hinge showing a variety of leucosome types, note the *in situ* patch leucosomes with garnet + quartz cores. (**b**) Contact between a massive non-foliated Al-poor gneiss and a strongly foliated aluminous gneiss. (**c**) Leucosomes parallel to, and slightly cross-cutting S1/2, minor folds are F3. (**d**) Thin leucosomes parallel to S1/2 and lensoid leucosomes parallel to F3 axial planes. Scale is 15 cm long.

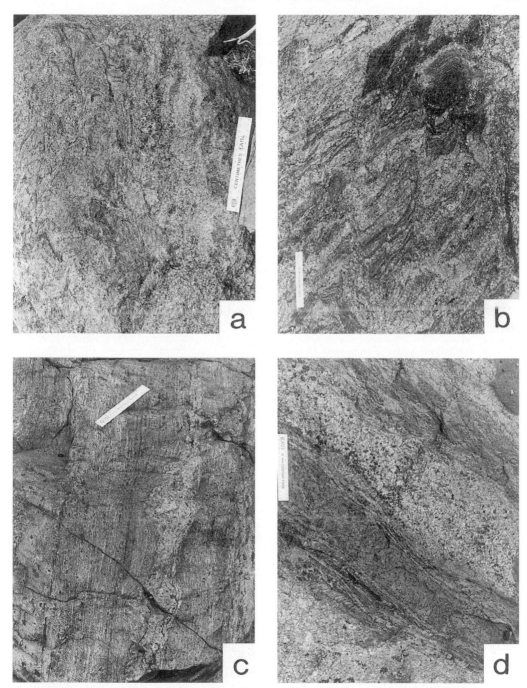

Fig. 6. Morphology of migmatites from F3 hinge domains (**a** & **b**) and S3 shear zone domains (**c** & **d**). (**a**) Wide K-feldspar-rich, leucosome dyke parallel to F3 axial plane containing schlieren and pieces of wall rock giving a diatexite-like appearance. (**b**) Part of the diffuse leucosome in a fold E (Fig. 3) showing *in situ* host rock remnants and schlieren, note melt in S3 planes (parallel to scale) is continuous with melt along S1/2 planes, this patch is diatexitic. (**c**) Pegmatite boudinaged in the centre of an S3 shear zone domain, note intense parallel S3 foliation and that some small patch leucosomes with orthopyroxene cores post-date S3. (**d**) Margin of an S3 shear zone showing screens of shear zone between leucosome veins oriented parallel to S3, note crenulated S1/2 inside the screen. Scale is 15 cm long.

to S3, and then along S1/2 planes (Fig. 6b) to accumulate in the synformal fold cores. A small melt-filled shear zone connects the synform to the larger antiformal closure to its SE, and magma flow sense in that shear is from the synform to the antiform. In the antiform a melanocratic layer is disrupted by the injection of granitic magma. A thick competent unit lies above the antiformal closure and leucosome has collected into a lens-shaped body beneath it. Melt has penetrated through the layer along a fracture to accumulate (forming a leucosome) in the F3 hinge above.

Sub-vertical dyke-like leucosomes oriented subparallel to S3 occur in some F3 fold hinge domains. Emplacement of these dykes post-dates F3, as some have enclaves containing F4 folds. Early phases of the dykes are plagioclase-rich and contain schlieren and wall-rock fragments. They are cut by later pink, K-feldspar-rich dykes containing fewer schlieren (e.g. right of fold B, Fig. 3). When these dykes occur in F3 fold closures containing thick, competent layers (e.g. fold A on Fig. 3), their geometry is more complex as many offshoots exist along the lithological layering, S1/2, and various fractures. Detailed cross-cutting relationships, and changes in texture and composition within the leucosomes, indicates that different batches of magma passed through the fracture array.

Small patch migmatites locally overprint the S1/2 foliation. Most have no preferred orientation, but some patches are elongated parallel to either S3 or S4. Many of the patches have cores consisting of garnet or orthopyroxene (typically as symplectitic intergrowths with quartz) and mantles of leucosome (Fig. 5a). In many cases cordierite and orthopyroxene in the host are partly replaced by biotite close to the patch leucosomes. Thus, the patch leucosomes formed *in situ*.

The S3 shear domains

At the shear zone margins, the S1/2 foliation is progressively steepened and overprinted by S3 crenulation zones 1 to 2 cm wide and spaced 5 to 15 cm apart. The crenulations become more closely spaced towards the interior of the shear zones until S1/2 can no longer be discerned; the foliation is then S3. There is a fundamental mineralogical and textural change accompanying this transition from S1/2 to S3. In most gneisses, the S1/2 foliation is principally defined by biotite, but in the wider S3 shear zones biotite is modally less abundant. The foliation in the biotite-poor rocks is defined by elongated quartz and cordierite grains, and by the elongation of

orthopyroxene or cordierite aggregates. Most of these rocks show replacement of cordierite by fine-grained biotite fringes, and orthopyroxene by small laths of biotite intergrown with plagioclase or quartz. This alteration produced 1 or 2 modal % biotite, and is most extensive on grain boundaries perpendicular to the S3 foliation. The shear zone domains are characterized by more abundant melanocratic layers devoid of quartz or feldspar, and more boudins of the quartz-rich cordierite-, or orthopyroxene-bearing lithologies than the F3 hinge domains.

In general, the S3 domains have a very strong foliation and banding and contain few pre- or syn-D3 migmatite structures; most leucosomes they contain are late- or post-D3. Some late D3 pegmatite veins are disrupted into boudin trains (Fig. 6c) and feldspar porphyroclasts, the sense of boudin offset and orientation of the associated elongation lineations (SE plunge) gives a NE-side-down (i.e. normal) movement sense and indicates later extensional reworking of S3 planes.

Two types of leucosome post-date the S3 foliation in the shear zones: patch types and veins. There are two varieties of *in situ* patch leucosome. In one type, cores of symplectitic garnet + quartz or orthopyroxene + quartz are surrounded by leucosome rims like those in the F3 hinge domains. Most patches are equant, but some have the leucosome part elongated in S3 or along S4. The other patch type is nebulitic and residual material is dispersed throughout.

The second type of post-S3 leucosome belong to the same set of subvertical post-D3 dykes present in the F3 fold hinge domains, but they have a different internal structure. The dykes are up to several metres wide and oriented parallel or sub-parallel to S3 (Fig. 6d). Dykes at the edges of S3 shear zones have intruded along, or between, S3 crenulation planes, although locally there are contacts that cross S3 crenulations. The margins of these dykes contain biotite flakes and biotite rich-schlieren which preserve the original orientation of S1/2 and the superimposed S3 and S4 foliations. The dykes may have been intruded during post-D3 extension. Locally the dykes are so numerous that only thin screens of shear zone wall rocks remain (shear zone H, Fig. 3) between dykes. Such places closely resemble the sheeted dyke and screen structure reported from the margins of the Wuluma Granite (Collins *et al.* 1989, p. 72, Fig. 8).

Late leucosome veins

The last two generations of leucosomes are not abundant, they cut across all structures and domains, including F4 folds, and have characteristic shallow dips. The older set comprises narrow (<20 cm), but laterally continuous veins of granitic pegmatite and aplite occurring in shallowly dipping en echelon arrays. The orientation of these veins is consistent with top-to-the-E movement. The younger set consist of small leucosome-filled tension gash arrays that imply a top-to-the NW (i.e. extension) movement sense after D4. The two vein arrays may represent a conjugate set formed during extension.

Geochemistry

The processes occurring during anatexis can be recognized by their geochemical signatures, provided that components which partition into the melt are plotted against those that partition into the residuum, or early crystallized phases. The major residual phases in the Wuluma migmatites are orthopyroxene, garnet and cordierite. Although biotite melted incongruently, some remains in the residuum and must be included. These four phases are the principal reservoir for (FeO + MgO) in the migmatites. The TiO_2 contained in biotite remains in the residuum after biotite breakdown as rutile or ilmenite, thus, the residuum is more fully represented by (FeO + MgO + TiO_2), but including TiO_2 has negligible effect. It is less clear which components could represent the melt fraction. In many cases SiO_2 is adequate, but in this case quartz is a major residual phase in some rocks. Alternatives are K_2O and Na_2O as they are partitioned into the melt. Melt could be represented by the compositions obtained from melting experiments on bulk compositions similar to the palaeosome at a temperature corresponding to that of the anatexis. Figure 7 shows a melt field obtained from 875–925 °C partial melting experiments on pelites and greywackes by Vielzeuf & Holloway (1988), Conrad et al. (1988), Patiño Douce & Johnson (1991), Beard et al. (1993) and Patiño Douce & Harris (1998).

The samples taken from the S3 shear zones have systematically higher (FeO + MgO), (also TiO_2, CaO, Cr, Sc and V), but systematically lower, SiO_2, K_2O and Na_2O (also Ba and Rb) contents (Table 1 and Fig. 7) than samples from the adjacent F3 fold hinges. Furthermore, the field of samples from the S3 shear zones lies farthest from the melt field, supporting the petrographic observation that the shear zone samples are more residual.

The samples from the F3 fold hinge domains belong to two groups. One includes those rocks

without leucosomes, and this field partially overlaps the samples from the S3 shear zone domains. The other group lies close to the experimental melt and leucosome fields, and came from fold closures where there was extensive injection of melt into the fold core. Samples EA34 and EA27 consist of disaggregated gneiss in a felsic matrix. Sample EA36 has high (FeO + MgO) and K_2O contents, which suggests that the host was first depleted by melt loss and subsequently injected

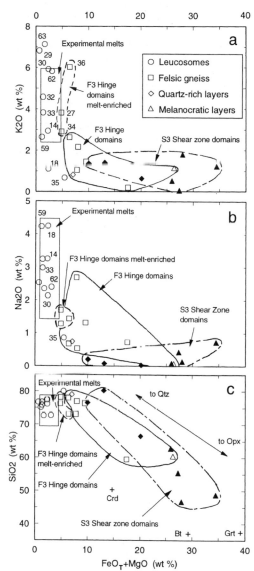

Fig. 7. Composition of the migmatites; (**a**) K_2O v. (FeO_T + MgO), (**b**) Na_2O v. (FeO_T + MgO), and (**c**) SiO_2 v. (FeO_T + MgO). All iron is treated as FeO and reported as FeO_T. Filled symbols represent samples from S3 shear zones.

Table 1. *Representative analyses of Wuluma migmatites*

(A)

Sample	Leucosomes										Felsic gneisses and melt		
	EA29	EA32	EA35	EA59	EA14	EA18	EA62	EA63	EA30	EA33	EA36	EA27	EA34
SiO_2 (wt%)	76.24	75.65	78.70	75.03	76.63	77.52	71.32	72.04	75.15	76.10	72.70	75.08	77.11
TiO_2	0.06	0.06	0.11	0.10	0.11	0.10	0.18	0.20	0.06	0.11	0.65	0.29	0.40
Al_2O_3	12.72	13.68	11.94	14.91	12.28	12.80	14.14	14.53	12.88	13.23	12.08	12.44	11.53
Fe_2O_3	0.58	1.08	3.03	0.62	1.97	2.01	2.44	1.45	1.64	0.92	4.80	3.30	2.98
MnO	0.01	0.01	0.05	0.01	0.06	0.05	0.05	0.01	0.02	0.01	0.04	0.03	0.03
MgO	0.14	0.36	2.63	0.57	0.45	0.56	0.80	0.34	0.85	0.50	2.12	1.80	2.15
CaO	0.31	1.21	0.65	1.81	1.41	1.88	1.25	1.31	0.27	1.55	0.12	0.43	0.64
Na_2O	2.51	3.23	0.84	4.22	3.22	4.25	2.35	2.33	2.11	2.95	1.43	1.67	1.26
K_2O	6.78	4.58	0.67	2.65	2.91	1.11	5.71	7.08	5.84	3.80	5.99	3.76	2.86
P_2O_5	0.07	0.07	0.04	0.06	0.09	0.06	0.09	0.08	0.08	0.07	0.06	0.06	0.05
LOI	0.23	0.14	0.46	0.34	0.22	0.15	0.39	0.47	0.28	0.30	0.22	0.32	0.42
Total	99.65	100.07	99.12	100.32	99.35	100.49	98.72	99.84	99.18	99.54	100.24	99.18	99.43
Rb (ppm)	111	61	11	22	36	9	133	149	91	54	121	83	55
Sr	62	64	17	124	85	71	119	146	58	78	40	42	33
Mg no.	32.34	39.76	63.22	64.55	31.15	35.56	39.37	31.71	50.65	51.84	46.66	51.93	58.83

(B)

Sample	Felsic gneisses					Quartz-rich layers					Melanocratic layers			
	EA22	EA16	EA15	EA19	EA42	EA24*	EA17	EA54*	EA44*	EA46*	EA37*	EA49*	EA61*	EA60
SiO_2 (wt%)	76.82	76.58	71.86	58.60	75.43	75.31	78.75	65.88	78.23	46.53	54.90	61.77	48.38	59.98
TiO_2	0.31	0.60	0.80	1.75	0.22	0.23	0.33	0.49	1.06	1.01	0.11	0.34	0.67	0.79
Al_2O_3	11.71	10.44	12.43	13.29	11.64	11.26	11.75	11.25	3.96	21.84	7.70	9.51	13.29	10.89
Fe_2O_3	3.67	6.42	6.68	14.45	5.06	3.85	3.83	16.48	8.43	15.80	17.55	18.52	19.94	14.90
MnO	0.03	0.03	0.18	0.19	0.11	0.08	0.03	0.40	0.05	0.10	0.71	0.26	0.27	0.14
MgO	3.02	3.82	1.70	4.27	3.31	6.40	3.56	5.14	5.20	13.62	11.32	8.85	16.24	12.77
CaO	0.08	0.32	1.74	6.47	0.14	0.01	0.22	0.40	0.03	0.16	8.00	0.28	0.41	0.04
Na_2O	0.66	1.32	2.64	0.71	0.52	0.20	0.74	0.03	0.08	0.14	0.42	0.09	0.72	0.08
K_2O	2.59	1.46	1.03	0.19	2.12	1.35	0.82	0.63	1.37	1.79	0.05	0.52	1.21	1.14
P_2O_5	0.04	0.06	0.08	0.14	0.04	0.03	0.05	0.05	0.04	0.13	0.07	0.03	0.03	0.03
LOI	0.59	0.33	0.01	<0.01	0.90	0.58	0.63	0.01	0.12	0.55	<0.01	<0.01	<0.01	0.03
Total	99.52	101.38	99.15	100.06	99.49	99.30	100.71	100.76	98.57	101.67	100.83	100.17	101.16	100.79
Rb (ppm)	61	34	26	18	54	62	34	25	85	138	2	26	62	60
Sr	24	12	62	109	19	2	15	2	2	4	11	4	15	2
Mg no.	61.97	54.09	33.51	36.92	56.44	76.70	64.80	38.18	54.99	63.06	56.09	48.62	61.73	62.93

All samples crushed in aluminium oxide ceramic mill. All major oxides (except Na_2O) and Sr were determined by X-ray fluorescence analysis at McGill University, Na_2O and Rb by instrumental neutron activation analysis at Université du Québec à Chicoutimi.
*Samples from S3 shear zone domains.

by a fractionated melt along its S1/2 foliation. These samples could be described as diatexitic.

The mineral assemblages indicate that all the sampled gneisses, quartz-rich and melanocratic rocks have melted and are residual to various degrees. Therefore, it is not possible to determine the composition of the palaeosome for the Ingula migmatites.

Some leucosomes are granitic, but others are granodioritic or even trondhjemitic (Fig. 7). Several studies (Barbey *et al.* 1996; Patiño Douce & Beard 1996; Patiño Douce & Harris 1998) have shown that trondhjemitic melts can form during the early stages of melting, when water is present to flux melting, or pressures are high, and plagioclase rather than muscovite breaks down. The high modal corderite and orthopyroxene, but low biotite, contents of the Wuluma migmatites, indicates that biotite dehydration melting, with (Al-rich felsic gneisses) or without (Al-poor felsic gneisses) aluminosilicate, reached a very advanced stage. Melt compositions obtained from biotite dehydration melting are potassic (e.g. Vielzeuf & Holloway 1988; Patiño Douce & Beard 1996). Therefore, it is unlikely that the plagioclase-rich leucosomes in the Wuluma migmatites represent primary partial melt compositions. The leucosome group shows a negative correlation between K_2O and Na_2O and a progressive increase in Rb/Sr ratio with K_2O (Fig. 8). Thus, the principal cause of compositional variation in the leucosome group is fractional crystallization of feldspar. Samples with low K_2O and Rb/Sr, but high Na_2O (samples EA14, EA18 and EA59) are of granodioritic and tonalitic/trondhjemitic composition and represent accumulations of early-crystallized plagioclase. In contrast, samples with low Na_2O, but high K_2O and Rb/Sr ratios (samples EA29, EA30; EA62 and EA63) represent fractionated granitic melts with more potassic compositions. Two granitic samples (EA32 and EA33) which are intermediate between these groups may be the least modified melt compositions. The small lensoid leucosome from an F3 fold axial plane (samples EA59) appear to be early crystallized material, whereas the large veins from F3 axial planes (EA29 and EA30) appear to be fractionated melts. The earlier phases (samples EA14 and EA18) of the subvertical dykes are rich in early crystallized plagioclase and later phases (EA62 and EA63) very fractionated.

Discussion

Partial melting is inferred to have occurred principally during the D3 transpressive deformation, since residual cordierite and idioblastic

orthopyroxene are parallel to S3. Minor melting continued during D4 deformation, and formed small *in situ* patch migmatites. Melting was, therefore, syntectonic, and melt segregation probably driven by applied tectonic stresses (e.g. McLellan 1988; Sawyer 1991, 1994; Stevenson 1989; Vigneresse *et al.* 1991). In contrast, the timing of the last stages of melt transport and accumulation is post-D3 to post-D4. This discrepancy in timing between melting and final melt accumulation arises because melt is mobile relative to the residuum (Sawyer 1994, 1996), and so can move in response to changes in deviatoric stress and pressure gradients, provided that temperatures remain above the solidus. According to thermal modelling (e.g. England & Thompson 1984) and thermo-chronological studies (e.g. Anovitz & Chase 1990; Brown & Dalmeyer 1996) rocks which reached >850 °C may take longer than 20 Ma to cool through the solidus.

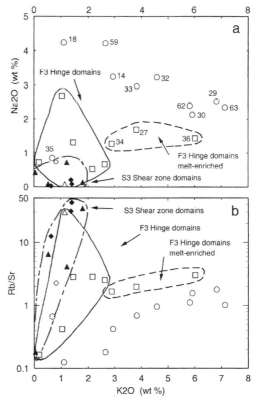

Fig. 8. Composition of the migmatites: (**a**) Na_2O vs K_2O, and (**b**) Rb/Sr vs K_2O for the migmatites. Note negative correlation between Na_2O and K_2O and positive correlation between Rb/Sr and K_2O for the leucosomes. Symbols as for Fig. 7.

SHRIMP U–Pb ages obtained from zircons taken from leucosomes in the section shown on Fig. 3 are: 1739 ± 10 Ma for the oldest syn-D3 leucosomes, 1728 ± 6 Ma for the most voluminous phase of D4 leucosomes, and 1712 ± 13 Ma for the late-stage, subhorizontal pegmatitic leucosomes. These ages are broadly coeval with the Wulma granite (1728 ± 3 Ma, Lafrance *et al.* 1995) and indicate that granulite facies melting and leucosome formation extended over a period of about 27 Ma, with the greatest volume of leucosomes formed between 1739 and 1728 Ma. Considering the errors, leucosome formation could have taken as little as 4 Ma, or as much as 50 Ma.

Melt could, therefore, be present and mobile for a considerable period of time in anatectic rocks if temperatures were high. This protracted history is recorded in the complexity displayed by leucosomes.

The Wuluma migmatites (excepting the leucosomes) are petrographically and geochemically residual; the greatest melt loss was from S3 shear zones. If the distribution of the S3 shear zones and the F3 fold hinges was due to the orientation of S1/2, as suggested by Lafrance *et al.* (1995), and not to a heterogeneous distribution of fertile rock types, then, shearing promoted either, more partial melting (due to variation in P, T or a_{H_2O}), or more complete melt extraction, in the shear zones. Plastic deformation of the solid matrix during contractional shearing may have led to a much lower volume of intergranular melt (i.e. smaller porosity) compared to the lower strain F3 hinge domains. High strain has destroyed most of the migmatite textures in the S3 shear zones; consequently it is not possible to determine whether melt extraction and movement was by the mechanisms of porous flow or magma/bulk flow (Sawyer 1996), also called melt-assisted granular flow (Rutter 1997). However, the preserved later (post-D3) migmatite structures indicate that melt transport was in veins formed parallel to S3 during D4 extension. Magma, or bulk, flow did not occur since there are no diatexites. This suggests that, although the rocks probably experienced 30% to 60% partial melting, based on the modal abundance of cordierite and orthopyroxene and the experimental results (Vielzeuf & Holloway 1988; Patiño Douce & Johnston 1991), the melt fraction in them at any one time was small (i.e. they were continually drained). During cooling of the S3 shear zones, a small melt fraction remained in the residual rocks and was preferentially located along grain boundaries perpendicular to S3, were it reacted with residual cordierite, orthopyroxene and garnet to form the late biotite.

Complex migmatite structures are preserved in F3 hinge domains. Most massive layers are thin (Fig. 5b), and melt was probably extracted via an interconnected networks of grain-edge channelways, since there are no leucosomes (or fractures) forming branching melt escape arrays of the type described by Weinberg (1998). However, such arrays existed in more anisotropic rocks, such as the gneisses. In these rocks, melt drained from the interconnected grain boundary networks into closely spaced arrays of centimetre-scale, lensoid channels generally oriented parallel to F3 (e.g. Fig. 5c). These are melt escape paths, since the leucosomes in them are accumulations of early crystallized phases and not melt. At this stage there was no significant residuum component preserved in the leucosomes, and none resemble diatexites. Melt flow was driven by stresses associated with progressive tightening of the F3 folds. Magmatic lineations are rare, but indicate that melt flow was parallel to fold hinges, as also noted by Collins & Sawyer (1996) elsewhere in the Arunta Inlier. The melt accumulated in whatever low pressure sites were available nearby, and began to crystallize there (i.e. evolve compositionally). Larger volumes of granitic magma form when several sources drained into the same site, commonly a fold hinge. In the Wuluma Hills, both synformal and antiformal fold closures accumulated melt, whereas Allibone & Norris (1992) report that in the Taylor Valley, melt accumulated preferentially in the antiforms. The morphology of these leucosomes is conspicuously different from those related to initial melt segregation; some are irregular, diffuse patches in hinges, others are elongated parallel to the anisotropy; typically they contain schlieren and resemble diatexites (Fig. 6b). Melt stayed in these sites only if they remained stable, if the sites were subject to deviatoric stresses their internal magma pressure rose, and the evolved melt in them was forced to move to a new, and structurally higher, lower pressure site. This stage of melt movement was in wide subvertical dykes along S3, F3 axial planes or along small shears. Magmatic erosion of the wall rocks gave rise to trains of schlieren and formation of flow structures, which resemble diatexites (Fig. 6a), at the stagnant dyke margins. Such tectonically driven melt movement occurred repeatedly during progressive deformation until the melt froze (i.e. the temperature declined below the solidus), or reached a site of long-term stability.

The anisotropy and competency contrast structure of the host changed with time. During progressive tightening, fold closures may become more competent as they lose melt, and eventually switch from being melt-loss to potential melt-accumulation sites (cf. Robin 1979) on the convex side of folded layers, or in fractures (e.g. Fig. 3). This is well illustrated by volumes of previously depleted rocks that were later injected by fractionated (i.e. K_2O-rich high Rb/Sr) melts.

When an accumulation of melt is subject to applied stresses the melt will become overpressured and dilate the pathway requiring the least expenditure of energy so as to migrate to lower pressure site. Where a strong planar anisotropy exists (i.e. lithological layering or foliation planes) this controls the orientation of the melt pathways, and hence the leucosomes (see Wickham 1987; Brown & Solar 1998b). Most leucosomes are, therefore, veins whose orientation is determined by the anisotropy plane that happens to be the most easily exploited within the local stress field. Simple subvertical leucosome veins predominate in the high strain S3 shear zone domains because there is only one uniform and very penetrative anisotropy (S3). In marked contrast, the low strain F3 domains offer several planes of strong anisotropy, hence leucosomes have complex geometries. Commonly, a migrating magma batch exploited both S1/2 and S3 foliations, the intersection of which, of course, parallels the fold hinges. Thus, although the Wuluma migmatites are morphologically diverse, most, irrespective of structural domain, are stromatic migmatites as defined by Brown & Rushmer (1997).

Both the S3 shear zone and F3 hinge domains are depleted and so must have contributed melt to nearby sinks, such as the Wuluma Pluton, or the smaller body on Fig. 2. Magma may not have transferred laterally from the S3 shear zone domains to the F3 hinge domains during contractional D3 deformation, but may have during subsequent extension, since the subvertical S3 planes are a viable path to higher crustal levels. However, as S3 shear zones grow, they may aid in evacuating the F3 hinge domains when penetrative S3 crenulation zones develop and establish a more direct pathway upwards. Since the S3 shear zones are more depleted and contain fewer leucosomes, they are more than just conduits for magma transport. They are also a significant source of granitic magma. When D4 extension occurred, they also became sites of melt accumulation.

Conclusions

In high-temperature anatectic regions, such as the granulite facies Wuluma migmatites, melt is present for long periods of time. Tectonic stresses during this time cause the melt to move from site to site in response to transient changes in pressure gradients resulting from the anisotropic nature of the rocks. The path and sequence of melt movement is preserved as multiple generations of cross-cutting leucosomes and by systematic compositional trends in the leucosomes. The geometry of these melt escape pathways (leucosome veins) is controlled by host rock anisotropy, notably foliations and lithological layering, and so varies between structural domains.

The general absence of diatexite migmatites, and the preservation of pre-migmatization fabrics, suggests that, during biotite dehydration melting, the melt fraction formed sufficiently slowly that it could be continually drained from the solid matrix during deformation, and collected in primary accumulation sites. Melt flow to secondary accumulation sites, caused by subsequent increments of deformation, is strongly chanellized along fewer and wider veins and can lead to wall rock erosion and the local development of diatexite-textured migmatites.

We thank A. Gorrie of Yambah Station for permission to work in the Wuluma Hills. This work has benifited greatly from many discussions in the field, with B. Lafrance, R. Vernon and P. Williams in the Arunta, and with S. Barbosa, G. Bergantz, M. Brown, I. Milord, G. Solar, A. Thompson and R. Weinberg in various other migmatite terranes. We are grateful to A. Castro for the invitation to participate in this volume, and to P. Barbey and M. Brown for very constructive reviews. The work was funded by a Natural Sciences and Engineering Research Council of Canada Collaborative Projects Grant 183274 to Sawyer and Australian Research Council grant A39230976 to R. H. Vernon, R. H. Flood and W. J. Collins.

References

ALLIBONE, A. H. & NORRIS, R. J. 1992. Segregation of leucogranite microplutons during syn-anatectic deformation: an example from the Taylor Valley, Antarctica. *Journal of Metamorphic Geology*, **10**, 589–600.

ANOVITZ, L. M. & CHASE, C. G. 1990. Implications of post-thrusting extension and underplating for P-T-t paths in granulite terranes: a Grenville example. *Geology*, **18**, 466–469.

BARBEY, P., BROUAND, M., LÊ FORT, P. & PECHER, A. 1996. Granite-migmatite genetic link: the example of the Manaslu granite and Tibetan Slab migmatites in central Nepal. *Lithos*, **38**, 63–79.

BARR, D. 1985. Migmatites in the Moines. *In*: ASHWORTH, J. R. (ed.) *Migmatites*. Blackie, Glasgow, 225–264.

BEARD, J. S., ABITZ, R. J. & LOFGREN, G. E. 1993. Experimental melting of crustal xenoliths from Kilbourne Hole, New Mexico and implications for the contamination and genesis of magmas. *Contributions to Mineralogy and Petrology*, **115**, 88–102.

BLACK, L. P. & SHAW, R. D. 1995. An assessment, based on U-Pb zircon data, of Rb-Sr dating in the Arunta Inlier, central Australia. *Precambrian Research*, **71**, 3–15.

BROWN, M. & DALLMEYER, R. D. 1996. Rapid Variscan exhumation and the role of magma in core complex formation: southern Brittany metamorphic belt, France. *Journal of Metamorphic Geology*, **14**, 362–380.

—— & RUSHMER, T. 1997. The role of deformation in the movement of granitic melt: views from the laboratory and the field. *In*: HOLNESS, M. B. (ed.) *Deformation-enhanced fluid transport*. Mineralogical Society Book Series, **8**. Chapman & Hall, London, 111–144.

—— & SOLAR, G. S. 1998a. Shear-zone systems and melts: feedback relations and self-organisation in orogenic belts. *Journal of Structural Geology*, **20**, 211–227.

—— & —— 1998b. Granite ascent and emplacement during contractional deformation in convergent orogens. *Journal of Structural Geology*, **20**, 1365–1393.

CLEMENS, J. C. & MAWER, C. K. 1992. Granitic magma transport by fracture propagation. *Tectonophysics*, **204**, 339–360.

COLLINS, W. J. & SAWYER, E. W. 1996. Pervasive granitoid magma transfer through the lower-middle crust during non-coaxial compressional deformation. *Journal of Metamorphic Geology*, **14**, 565–579.

—— & SHAW, R. D. 1995. Geochronological constraints on orogenic events in the Arunta Inlier; a review. *Precambrian Research*, **71**, 315–346.

—— & TEYSSIER, C. 1989. Crustal-scale ductile fault systems in the Arunta Inlier, central Australia. *Tectonophysics*, **158**, 49–66.

——, FLOOD, R. H., VERNON, R. H. & SHAW, S. E. 1989. The Wuluma granite, Arunta Block, central Australia: an example of in situ, near-isochemical granite formation in a granulite facies terrane. *Lithos*, **23**, 63–83.

CONRAD, W. K., NICHOLLS, I. A. & WALL, V. J. 1988. Water-saturated and -undersaturated melting of metaluminous and peraluminous crustal compositions at 10 kb: evidence for the origin of silicic magmas in the Taupo Volcanic Zone, New Zealand and other occurrences. *Journal of Petrology*, **29**, 765–803.

CUNEY, M. & BARBEY, P. 1982. Mise en evidence de phenomenes des cristallisation fractionee dans les migmatites. *Comptes Rendus de l'Academie des Sciences, Paris*, **295**, 37–42.

D'LEMOS, R. S., BROWN, M. & STRACHAN, R. A. 1992. Granite magma generation, ascent and emplacement within a transpressional orogen. *Journal of the Geological Society, London*, **149**, 487–490.

ENGLAND, P. C. & THOMPSON, A. B. 1984. Pressure-temperature-time paths of regional metamorphism I. Heat transfer during the evolution of

regions of thickened continental crust. *Journal of Petrology*, **25**, 894–928.

FROST, B. R. & CHACKO, T. 1989. The granulite uncertainty principle: limitations on thermobarometry in granulites. *Journal of Geology*, **97**, 435–450.

GREENFIELD, J. E., CLARKE, G. L. & WHITE, R. W. 1996. A sequence of partial melting reactions at Mt. Stafford, central Australia. *Journal of Metamorphic Geology*, **16**, 363–378.

HAND, M. & DIRKS, P. H. G. M. 1992. The influence of deformation on the formation of axial-planar leucosomes and the segregation of small melt bodies within the migmatitic Napperby Gneiss, Central Australia. *Journal of Structural Geology*, **14**, 591–604.

LAFRANCE, B., CLARKE, G. L., COLLINS, W. J. & WILLIAMS, I. S. 1995. The emplacement of the Wuluma granite: melt generation and migration along steeply dipping extensional fractures at the close of the Late Strangways orogenic event, Arunta Block, central Australia. *Precambrian Research*, **72**, 43–67.

McLELLAN, E. L. 1988. Migmatite structures in the Central Gneiss Complex, Boca de Quadra. *Journal of Metamorphic Geology*, **6**, 517–543.

PATIÑO DOUCE, A. E. & BEARD, J. S. 1996. Effects of P, $f(O_2)$ and Mg/Fe ratio on dehydration melting of model metagraywackes. *Journal of Petrology*, **37**, 999–1024.

—— & HARRIS, N. 1998. Experimental constraints on Himalayan anatexis. *Journal of Petrology*, **39**, 689–710.

—— & JOHNSTON, A. D. 1991. Phase equilibria and melt productivity in the pelitic system: implications for the origin of peraluminous granitoids and aluminous granulites. *Contributions to Mineralogy and Petrology*, **107**, 202–218.

PETFORD, N. 1996. Dykes or diapirs? *Transactions of the Royal Society of Edinburgh: Earth Sciences*, **87**, 105–114.

——, KERR, R. C. & LISTER, J. R. 1993. Dike transport of granitoid magmas. *Geology*, **21**, 845–848.

ROBIN, P.-Y. F. 1979. Theory of metamorphic segregation and related processes. *Geochimica et Cosmochimica Acta*, **43**, 1587–1600.

RUTTER, E. H. 1997. The influence of deformation on the extraction of melts: A consideration of the role of melt-assisted granular flow. *In*: HOLNESS, M. B. (ed.) *Deformation-enhanced Melt Segregation and Metamorphic Fluid Transport*. Mineralogical Society Book Series, **8**. Chapman and Hall, London, 82–110.

SAWYER, E. W. 1987. The role of partial melting and fractional crystallization in determining discordant migmatite leucosome compositions. *Journal of Petrology*, **28**, 445–473.

—— 1991. Disequilibrium melting and the rate of melt-residuum separation during migmatisation of mafic rocks from the Grenville Front, Quebec. *Journal of Petrology*, **32**, 701–738.

—— 1994. Melt segregation in the continental crust. *Geology*, **22**, 1019–1022.

—— 1996. Melt segregation and magma flow in migmatites: implications for the generation of granite magmas. *Transactions of the Royal Society of Edinburgh: Earth Sciences*, **87**, 85–94.

—— 1998. Formation and evolution of granite magmas during crustal reworking: the significance of diatexites. *Journal of Petrology*, **39**, 1147–1167.

SHAW, R. D., LANGWORTHY, A. P., OFFE, L. A., STEWART, A. J., ALLEN, A. R. & SENIOR, B. R. 1979. *Geological report on 1:100,000-scale mapping of the southeastern Arunta Block, Northern Territory*. Bureau of Mineral Resources Australia, record 1979/47, BMR Microform MF, **133**.

——, STEWART, A. J. & BLACK, L. P. 1984. The Arunta Inlier: a complex ensialic mobile belt in central Australia. Part 2: Tectonic history. *Australian Journal of Earth Sciences*, **31**, 457–484.

STEVENSON, D. J. 1989. Spontaneous small scale melt segregation in partial melts undergoing deformation. *Geophysical Research Letters*, **66**, 1067–1070.

STEWART, A. J., SHAW, R. D. & BLACK, L. P. 1984. The Arunta Inlier: a complex ensialic mobile belt in central Australia. Part 1: stratigraphy, correlations and origin. *Australian Journal of Earth Sciences*, **31**, 445–455.

VIELZEUF, D. & HOLLOWAY, J. R. 1988. Experimental determination of the fluid-absent melting relations in the peltic system. *Contributions to Mineralogy and Petrology*, **98**, 257–276.

—— & MONTEL, J. M. 1994. Partial melting of metagreywackes. Part I. Fluid-absent experiments and phase relationships. *Contributions to Mineralogy and Petrology*, **117**, 375–393.

VIGNERESSE, J. L., CUNEY, M. & BARBEY, P. 1991. Deformation assisted crustal melt segregation and transfer. *In: Geological Association of Canada/ Mineralogical Association of Canada Abstracts volume*, **16**, A128.

WEINBERG, R. F. 1996. Ascent mechanism of felsic magmas: news and views. *Transactions of the Royal Society of Edinburgh: Earth Sciences*, **87**, 95–103.

—— 1998. Mesoscale pervasive felsic magma migration: alternatives to dyking. *Lithos*, **46**, 393–410.

WICKHAM, S. M. 1987. The segregation and emplacement of granitic magma. *Journal of the Geological Society, London*, **144**, 281–297.

WINDRIM, D. P. & MCCULLOCH, M. T. 1986. Nd and Sr isotopic systematics of central Australian granulites: chronology of crustal development and constraints on the evolution of the lower crust. *Contributions to Mineralogy and Petrology*, **94**, 289–303.

ZHAO, J.-X. & BENNETT, V. 1995. SHRIMP U-Pb zircon geochronology of granites in the Arunta Inlier, central Australia: implications for Proterozoic crustal evolution. *Precambrian Research*, **71**, 17–43.

—— & MCCULLOCH, M. T. 1995. Geochemical and Nd isotopic systematics of granites from the Arunta Inlier, central Australia: implications for Proterozoic crustal evolution. *Precambrian Research*, **71**, 265–299.

Partial melting and *P–T–t* evolution of LP/HT metamorphic terranes: an example from the Svecofennian K-feldspar–poor leucosome migmatite belt, Southern Finland

H. MOURI* & K. KORSMAN

Geological Survey of Finland, PO Box 96, FIN-02151 Espoo, Finland

*Present address: Department of Geology and Geophysics, University of Minnesota, Mineapolis, MN55455, USA (e-mail: mouri001@tc.umn.edu)

Abstract: The migmatites with Kfs-poor leucosomes from the Svecofennian domain of southern Finland are characterized by a suite of anastomosing leucosome veins and patches characterized by low and varying abundance of K-feldspar with or without cordierite. These leucosomes are generated by *in situ* melting reactions at different $P–T–a$H$_2$O conditions. Microtextural analysis in conjunction with THERMOCALC calculations and geothermobarometry show that these rocks were metamorphosed under granulite facies conditions at 700–750 °C, 4–5 kbar and aH$_2$O = 0.4–0.7. The formation of cordierite coronas around garnet and the late crystallization of andalusite suggest that the final stage of the $P–T$ history was characterized by decompression and cooling to within the andalusite stability field, estimated at 500–650 °C and 3–4 kbar. U–Pb and Sm–Nd conventional analyses of monazite and garnet, respectively, from different parts of these migmatites (mesosomes and leucosomes) indicate that, within error limits, all leucosomes were formed at about 1878 Ma during a single tectonometamorphic event. Since there is no evidence of an earlier high-grade metamorphic event in this area, it is assumed that this is the approximate age of peak metamorphism and partial melting.

Partial melting processes, formation of migmatites and generation of large volumes of granitoids in high-grade metamorphic terranes and their role in crustal evolution are the subjects that continuously capture the attention of a broad spectrum of earth scientists. These subjects include the following.

(1) The genetic link between migmatites and granitoids at a regional scale. This has been an important subject of discussion since the work of Sederholm (1923). Based on experimental data (e.g. Le Breton & Thompson 1988; Vielzeuf & Holloway 1988), *in situ* melting can generate up to 40% of melt within the temperature range 650–800 °C in pelitic, quartzo-feldspathic and mafic rocks. Melting of crustal material is considered to generate most of the peraluminous granitoids (e.g. Barbarin 1999). According to Sawyer (1995) granitic plutons may be considered as an extension of the melting process giving rise to large volumes of diatexitic migmatites. However, according to Ashworth (1985), the hypothesis of *in situ* melting is valid only if the proportion of mesosome dominates. Leucosome-dominated rocks seem to be essentially the locus of pervasive granitic injection.

(2) The role of fluid composition in the formation of migmatites and the effect of water activity on the chemical composition of leucosomes is another matter of debate. According to the recent experimental data of Patiño Douce & Harris (1998) trondhjemitic leucosomes instead of granitic leucosomes are formed by melting of metapelitic rocks under water saturated conditions at $P \approx 6$ kbar and $T \leqslant 750$ °C.

(3) Another subject is the origin of leucosomes in layered and stromatic migmatites at a small scale. Two hypotheses have been proposed (e.g. Sawyer 1991): (i) anatectic origin related to dehydration melting reactions during high temperature metamorphism and (ii) tectonic origin, or subsolidus reaction related to physical processes such as deformation.

Migmatites have been considered for many years as good candidates for the study of restite-magma separation at a small scale (*in situ* melting), even if the melt product is frozen locally to form layered and small-scale leucosome veins rather than being efficiently extracted to become a granite pluton.

In this paper we show an example of *in situ* partial melting under low-pressure/

From: CASTRO, A., FERNÁNDEZ, C. & VIGNERESSE, J. L. (eds) *Understanding Granites: Integrating New and Classical Techniques.* Geological Society, London, Special Publications, **168**, 239–253. 1-86239-058-4/99/$15.00 © The Geological Society of London 1999.

high-temperature and relatively high aH_2O conditions giving rise to large outcrops of migmatites with less than 50% of leucosomes of trondhjemitic–tonalitic compositions (Kfs-poor leucosomes) from the Southern Finland Svecofennian domain. The study aims at establishing the P–T evolution of these migmatitic rocks, the processes of melting, the relationship between the different leucosome generations observed on the field and the time scale of partial melting using U–Pb and Sm–Nd conventional methods on monazite and garnet, respectively.

Geological setting

Details on the geology of the Svecofennian domain has been published in more detail by

Korsman *et al.* (1999). They are only summarized in this paper.

The Kfs-poor leucosome migmatite belt, so-called Trondhjemite–Tonalite Migmatite Belt, belongs to the Central Finland Island Arc Complex (Fig. 1). This belt occurs as a narrow zone to the south and west of the Granitoid Complex of Central Finland, and is bounded in the south by the Granite Migmatite belt (Fig. 1). According to Korsman *et al.* (1999) this belt was caused by collision of the Svecofennian domain with the Archaean continent. The belt was intruded by syn- to post-orogenic granitoids interpreted as the product of interaction and mixing of mantle- and crustal-derived partial melts (Lahtinen & Huhma 1996). In addition, rare MORB-type ultramafic volcanic rocks,

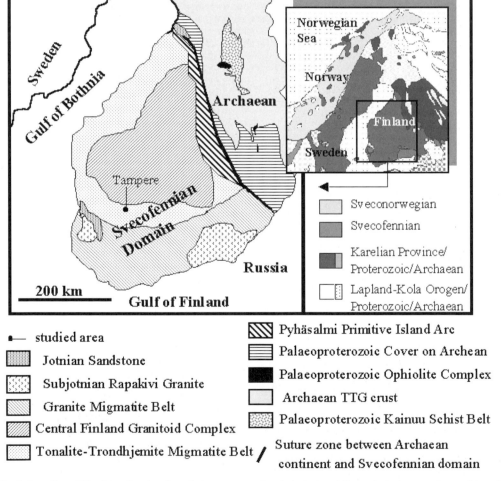

Fig. 1. Location of the Svecofennian domain in a more regional context and the main tectono-metamorphic units of this domain in southern Finland, after Korsman *et al.* (1999).

probably older than 1.91 Ga, and mafic and ultramafic cumulates dated at 1.89 Ma (Peltonen 1995) are found in some places within the belt. An attempt of estimation of the age of metamorphism in the studied migmatite belt has been made by the study of the relationship between deformation, magmatism and metamorphism (Korsman *et al.* 1999) and considered to be at about 1885 Ma. However, no direct dating and detailed study of the metamorphic events and partial melting processes have been previously undertaken in this area.

Field relationships and petrology

In this paper we use the nomenclature proposed for migmatites by Mehnert (1968) extended later by Johannes & Gupta (1982), the mineral abbreviations used are from Kretz (1983). The migmatites studied here contain less than 50% leucosomes and exhibit a stromatic banded structure in which it is possible to distinguish dark mesosomes with layers, veins and patches of leucosomes (Fig. 2a–c). These migmatites are characterized by different generations of anastomosing leucosomes with a slight increase in the proportion of K-feldspar from Kfs-poor (L1) through mildly Kfs-enriched (L2) and (L3) to granitic composition (L-granite).

Petrography

The mesosomes are predominantly composed of bitotite, quartz (Table 1) together with plagioclase sillimanite, cordierite, garnet and rare K-feldspar and late andalusite. Biotite occurs

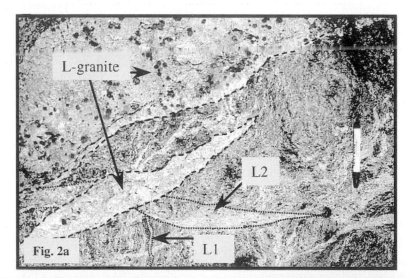

Fig. 2. (**a**) Photomicrograph showing field relationships between the different types of leucosomes (L1, L2, L-granite) and the mesosome. (**b**) Hand specimen of migmatite with light layered leucosome and dark biotite-rich mesosome with porphyroblasts of garnet. (**c**) Hand specimen of the discordant-leucosome vein L2 showing the fragments of dark mesosome.

Table 1. *Modal proportions of different phases in the mesosmoes and leucosomes of the studied migmatites*

	% Kfs	% Pl	% Bt	% Qtz	% Grt	% Sil
Mesosome	1–5	14–29	34–36	34–39	1–2	1–3
Layered leucosome (L1)	<1	40–45	4–6	44–46	<1	0
Leucosome vein (L2)	10–11	40–42	21–23	23–25	<1	<1
Crd leucosome (L3)	13–15	36–38	18–20	26–27	0	0
L-granite	42–44	23–25	2–4	26–28	0	<1

either with sillimanite in the main foliation plane or randomly distributed in the matrix in equilibrium with cordierite, sillimanite, plagioclase and quartz. Plagioclase is moderately abundant (14–29 vol.%). It has been observed as: (i) rare and small inclusions in garnet, (ii) in the matrix associated with cordierite, (iii) in some cordierite coronas and (iv) along the main foliation with sillimanite and biotite. Garnet occurs as rare porphyroblasts (about 3 mm diameter) and sometimes contains inclusions of quartz, sillimanite, biotite and plagioclase. Garnet is often isolated from sillimanite–biotite assemblages by well-developed coronas of cordierite in equilibrium with plagioclase. Cordierite occurs also in the matrix as large crystals containing inclusions of sillimanite, biotite, quartz and rare plagioclase. K-feldspar is rare and only observed in the matrix in textural equilibrium with plagioclase and quartz.

Different types of anastomosing leucosomes are identified on the basis of field observations (Fig. 2a–c).

Layered leucosomes (L1) are the most abundant, it occurs as thin layers (1–10 cm) parallel (Fig. 2a, b) to the dark mesosome characterized by very low abundance of K-feldspar (⩽1 vol.%). The main minerals in decreasing order of abundance (Table 1) are: quartz, polygonal plagioclase, biotite and rounded fragmented garnet observed mainly at the boundary between leucosomes and mesosomes.

Discordant leucosome veins (L2) occur as veins ranging in thickness from a few centimetres to less than 1 m, cutting across L1 and the mesosome (Fig. 2a). These leucosomes contain more K-feldspar (10 vol.%) than L1 together with plagioclase, quartz and biotite (Table 1). In some places, these leucosomes contain small rounded crystals of garnet (less than 1 mm in diameter) together with biotite, late muscovite and minor sillimanite occurring as small dark patches (Fig. 2c). These patches are probably fragments of mesosome, dragged mechanically by the leucosome.

Cordierite-bearing leucosome (L3) occurs as small patches overprinting the main foliation. It is characterized by a slightly higher Kfs content

(12–15 vol.%) than L1 and L2. This Kfs occurs as an interstitial phase together with cordierite, polygonal plagioclase and quartz.

Granitic leucosomes (L-granite) are rare and occur as wide bands (c. 50 cm) discordant with respect to the country-rock structures and cutting sharply across mesosomes and leucosomes (L1 and L2) (Fig. 2a). The texture is pegmatitic and comprises of large crystals of K-feldspar (44 vol.%) (Table 1), euhedral quartz, myrmekitic plagioclase and very rare prisms of biotite and sillimanite. All leucosomes show similar microstructures, characterized by euhedral crystals of plagioclase with crystal faces against quartz. In addition, the discordant leucosome vein (L2) is characterized by zoning of plagioclase. These features suggest that they crystallized from melt (e.g. Vernon & Collins 1988). However such an anatectic origin does not preclude melt segregation to be tectonically controlled in a way described by Sawyer & Barnes (1988).

Mineral chemistry

Representative analyses of the minerals from the mesosomes and the different types of leucosomes are listed in Table 2.

Biotites in mesosomes and leucosomes are compositionally heterogeneous (Table 2), X_{Fe} (Fe/Fe + Mg) ranges between 0.50 and 0.73 and TiO_2 between 0 and 4.16 wt%. Biotites in leucosomes are relatively richer in iron ($X_{Fe} = 0.60–0.73$) than biotites in the mesosomes ($X_{Fe} = 0.50–0.60$), whereas Al_2O_3 in all samples shows a similar range between 18 and 20 wt%. Small but significant differences in biotite composition are found in the mesosome and the discordant leucosome vein (L2). Biotite inclusions in garnet have slightly lower X_{Fe} (0.52–0.54) than biotite in the matrix ($X_{Fe} = 0.56–0.58$) and biotite in the main foliation plane ($X_{Fe} = 0.59–0.60$). Biotite from the discordant leucosome vein (L2) is also heterogeneous in composition: biotite in the small fragments (see Fig. 2c) has similar X_{Fe} (0.57–0.59) to biotite in the mesosome, whereas biotite in the matrix with plagioclase and quartz

Table 2. Representative chemical compositions of biotite, garnet and cordierite in the mesosome and the adjacent leucosomes

	Bt mesosomes		Bt-L1		Bt-L2		Grt mesosome		Grt leucosomes		Crd mesosomes		Crd-L3	
	Bt(f)	Bt(m1)	Bt(m2)	Bt(m2)	Bt/grt	Bt(m2)	Core	Rim	Core	Rim	Crd	Corona		
SiO$_2$	34.94	35.26	34.00	34.81	34.39	34.92	36.82	36.53	36.93	37.40	47.92	48.31	48.51	48.70
TiO$_2$	2.42	0.27	4.00	4.16	0.8	3.28	0.02	0.03	0.00	0.02	0.00	0.02	0.01	0.00
Al$_2$O$_3$	19.72	21.34	19.00	19.14	20.79	18.51	20.53	20.50	20.82	20.75	31.70	31.70	31.83	33.19
Cr$_2$O$_3$	0.1	0.02	0.00	0.16	0.06	0.01	0.00	0.06	0.01	0.01	0.00	0.00	0.00	0.00
MgO	7.94	8.56	6.00	6.52	8.43	7.38	4.04	2.27	3.64	2.88	6.04	6.58	6.89	7.61
Fe$_2$O$_3$	—	—	—	—	—	—	0.50	0.06	0.91	0.00	—	—	—	—
FeO	20.85	20.98	22.00	22.21	20.98	21.11	34.22	37.02	34.98	37.07	11.53	10.85	9.92	9.03
MnO	0.09	0.04	0.00	0.05	0.1	0.17	1.23	1.29	1.06	1.21	0.24	0.14	0.20	0.01
CaO	0.01	0	0.00	0.01	0.02	0.01	1.03	0.89	1.15	0.91	0.01	0.00	0.03	0.000
BaO	0	0.07	0.00	0.39	0	0	—	—	—	—	—	—	—	—
Na$_2$O	0.22	0.3	0.00	0.13	0.2	0.17	0.01	0.03	0.01	0.01	0.27	0.25	0.38	0.20
K$_2$O	9.44	9.07	9.00	9.35	9.05	9.81	0.00	0.01	0.03	0.01	0.01	0.15	0.00	0.00
Sum	95.73	95.91	94.00	96.93	94.82	95.37	98.40	98.69	99.54	100.27	97.72	98.00	97.78	98.75
X_{Fe}	0.596	0.579	0.66	0.656	0.583	0.616	0.825	0.902	0.844	0.878	0.51	0.48	0.44	0.40

Bt(f), biotite in the foliation with sillimanite; Bt(m1), biotite in the matrix with sillimanite; Bt(m2), biotite in the matrix with plagioclase and quartz; Bt/grt, biotite in the matrix with sillimanite + cordierite Bt(m2), biotite with garnet in the small fragments of mesosome within L2.

Table 3. *Average* P–T *calculation using* THERMOCALC *(Powell & Holland 1988) for the mesosome and the discordant leucosome vein (L2) assembalges*

aH_2O	Mesosome without andalusite (peak conditions)		Mesosome with andalusite (retrograde conditions)		Leucosome vein (L2) (peak conditions)	
	T (°C)	P (kbar)	T (°C)	P (kbar)	T (°C)	P (kbar)
1	799 ± 54	5.4 ± 0.5	High fit		794 ± 55	5.5 ± 1.5
0.9	784 ± 52	5.2 ± 0.5	High fit		770 ± 53	5.4 ± 1.5
0.8	771 ± 51	5.1 ± 0.5	High fit		767 ± 52	5.2 ± 1.5
0.7	756 ± 49	5.0 ± 0.5	High fit		754 ± 50	5.1 ± 1.5
0.6	743 ± 47	4.7 ± 0.5	631 ± 23	3.6 ± 0.6	741 ± 48	5.0 ± 1.5
0.5	726 ± 45	4.5 ± 0.4	631 ± 21	3.5 ± 0.6	725 ± 46	4.8 ± 1.5
0.4	706 ± 43	4.2 ± 0.4	629 ± 19	3.4 ± 0.4	705 ± 44	4.6 ± 1.4
0.3	679 ± 40	3.9 ± 0.4	626 ± 17	3.4 ± 0.4	679 ± 41	4.4 ± 1.4
0.2	642 ± 36	3.6 ± 0.4	619 ± 17	3.3 ± 0.4	644 ± 38	4.0 ± 1.4
0.1	584 ± 33	3.1 ± 0.4	604 ± 17	3.3 ± 0.4	587 ± 33	3.4 ± 1.4

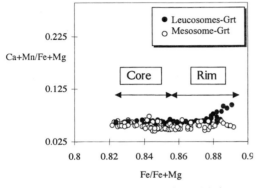

Fig. 3. Variation of X_{Fe} v. Ca + Mn/Fe + Mg in garnets from the mesosome and the discordant leucosome vein (L2).

has higher X_{Fe} (0.60–0.62). Garnet in both mesosome and leucosomes is almandine-rich with X_{Fe} ranging between 0.82 and 0.90 (Table 3). In both cases garnet is characterized by an unzoned core with slight enrichment in Fe towards the rim and corresponding depletion in the Mg content (Fig. 3). Such patterns are typical of garnet in high-grade rocks and can be interpreted as indicating that garnets were at one time homogeneous, and that subsequent diffusion at the rim has modified the composition during retrogression (Spear 1993). Garnet from the leucosomes is characterized by a sharp increase in Mn on approaching the edge (Fig. 3). This has been interpreted as a back diffusion (e.g. Selverstone & Hollister 1980). The similarity of garnets compositions is probably the result of homogenization at high temperature conditions. Cordierite is compositionally heterogeneous (Table 2); cordierite coronas around garnet have higher X_{Fe} (0.48–0.51) than cordierite in

the matrix and cordierite in the leucosome patches (L3) (X_{Fe} = 0.40–0.45). Plagioclases in the mesosome, the layered leucosome (L1) and the L-granite are homogeneous in composition, while in the discordant vein (L2), unzoned as well as normally zoned plagioclase are present. In the mesosome, L1 and L2 leucosomes, X_{an} varies over a wide range (0.18–0.34). Plagioclase from the L-granite is more sodic, with X_{an} of 0.16–0.18. It is worth noting that plagioclase inclusions in garnet have very low X_{an} (0.05).

Reaction textures

Textures observed in the mesosome and formation of the different generations of leucosomes can be explained by progressive partial melting reactions involving increasing temperature and decompression. The early melting reaction involves mainly plagioclase and quartz. However, a tiny amount of garnet requires biotite and sillimanite as a reactant of the melting reaction which, therefore should be as follows:

$$Pl + Qtz + Bt + Sil \Rightarrow Grt + melt. \quad (1)$$

The presence of rare remnant sillimanite, biotite, quartz and albite-rich plagioclase inclusions within garnet support this reaction. K_2O released from biotite is distributed between plagioclase and eventually K-feldspar within the leucosomes. However, the fact that garnet is not abundant in the studied migmatites (1–2 vol.%), and that the leucosomes are Kfs poor, and rich in plagioclase and quartz, we suggest that melting reaction (1) involves mainly plagioclase and quartz with minor biotite. Although biotite is considered as minor phase during this melting stage, K-feldspar is one of the expected products, which has not been found in equilibrium with garnet in the mesosome; it is present mainly in the leucosomes in different proportions (Table 1). This could be explained by removal of the K component or mobilization at the scale of the thin-section and crystallization within the melt fraction (e.g. Ashworth 1985).

The melt formed by reaction (1) could correspond to the composition of the layered leucosome (L1) and the discordant vein (L2) since garnet is encountered in both leucosomes. Conditions of formation of L1 and L2 and the significance of the difference in K-feldspar content in these two leucosomes will be presented later in the discussion section.

Cordierite coronas around garnet could be explained by the breakdown of garnet, sillimanite and quartz and the following reaction is proposed:

$$Grt + Sil + Qtz + (Bt1) \Rightarrow Crd + (Bt2). \quad (2)$$

It has been noted that cordierite in the matrix of the mesosome and in the leucosome patches (L3) is different in composition and texture than cordierite in the coronas. This cordierite could be a peritectic phase resulting by the incongruent melting reaction of biotite during the decompression stage; the following reaction is proposed:

$$Bt + Sil + Qtz \rightarrow Crd + melt. \quad (3)$$

The melt composition in equilibrium with cordierite may correspond to the composition of L3 containing higher content of K-feldspar (12–15 vol.%) than L1 and L2.

Paragenetic analysis

In order to discuss the metamorphic evolution of the studied migmatites, all the observed assemblages have been represented within the appropriate NKFMASH system. To interpret the observed melting reactions and to understand the relationships between leucosomes and mesosomes, bulk compositions of L1, L2, L3 and L-granite were projected on the same compatibility diagram as well as the mesosome assemblages (Fig. 4). In first approximation, water is considered in excess together with Pl, Sil and Qtz and all these phases are used as projection phases to reduce the representative system NKFMASH to KFM. The rectangular diagram K/F + M v. F/F + M (Fig. 4) shows the following. (i) All the observed reaction textures (1), (2) and (3) correspond to continuous equilibria within the NKFMASH system. These equilibria are represented by triangles within the compatibility

diagram and correspond to different P–T domains separated by univariant reactions. (ii) L1, L2 and L3 compositions preclude equilibrium between mesosome assemblages and L-granite. This may indicate that this granite band is formed at different P–T–aH_2O conditions either from the same source as L1, L2 and L3 or it represents an injection of melt from an external source. This last hypothesis is consistent with the sharp contacts of this band observed in the field (Fig. 2a). Therefore, we suggest that there is no genetic link between this granite-band and the mesosome assemblages. (iii) Garnet is in equilibrium with L1 and L2, whereas cordierite with low X_{Fe} is in equilibrium with L3. Therefore, L1 and L2 can be interpreted as resulting from the same reaction (1) occurring at the same P–T conditions, but probably variable aH_2O, whereas L3 is formed by reaction (3), (iv) Garnet and cordierite coronas are not in equilibrium with K-feldspar, which is consistent with the textural observations.

P–T path derived from the paragenetic analyses

Bulk rock compositions of the studied migmatites are close to those of Vielzeuf & Holloway

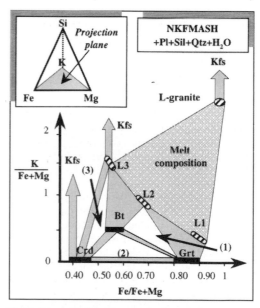

Fig. 4. Compatibility diagram in the system NKFMASH projected from water, quartz, plagioclase and sillimanite showing the phase relationship between the mesosome parageneses and the leucosome bulk rock compositions (L1, L2, L3 and L-granite) and the interpretation of the observed reaction textures (1, 2 and 3).

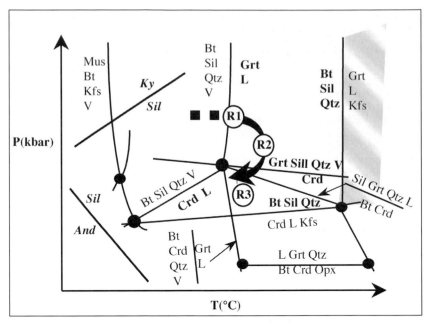

Fig. 5. Vielzeuf & Holloway (1988) $P–T–X$ pseudosection and infered $P–T–t$ path.

(1988). Therefore, the pseudosection for an intermediate X_{Mg} proposed by these authors can be used to draw the $P–T$ path followed by the migmatites (Fig. 5). The divariant assemblages are represented by single lines for clarity (Vielzeuf & Holloway 1988). The observed textures can be explained by a decompression $P–T$ path from peak conditions determined by the melting reaction (1). The $P–T$ domain of the studied migmatites is located on the low temperature side of the fluid-absent melting reaction Bt + Sil + Qtz $\Rightarrow$ Grt + Liq + Kfs (Fig. 5) since no K-feldspar has been observed in equilibrium with garnet. The $P–T$ position of this reaction is about 850 °C at 6 kbar (Vielzeuf & Holloway 1988), although the temperature depends upon the occurrence and composition of plagioclase and to a lesser extent on the X_{Mg} of the rock. The addition of Na to the KFMASH system shifts melting curves to lower temperatures. For common plagioclase-bearing metapelitic rocks, the above reaction is located at about 750 °C between 6 and 10 kbar (Le Breton & Thompson 1988). Metamorphic cordierite is usually not stable at pressures greater than 6 kbar in most common metapelites (e.g. Aranovich & Podlesskii 1983). Therefore, cordierite replacing garnet (reaction 2) indicates that the peak event was followed by decompression. Moreover, reaction (3) has a relatively flat slope (Vielzeuf & Holloway 1988) at about 4 kbar between 700 and 750 °C. Therefore, cordierite-bearing

leucosome (L3) could have been formed during the decompression stage.

Estimation of the metamorphic conditions

The $P–T–aH_2O$ conditions of the study rocks have been constrained by combining THERMO-CALC 2.5 calculation (Powell & Holland 1988) with the available classical geothermometers and the experimental data. Given the above discussion, calculations have been performed at variable aH_2O. The results are shown in Tables 3 and 4 and Fig. 6.

Peak metamorphic conditions

In order to estimate the peak metamorphic conditions in the mesosome THERMOCALC calculations were performed for the Bt–Grt–Sil–Crd–Pl–Kfs assemblages. Calculations of an average $P–T$ at different aH_2O yield an independent set of reactions in all cases (Table 3); from $aH_2O = 0–1$, av. $P–T = 3–5.5$ kbar, and 600–800 °C.

To constrain the peak metamorphic temperature an attempt of calculation was made on the basis of the Fe–Mg cation exchange between core compositions of garnet and biotite using different calibrations of Ferry & Spear (1978), Hodges & Spear (1982), Perchuk & Lavrent'eva (1983), Williams & Grambling (1991). The results obtained (Table 4) using the formulation of

Table 4. *Temperature estimates at 4 and 5 kbar in the mesosome and the adjacent discordant leucosome vein (L2) using Grt–Bt calibration of: 1, Ferry & Spear (1978); 2, Hodges & Spear (1982); 3 and 4, Perchuk & Lavrent'eva (1983); 5, Williams & Grambling (1991)*

Grt–Bt (core–core)		Grt–Bt (rim–rim)
Mesosome	at 4 kbar	
1 & 2 & 5 $\Rightarrow T = 710$–770 °C		$T = 491$–565 °C
3 & 4 $\quad \Rightarrow T = 682$–693 °C		
	at 5 kbar	
1 & 2 & 5 $\Rightarrow T = 715$–775 °C		
3 & 4 $\quad \Rightarrow T = 680$–686 °C		$T = 494$–557 °C
Discordant leucosome vein (L2)	at 4 kbar	
1 & 2 & 5 $\Rightarrow T = 700$–760 °C		$T = 508$–573 °C
3 & 4 $\quad \Rightarrow T = 668$–676 °C		
	at 5 kbar	
1 & 2 & 5 $\Rightarrow T = 702$–763 °C		
3 & 4 $\quad \Rightarrow T = 665$–669 °C		$T = 527$–568 °C

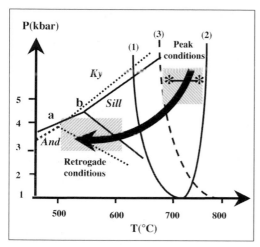

Fig. 6. THERMOCALC results for the peak and retrograde conditions: (1) metapelite solidus, (2) Bt + Sill + Qtz + Plg − Grt + Kfs + L after Le Breton & Thompson (1988), (3) Qtz–Ab–H$_2$O after Clemens & Vielzeuf (1987), *– – ––* – range of solidi temperatures for melting of plagioclase (An$_0$–An$_{80}$) + quartz at 5 kbar after Johannes (1978), (a) AS-triple point of Holloway (1971), (b) data set of Powell & Holland (1988).

Ferry & Spear (1978), and Hodges & Spear (1982) gave values in the range of 710–775 °C at 4.5–5 kbar, whereas, the formulations of Perchuk & Lavrent'eva (1983), and Williams & Grambling, (1991) gave lower temperatures (< 700 °C). Temperatures lower than 700 °C at such pressures are not considered realistic. According to Chipera & Perkins (1988) those calibrations of the garnet--biotite thermometer which use solely Fe–Mg partition data yield more precise results than those that incorporate the effect of other

components. However, the more recent reformulation of the garnet–biotite thermometer by Bhattacharya *et al.* (1992) also yields results lower than 700 °C. Such low values are probably the result of not considering the effect of Ti in biotite. Based on these results, we conclude that the most likely temperature range is 700–750 °C at 4–5 kbar, implying aH$_2$O in the range 0.4–0.7 (Table 3). Such aH$_2$O conditions are consistent with the recent data of Aranovich & Newton (1997). These results are consistent with the experimental data of Patiño Douce & Harris (1998) who have shown that leucosomes of trondhjemitic compositions could be generated at $T \approx 750$ °C and $P \approx 6$ kbar by melting of metapelitic rocks at high aH$_2$O.

Retrograde conditions

Using andalusite instead of sillimanite and the rim compositions instead of the core compositions, average P–T calculations using THERMO-CALC yield a complete set of independent reactions only for a domain in which aH$_2$O is between 0.1 and 0.6. The average P–T conditions are of 3–3.6 kbar and 619–631 °C (Table 4). These results are consistent with garnet–biotite thermometry ranging between 500 and 630 °C at 3.5–4 kbar.

Age determinations

The approach adopted here was to date monazite and garnet from the mesosomes and the adjacent leucosomes. The migmatites contain both monazite and garnet, while the L-granite (sample 45) yielded only monazite. The samples used for garnet dating are of two types: (1) one is a mesosome-L1 pair (sample 46) and (2) one mesosome–L1–L2 triplet (sample 51) (Tables 5 and 6). It is worth noting that cordierite-

Table 5. *Sm–Nd data on garnet–whole rock from the mesosome and the leucosomes of the studied migmatites*

Sample	Sm (ppm)	Nd (ppm)	$^{147}Sm/^{144}Nd$	$^{143}Nd/^{144}Nd \pm 2\sigma$	$\varepsilon_{Nd}(1900)$	T (grt-wr) (Ma)
Sample 46A-B: mesosome–L1 pair						
46B = mesosome						
Whole rock	6.29	38.79	0.0980	0.511382 ± 11	-0.5	
Garnet	0.773	0.481	0.9744	0.522300 ± 230		1893 ± 40
46A = Leucosome (L1)						
Whole rock	4.97	28.40	0.1057	0.511430 ± 10	-1.4	
Garnet	0.6	0.6	0.5146	0.516463 ± 26		1871 ± 14
Sample 51A-B: mesosome–L1–L2 triplet						
51-1A = mesosome						
Whole rock	3.83	23.25	0.0997	0.511408 ± 12	-0.4	
Garnet	0.10	0.093	0.6822	0.518603 ± 60		1877 ± 18
51-3B = L1						
Whole rock	5.55	31.90	0.1051	0.511450 ± 10	-0.9	
Garnet	1.32	2.25	0.3551	0.514543 ± 32		1880 ± 23
51-1B = L2						
Whole rock	7.65	37.49	0.1233	0.511663 ± 10	-1.2	
Garnet	0.85	0.62	0.8220	0.520134 ± 33		1843 ± 11

Table 6. *U–Pb data on monazite from the mesosome and the leucosomes of the studied migmatites*

Sample	Measured								Age (Ma)		
	wt (mg)	U (ppm)	Pb (ppm)	$^{206}Pb/^{204}Pb*$	$^{208}Pb/^{206}Pb*$	$^{206}Pb/^{238}U*$	$^{207}Pb/^{235}U$		$^{206}Pb/^{238}U$	$^{207}Pb/^{235}U$	$^{207}Pb/^{206}Pb$
Mesosome–L1 pair											
46B = mesosome	1.2	3600	4323	87106	3.0	0.33973	5.3822		1885	1882	1878
46A = L1	0.9	4720	5021	93335	2.5	0.34033	5.3931		1888	1884	1879
Mesosome–L1–L2 triplet											
51-1A = mesosome	1.4	4752	4303	67232	2.0	0.33830	5.3602		1879	1879	1879
51-3B = L1	1.2	4984	4545	28540	2.0	0.33704	5.33512		1872	1874	1877
51-1B = L2	1.8	6718	5915	27165	1.9	0.34332	5.4381		1903	1891	1878
L-granite											
45	2.5	1.69%	1.11%	295146	1.1	0.34596	5.4818		1915	1898	1879

*Corrected for mass fractionation (0.1%/a.m.u.), blank (0.2 ng Pb) and common lead (Stacey & Kramers 1975).
2σ-error estimates are 1.5% for Pb/U ratio in monazite.

bearing leucosome (L3) is not dated because of its occurrence as a small and rare patch impossible to separate exclusively from the mesosome.

About 2 kg of each sample were crushed and separated into magnetic and non-magnetic fractions. Garnet and monazite were then separated by heavy liquids and by magnetic separation using a Frantz isodynamic. The final purification was achieved by handpicking with a binocular microscope.

Sm–Nd method

In order to avoid problems of contamination of garnet isotopic composition by microscopic inclusions of monazite and zircon, leaching of garnet fractions was performed following the stepwise dissolution method of DeWolf *et al.* (1996). Garnet fraction was then dissolved in two Savillex teflon beakers in order to improve the dissolution. After careful evaporation of fluorides (with $HClO_4$ and HNO_3) the residue was dissolved in 6 N HCl and finally the solutions were combined and the mixed $^{149}Sm–^{150}Nd$ spike was added into a clear solution (no aliquoting). Measurements were made in a dynamic mode on a VG Sector 54 mass spectrometer using triple filaments at the laboratory of Isotope Geology of The Geological Survey of Finland (GSF). Estimated error in $^{147}Sm/^{144}Nd$ is 0.4%, $^{143}Nd/^{144}Nd$ ratio is normalized to $^{146}Nd/^{144}Nd$ 0.7219 and average value for La Jolla standard was $^{143}Nd/^{144}Nd = 0.511852 \pm 12$ ($n = 20$, all errors quoted in this paper are 2σ). For many garnet analyses the precision of the $^{143}Nd/^{144}Nd$ is not very good due to low abundance of Nd. This is partly due to analytical procedure in GSF, which involves triple filament analyses of Nd+ ions. Nevertheless, some of the analyses allow garnet–whole rock dating with a precision of 10–20 Ma.

U–Pb method

Chemical treatment and separation of U and Pb from about 0.5 mg of monazite were done following the method described by Krogh (1973). The mass spectrometry and the isotope measurement of U and Pb were described by Huhma (1986).

Results and interpretation

Sm–Nd on garnet–whole rock. Sm and Nd concentrations and Nd isotopic ratios are given in Table 5. In all samples $^{147}Sm/^{144}Nd$ ratio of garnet is higher than that from the whole rock. $^{147}Sm/^{144}Nd$ ratio in garnet from L1 (sample 51-3B) is slightly lower than in garnet from the adjacent mesosome (51-1A) and L2 (sample 51-1B), respectively. This is likely due to the persistence of some monazite inclusions in spite of leaching since also the concentrations of Sm and Nd are slightly high (Table 5).

The absolute garnet ages vary from 1843 ± 11 to 1893 ± 40 Ma. The data indicate however, that two marginally different age groups may be present. The pair mesosome-L1 (samples 51-1A/ 51-3B, and 46A/46B) have an age ranging between 1871 ± 14 and 1893 ± 40 Ma and the leucosome-vein (L2) (51-1B) has an age of 1843 ± 11 Ma which is slightly younger than the age of mesosomes and layered leucosome (L1) age. Numerical analysis of these two groups of ages suggest that they are statistically different, but caution must be exercised in translating this result into geological significance, especially when the youngest age is defined by only one analysis.

Depending on the closure temperature (T_c) of Sm–Nd system, the obtained ages could be interpreted either as peak or cooling ages. This important aspect of garnet Sm–Nd studies is still debated (e.g. Guangaly *et al.* 1998 and references therein). According to Patchett & Ruiz (1987) garnets give Sm–Nd ages that reflect cooling events, whereas Burton & O'Nions (1991) show examples of garnets that crystallized during the prograde stage. Uncertainty on the absolute closure temperature (T_c) arises because it is known to be dependent upon a number of factors such as diffusion rate, cooling rate, grain size, crystal geometry, composition and thermal history experienced by the rock. Therefore, garnet is unlikely to possess a unique closure temperature.

In the case of the studied migmatites, garnet was formed under granulite-facies conditions. If the T_c of Sm–Nd system is higher than 800 °C (e.g. Cohen *et al.* 1988; Mezger *et al.* 1989), then the first Sm–Nd age group ranging between 1871 ± 14 and 1893 ± 40 Ma could reflect the peak metamorphic event, since the T_c is higher than the estimated peak conditions (700–750 °C). This age therefore could represent the time at which partial melting processes and formation of garnet, L1 and L2 by reaction (1) have occurred. By contrast, the Sm–Nd age of 1843 ± 11 Ma obtained for L2 could be either an artifact due to the small grain size and scarcity of garnet in this leucosome or because of Sm–Nd fractionation resulting from garnet dissolution during the leaching processes. Therefore, the difference in age between the mesosome and the leucosome may not be realistic.

U–Pb on monazite. In order to constrain the age of the metamorphism in the studied migmatites, U–Pb dating on monazite was performed and the results are presented in Table 6. Monazite from both mesosomes and all the leucosomes yielded practically the same age of 1878.5 ± 1.5 Ma (average).

The monazites in the studied migmatites are often observed in association with biotite–sillimanite assemblages in the main foliation and as inclusions in garnet and, therefore, they grew before or at the same time as garnet formed by the reaction (1) at peak conditions. Thus we interpret the monazite ages as a peak metamorphic age. This requires that the T_c for these monazites is a bit higher than the peak metamorphic conditions of 750 °C reached by the migmatites as it was suggested by Spear & Parrish (1996). This is consistent with the previous studies which show that monazite may preserve ages corresponding to peak temperatures up to granulite-facies conditions (DeWolf *et al.* 1993), and may even record prograde growth ages (Vry *et al.* 1996).

Discussion

P–T evolution

Calculated *P–T* conditions combining THERMO-CALC calculations, conventional geothermometry and experimental data agree with the metamorphic conditions and the *P–T* path deduced from the paragenetic analysis. The absence of early muscovite suggests that the stability field of muscovite + quartz has been exceeded. The *P–T* diagrams (Figs 5 and 6) show that the peak metamorphic conditions are within the stability domain of sillimanite, and above the H_2O-saturated pelite solidus (curve 1 in Fig. 6). This domain is located within the range of solidus temperatures for the melting of plagioclase + quartz at 5 kbar (Johannes 1978). These results indicate that the studied migmatites were metamorphosed under granulite-facies conditions at 700–750 °C, 4–5 kbar and aH_2O of 0.4–0.7. Formation of cordierite coronas around garnet and late crystallization of andalusite suggest that the final stage of the $P–T$ history is characterized by decompression and cooling to the andalusite stability field estimated at 500–650 °C and 3–4 kbar.

Isotopic data on garnet–whole rock and monazite suggest that the migmatites were metamorphosed at c. 1878 Ma during a single tectonometamorphic event. Since there is no evidence of any earlier high-grade metamorphic event in this area, it is assumed that 1878 Ma is the approximate age of the peak metamorphic event. This is consistent with the ion probe dating of zircon rims of metamorphic origin from these migmatites (Mouri et al. 1999). The fact that both monazite and garnet in all samples gave the same age within error indicates that the partial melting and the formation of different leucosome generations occurred within a negligible interval of time. Partial melting probably started at peak conditions at the same time as garnet formation, and continued during the decompression stage simultaneously with the formation of cordierite.

Melting processes and the origin of Kfs-poor leucosomes

Starting dehydration melting temperatures are widely scattered because melting of rocks may involve many phases. This is made more complex by the absence or presence of fluid, its nature and the effect of fO_2. Furthermore, it has been shown that melt compositions are mainly controlled by source composition (e.g. Patiño Douce & Beard 1995). However, more recently, Patiño Douce & Harris (1998) have shown that changing the conditions under which melting takes place can have effects on melt composition that are comparable in magnitude with the effects of changing source composition. The depletion in potassium content of leucosomes reflects the preferential destabilization of plagioclase relative to mica with increasing H_2O activity (e.g. Whitney & Irving 1994; Patino Douce & Harris 1998). As a consequence, the residual material (mesosome) is rich in mica and quartz, but depleted in plagioclase comparing to the

leucosomes which is a feature observed in the migmatites studied here (Table 1).

Therefore this material still constitutes fertile magma sources and could generate further significant volumes of granitic melts by dehydration melting at temperatures higher than 850 °C (e.g. Patiño Douce & Johnston 1991). The different generations of leucosomes observed in the field can be interpreted as different $P–T–aH_2O$ stages of in situ progressive melting. Variation in potassium content in these leucosomes is probably related to variation in aH_2O during partial melting. According to Patiño Douce & Harris (1998), melting of metasedimentary rocks under water saturated conditions at $T = 750$ °C and $P = 6$ kbar gives rise to melts of trondhjemitic compositions, whereas water under-saturated melting generates melts of granitic composition for identical $P–T$ and source composition.

In order to check this hypothesis in the present natural system, bulk compositions of leucosomes (L1, L2 and L3) and mineral compositions from the mesosome were projected on the same compatibility diagram (Fig. 7). In this projection, biotite is considered to be in excess instead of water (H_2O) and is used as a projection phase together with Qtz, Sil and Pl. Therefore, NKFMASH system is reduced to H–K–FeMg and represented by the rectangular diagram $H/K + FM$ v. $K/K + FM$ (Fig. 7). A water content of 2 wt%, corresponding to $P–T$ con-

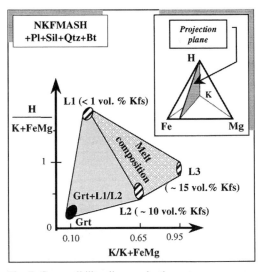

Fig. 7. Compatibility diagram in the system NKFMASH projected from biotite, quartz, plagioclase and sillimanite showing the relationships between garnet and leucosomes bulk rock composition.

ditions of 6 kbar and 750 °C, according to the experimental data of Patiño Douce & Harris (1998), has been used in order to account for the H_2O released during crystallization. This projection shows the following. (i) The layered leucosome (L1) generated by reaction (1) is projected at $H/K + FM > 1$, whereas, the discordant (L2) and the Crd-bearing (L3) leucosomes are projected at $H/K + FM < 1$. Therefore, we suggest that L1 (1 vol.% of Kfs) is formed under water saturated conditions by melting reaction (1) involving mainly Pl + Qtz. Whereas L2 (10 vol.% Kfs) and L3 (15 vol.% of Kfs) could have been formed under water under-saturated conditions by melting reaction involving mainly Bt + Qtz (reactions 1 and 3, respectively). (ii) L1 and L2 compositions are in equilibrium with garnet and preclude equilibrium between this phase and (L3). This also suggests that the same garnet producing reaction (1) formed L1 and L2 under the same $P-T$ conditions close to the thermal peak but at different aH_2O conditions. By contrast, cordierite-bearing leucosomes (L3) with relatively high proportion of K-feldspar (15 vol.%) are interpreted to have formed during the decompression stage and at relatively low aH_2O.

The origin of the L-granite remains obscure. On the basis of the phase relationships and field observations, this leucosome is more likely to be an injection from an external source at peak metamorphic conditions (1878 Ma). According to Patiño Douce & Harris (1998), migmatites with trondhjemitic leucosomes may represent the earliest appearance of melt during prograde metamorphism of a thickening orogenic belt. This is consistent with the studied rocks since they are characterised by Kfs-poor leucosomes and represent the oldest high grade metamorphic rocks occurring in the Svecofennian domain within a crust of a maximum thickness of about 60 km (Korsman et al. 1999). The low-pressure, high-temperature metamorphism is probably related to extensive magma under-plating caused by the collision between the Archaean continent and the Svecofennian domain and could explain the large volumes of syn-orogenic magmas. However, the variability of the bulk leucosome compositions described in the studied migmatites is inconsistent with this large production of homogeneous magmas. The partial melting observed in the studied area does not seem to be the source of the granitic magmas. Nevertheless, late leucosome crosscutting migmatites are of granitic composition. As they originate from higher P and T conditions, this suggests that extraction of homogeneous magmas took place at deeper level.

Conclusions

From the field relationships, petrographic observations and isotope geology we conclude that the Southern Finland Svecofennian Kfs-poor leucosomes migmatites have experienced beginning of melting at peak conditions of about 700–750 °C and 4–5 kbar followed by decompression and cooling to the andalusite stability field. Melting reactions produced melts of variable compositions mainly with respect to K-feldspar contents. The first leucosomes (L1) are characterized by very low K-feldspar content and high plagioclase + quartz content. Whereas the last leucosomes (L3) are scarce and have slightly higher abundance of K-feldspar. This variation in composition could be explained by the decrease in water activity conditions gradually from the peak metamorphism. The whole melting process has taken place during a single tectonometamorphic event dated at c. 1878 Ma.

This work has been funded by the Academy of Finland. I.I.M. would like to acknowledge the staff of the Departments of Petrology, Mineralogy and Isotope Geology in GSF-Espoo for discussions, assistance in the field and with mineral separation and chemistry, special thanks are due to H. Huhma for isotope measurements and discussions on the isotopic part of this manuscript, M. Guiraud for useful comments and numerous discussions, J. R. Kienast for supporting partly the microprobe analyses and discussions on the earlier version of the paper, P. Sorjonen-Ward for language correction, A. Castro, G. Rogers, G. Watt and an anonymous reviewer are thanked for their fruitful comments which helped to improve the content of the paper.

References

ASHWORTH, J. R. 1985. Introduction. In: ASHWORTH, J. R. (ed.) Migmatites. Blackie, Glasgow, 1–35.

ARANOVICH, L. Y. & NEWTON, R. C. 1997. Phase equilibrium constraints on the granulite fluid composition. In: HÖLTTÄ, P. (ed.) Mineral equilibria and databases; meeting of the Mineral Equilibria Working Group of the International Mineralogical Association (IMA). Geological Survey of Finland, Abstract guide, 46, 9.

—— & PODLESSKII, K. K. 1983. The cordierite–garnet–sillimanite–quartz equilibrium: experiments and applications. In: SAXENA, S. K. (ed.) Kinetics and equilibrium in mineral reactions: Advances in physical geochemistry. Springer-Verlag, New York, 173–198.

BARBARIN, B. 1999. A review of the relationships between granitoids types, their origins and their geodynamic environments. Lithos, 46, 605–626.

BHATTACHARYA, A. & MOHANTY, L., MAJI, A., SEN, S. K. & RAITH, M. 1992. Non ideal mixing in the phlogopite-annite binary: constraints from experimental data on Mg–Fe partitioning and a reformulation of the biotite–garnet geotherm-ometer. *Contributions to Mineralogy and Petrology*, **111**, 87–93.

BURTON, K. W. & O'NIONS, R. K. 1991. High-resolution garnet chronometry and the rates of metamorphic processes. *Earth and Planetary Sciences Letters*, **107**, 649–671.

CHIPERA, S. J. & PERKINS, D. 1988. Evaluation of biotite–garnet geothermometers; application to the English River subprovince, Ontario. *Contributions to Mineralogy and Petrology*, **98**, 40–48.

CLEMENS, J. D. & VIELZEUF, D. 1987. Constraints on melting and magma production in the crust. *Earth and Planetary Science Letters*, **86**, 287–306.

COHEN, A. S., O'NIONS, R. K., SIEGENTHALER, R. & GRIFFIN, W. L., 1988. Chronology of the pressure-temperature history recorded by granulite terrane. *Contributions to Mineralogy and Petrology*, **98**, 303–311.

DEWOLF, C. P., BELSHAW, N. S. & O'NIONS, R. K. 1993. A metamorphic history from micron-scale $^{207}Pb/^{206}Pb$ chronometry of Archean monazite. *Earth and Planetary Sciences Letters*, **120**, 207–220.

——, ZEISSLER, C. J., HALLIDAY, A. N., MEZGER, K. & ESSENE, E. J. 1996. The role of inclusions in U–Pb and Sm–Nd garnet geochronology; stepwise dissolution experiments and trace uranium mapping by fission track analysis. *Geochimica et Cosmochima Acta*, **60**, 121–134.

FERRY, J. M. & SPEAR, F. S. 1978. Experimental calibration of the partitioning of Fe and Mg between biotite and garnet. *Contributions to Mineralogy and Petrology*, **66**, 113–117.

GUANGALY, J., TIRONE, R. & HERVIG, R. L. 1998. Diffusion kinetics of Samarium and Neodymium in garnet and a method for determining rates of rocks. *Sciences*, **281**, 805–807.

HODGES, K. V. & SPEAR, F. S. 1982. Geothermometry, geobarometry and the Al_2SiO_3 triple point at Mt Moosilauke, New Hampshire. *American Mineralogist*, **67**, 1118–1134.

HUHMA, H. 1986. Sm–Nd, U–Pb and Pb–Pb isotopic evidence for the origin of the Early Proterozoic Svecokarelian crust in Finland. *Geological Survey of Finland, Bulletin*, **337**, 52.

JOHANNES, W. 1978. Melting of plagioclase in the system Ab–An–H_2O and Qz–Ab–An–H_2O at P H_2O = 5 kbars, an equilibrium problem. *Contributions to Mineralogy and Petrology*, **66**, 295–303.

—— & GUPTA, L. N. 1982. Origin and evolution of a migmatite. *Contributions to Mineralogy and Petrology*, **79**, 114–123.

KORSMAN, K., KORJA, T., PAJUNEN, M., VIRRANSALO, P. & THE GGT/SVEKA WORKING GROUP, 1999. The GGT/SVEKA Transect—Structure and Evolution of the Continental Crust in the Palaeoproterozoic Svecofennian Orogen in Finland. *International Geology Review*, **41**, 287–333.

KRETZ, R. 1983. Symbols for rock-forming minerals. *American Mineralogist*, **68**, 277–279.

KROGH, T. E. 1973. A low-contamination method for hydrothermal decomposition of zircon and extraction of U and Pb for isotopic age deter-minations. *Geochimica et Cosmochimica Acta*, **37**, 485–494.

LAHTINEN R. & HUHMA H. 1996. Isotopic and geochemical constraints on the evolution of the 1.93–1.79 Ga Svecofennian crust and mantle in Finland. *Precambrian Research*, **82**, 13–34.

LE BRETON, N. & THOMPSON, A. B. 1988. Fluid-absent (dehydration) melting of biotite in metapelites in the early stages of crustal anatexis. *Contributions to Mineralogy and Petrology*, **99**, 226–237.

MEHNERT, K. R. 1968. *Migmatites and the origin of granitic rocks*. Elsevier, Amsterdam.

MEZGER, K., HANSON, G. N. & BOHLEN, S. R. 1989. U–Pb systematics of garnet: dating the growth of garnet in the Late Archean Pikwitonei granulite domain at Cauchon and Natawahuna Lakes, Manitoba, Canada. *Contributions to Mineralogy and Petrology*, **101**, 136–148.

MOURI, H., KORSMAN, K. & GUIRAUD, M. 1999. Contribution of CL imaging to the understanding of the zircon U–Pb ion probe ages and geological implications: example from the LP/HT Kfs-poor leucosomes migmatites from Tampere area (southern Finland). *Terra Abstracts II*, **4**, 131.

PATIÑO DOUCE, A. E. & BEARD, J. S. 1995. Dehydra-tion-melting of biotite gneiss and quartz amphi-bolite from 3 to 15 kbar. *Journal of Petrology*, **36**, 707–738.

—— & HARRIS, N. W. 1998. Experimental constraints on the Himalayan Anatexis. *Journal of Petrology*, **39**, 689–710.

—— & JOHNSTON, A. D. 1991. Phase equilibria and melt productivity in the pelitic system: implica-tions for the origin of peraluminous granitoids and granulites. *Contributions to Mineralogy and Petrology*, **107**, 202–218.

PATCHETT, P. J. & RUIZ, J. 1987. Nd isotopic ages of crust formation and metamorphism in the Pre-cambrian of eastern and southern Mexico. *Contributions to Mineralogy and Petrology*, **96**, 523–528.

PELTONEN, P. 1995. Petrogenesis of ultramafic rocks in the Vammala nickel belt; implications for crustal evolution of the early Proterozoic Svecofennian arc terrane. *Lithos*, **34**, 253–274.

PERCHUK, L. L. & LAVRENT'EVA, I. V. 1983. Experimen-tal investigation of exchange equilibria in the system cordierite–garnet–biotite. *In*: SAXENA, S. K. (ed.) *Kinetics and equilibrium in mineral reactions*. Springer, 199–239.

POWELL, R. & HOLLAND, T. J. B. 1988. An internally consistent dataset with uncertainties and corre-lations; 3. Applications to geobarometry, worked examples and a computer program. *Journal of Metamorphic Geology*, **6**, 173–204.

SAWYER, E. W. 1991. Desequilibrium melting and the rate of melt-residuum separation during migma-tisation of mafic rocks from Grenville Front, Quebec. *Journal of Petrology*, **32**, 701–738.

—— 1995. Melt segregation or magma mobility—the difference between migmatites and granites. *USGS Circular*, **1129**, 133–134.

—— & BARNES, S.-J. 1988. Temporal and com-positional differences between subsolidus and anatectic migmatite leucosomes from the Quetico metasedimentary belt, Canada. *Journal of Metamorphic Geology*, **6**, 437–450.

SEDERHOLM, J. J. 1923. *On migmatites and associated Pre-Cambrian rocks of south-western Finland. Part 1. The Pellinge region.* Bulletin de la Commission Géologique de Finlande. **58**.

SELVERSTONE, J. V. & HOLLISTER, L. S. 1980. Cordierite-bearing granulites from the Coast Ranges, British Columbia; P–T conditions of metamorphism. *Canadian Mineralogist*, **18**, 119–129.

SPEAR, F. S. 1993. *Metamorphic phase equilibria and pressure-temperature-time paths.* Mineralogical Society of America, Book Crafters, Inc., Chelsea, MI.

—— & PARRISH, R. R. 1996. Petrology and cooling rates of the Valhalla Complex, British Columbia. *Canadian Journal of Petrology*, **37**, 733–765.

VERNON, R. H. & COLLINS, W. J. 1988. Igneous microstructures in migmatites. *Geology*, **16**, 1126–1130.

VIELZEUF, D. & HOLLOWAY, J. R. 1988. Experimental determination of the fluid-absent melting relations in the pelitic system; consequences for crustal differentiation. *Contributions to Mineralogy and Petrology*, **98**, 257–276.

VRY, J., COMPSTON, W. & CARTWRIGHT, I. 1996. SHRIMP II dating of zircons and monazites; reassessing the timing of high-grade metamorphism and fluid flow in the Reynolds Range, northern, Arunta Block, Australia. *Journal of Metamorphic Geology*, **14**, 335–350.

WHITNEY, D. L. & IRVING, A. J. 1994. Origin of K-poor leucosomes in a metasedimentary migmatite complex by ultrametamorphism, syn-metamorphic magmatism and subsolidus processes. *Lithos*, **32**, 173–192.

WILLIAMS, M. L. & GRAMBLING, J. A. 1991. Manganese, ferric iron, and the equilibrium between garnet and biotite. *American Mineralogist*, **75**, 886–908.

Evidence of magmatic hybridization related with feeding zones: the synkinematic Guitiriz granitoid, NW Iberian Massif

M. MENÉNDEZ & L. A. ORTEGA

Departamento de Mineralogía y Petrología, Universidad del País Vasco, Aptdo. 644, Bilbao 48080, Spain (e-mail: npbmemam@lg.ehu.es)

Abstract: The hybridization zone of Guitiriz (NW Iberian Massif) displays numerous features resulted from the interaction of two contemporaneously emplaced felsic and intermediate magmas. A combined study, including field geology, petrography, chemistry and Sr isotopic data, reveals a simple two-component mixing between a mantle-derived magma with an initial $^{87}Sr/^{86}Sr$ isotopic ratio <0.7042 and a crustal-derived component with an initial Sr isotopic ratio >0.7104. Both magmas rose through a common feeding zone developed in an extensional regime in which favourable conditions for magma mixing occurred. Calculated emplacement temperature and pressure for the Guitiriz hybridization zone are 700 °C and 2.5 kbar, respectively, and are comparable with those estimated from regional metamorphism. Magmatic epidote occurrence, exhibiting textural and compositional characteristics suggestive of magmatic origin, does not require high-pressure granite solidification. Its presence may be related to a rapid magma transport through the feeding zone that favoured the preservation of epidote.

Hybridization between contemporaneous magmas is a widespread petrogenetic process which ranges from complete and homogeneous mixing where the identity of the different magmas is not apparent, to magma mingling where original components maintain much of their individual characteristics. Many authors had invoked hybridization to account for observed field, petrographic features (Vernon *et al.* 1988; Didier & Barbarin 1991 and references therein; Hibbard 1995) chemical (Zorpi *et al.* 1989; D'Lemos 1996) and isotopic features (Moreno Ventas *et al.* 1995; Castro *et al.* 1995; Janousek *et al.* 1995; Galán *et al.* 1996). When the involved magmas come from completely different sources, mantle and crust, the resultant features of the hybrid rocks are very remarkable. Usually, the large differences in magma chemistry and in solidus temperatures result in extensive disequilibrium and ensure the preservation of many interaction features (D'Lemos 1996).

Mixing operates at various stages of the granite history, from the initial injection of a mafic-intermediate magma into a felsic one, to the ascent and final emplacement of the pluton. Recent models for granite pluton assembly infer that magma is transported from the source through conduits or dykes (Koyaguchi 1985; Clemens & Mawer 1992; Castro *et al.* 1995). The ascent through conduits appears to be an efficient mechanism for hybridization to occur (Koyaguchi 1985).

Here below we present detailed field, petrographic, chemical and isotopic features of a hybridization area geographically related with a feeding zone of the Guitiriz granitoid (NW of the Iberian Massif). There, it seems that mixing occurred between contemporaneous magmas from two different sources, mantle and crust. Particular emphasis was devoted to the study of magmatic epidote, *P–T* conditions of magma emplacement and geochemical evidence of compositional changes as a result of hybridization.

Geological setting

The Guitiriz granitoid intruded Precambrian metavolcanic rocks and Ordovician–Silurian schists of the northern Ollo de Sapo domain. This domain extends more than 200 km, trending N–S, along the eastern boundary of the Central Iberian zone of the Iberian Massif (Julivert *et al.* 1974) (Fig. 1). Three main deformation phases form part of a progressive deformation process at middle to shallow crustal levels in response to

From: Castro, A., Fernández, C. & Vigneresse, J. L. (eds) *Understanding Granites: Integrating New and Classical Techniques.* Geological Society, London, Special Publications, **168**, 255–272. 1-86239-058-4/99/$15.00 © The Geological Society of London 1999.

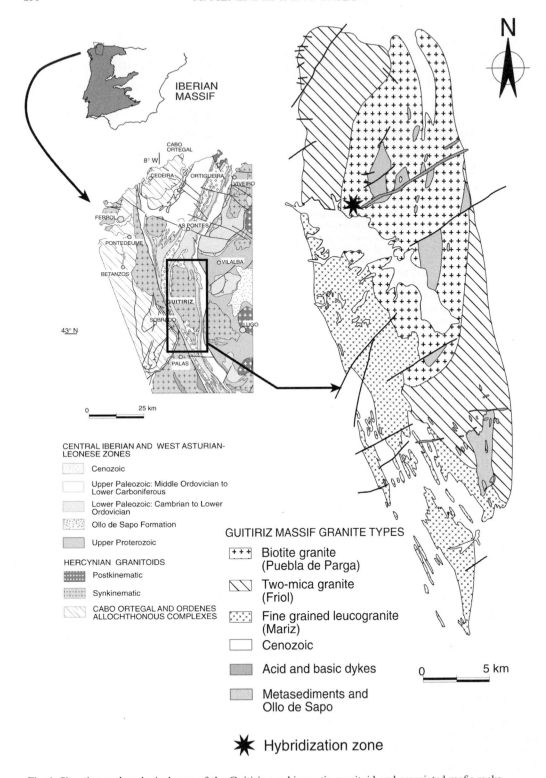

Fig. 1. Situation and geological map of the Guitiriz syn-kinematic granitoid and associated mafic rocks.

Variscan continental collision and associated crustal thickening (Díez Balda *et al.* 1991). The Guitiriz granite intrusion is synkinematic with respect to D3 (Aranguren *et al.* 1996) which corresponds to a retrometamorphic phase. Minimum conditions accomplished during D2 (Aranguren *et al.* 1996; Martínez *et al.* 1996) reached amphibolite facies conditions at >600 °C; 5 kbar (Arenas 1991; Martínez *et al.* 1996).

The Guitiriz granitoid is bounded by two major, roughly N–S-trending, ductile shear zones (Iglesias & Choukroune 1980; Aranguren *et al.* 1996): the Valdoviño fault to the west and the Vivero fault to the east (Fig. 1). The Valdoviño fault is a sinistral strike-slip fault whereas the Vivero fault is a dextral, trastensional shear zone.

The Guitiriz massif is composed of three main types of granitic rocks (Fig. 1): (i) a biotite granite with K-feldspar porphyrocrysts (Puebla de Parga facies), (ii) a medium to coarse-grained equigranular two mica granite (Friol facies), and (iii) a fine-grained two-mica leucogranite (Mariz facies).

These three types of granite have been classically ascribed to two different petrogenetic series (Capdevila *et al.* 1973; Corretgé 1983): (i) Friol and Mariz to a series of syn- to late-kinematic two-mica granitoids, usually peraluminous and albite/or K-feldspar rich, with evidence of an anatectic origin and usually high Sr_i ratios; (ii) the Puebla de Parga facies would belong to the series of syn- to post-kinematic biotite-rich granitoids with calc-alkaline tendency, sometimes associated with intermediate-basic to ultrabasic rocks and generally showing lower Sr_i ratios.

Yet, detailed structural and anisotropy of magnetic susceptibility studies (Aranguren 1994; Aranguren *et al.* 1996) have shown that the granitoids of the Guitiriz massif may have emplaced during a single intrusive episode. Moreover, petrological and geochemical data (Bellido *et al.* 1990) suggest a co-genetic relationship between the three granite facies distinguished. The same authors proposed an intrusion age of *c.* 319 Ma (whole rock Rb/Sr) for the massif. Some new exposures and road-cuts have allowed for the study of mafic–intermediate rocks not described in previous works. One of the outcrops shows about 100 m of exposure of clean rock where most of the photographs and samples have been taken for this study (Fig. 2). The host rock of the mafic–intermediate bodies is the Puebla de Parga facies of biotite granite with K-feldspar porphyrocrysts. A great amount of rocks at the outcrop show features intermediate between those of the granite and the mafic intrusive (Fig. 2).

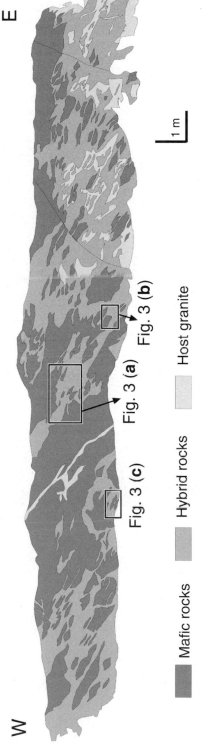

Fig. 2. Schematic section across the Guitiriz mixing zone. Boxes show locations of photographs in Fig 3.

Field relationships and petrography

Commonly, the mafic–intermediate bodies, ranging in size from a few centimetres to tens of metres (Fig. 3a), have an ellipsoidal, flattened shape and fine-grained equigranular texture. Double enclaves with a more mafic inner region are not uncommon. The spatial arrangement of mafic–intermediate masses within the granite suggests a dyke-like intrusion process. The contacts between the mafic–intermediate bodies and the host granite are generally sharp and lobate and, more rarely, gradational (Fig. 3b, c). Isolated K-feldspar crystals from the host granite are often incorporated into the mafic–intermediate bodies. Some of the feldspars show little or no modification but others have partially lost their original morphology. This suggests a magmatic state of the mafic–intermediate rock to allow the incorporation of the feldspar crystals. Local back-veining of the mafic rocks by host granitoid is indicative of synplutonic injection of mafic–intermediate

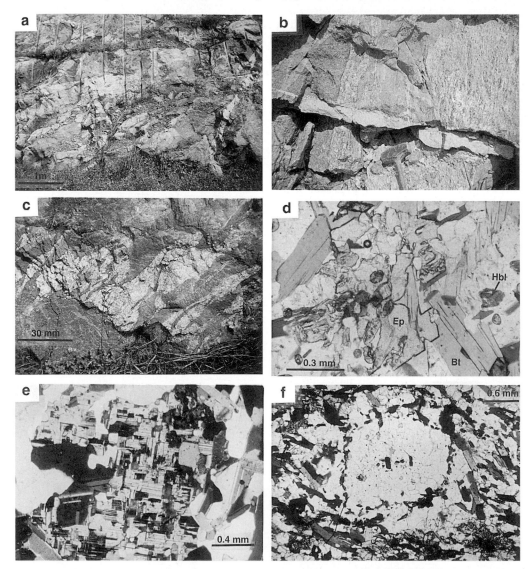

Fig. 3. Some field and petrographic relationships of the magma mixing zone. (**a**) Field relationships between felsic, mafic and hybrid rocks. (**b**) Hybrid zone with flow structures. Felsic dyke cross cutting the hybridization zone. (**c**) Lobate contacts between mafic rocks and granite host. (**d**) Magmatic epidote. (**e**) Boxy cellular plagioclase. (**f**) Quartz xenocrysts.

magma into the granitic one. These veins are discontinuous indicating that the host granite was mobile during their emplacement (Hibbard 1995).

Petrographically, the more mafic rocks are tonalitic in composition and fine-grained with micro-porphyritic texture due to the presence of epidote, titanite and some plagioclase phenocrysts. The partially recrystallized mesostasis is almost quartzofeldspathic with hornblende, minor allanite, apatite and zircon, along with minor amounts of secondary chlorite and muscovite.

Biotite commonly occurs as subhedral crystals defining the magmatic foliation. The alignment of biotite is parallel to the foliation in the host granite, suggesting as well that the two magmas were not fully crystallised at the time of interaction. Hornblende occurs as subhedral crystals usually associated with biotite, epidote or titanite. Titanite crystals are euhedral when included in epidote or biotite, but may be completely skeletal when in contact with the matrix. Plagioclase phenocrysts are euhedral to subhedral, exhibiting optical zonation (An_{38} to An_{31} from core to rim), strong twinning and inclusions of apatite, amphibole, sphene and biotite, whereas the finer-grained plagioclase of the mesostasis (An_{36}) does not show optical zonation.

Coarse hexagonal epidote crystals are euhedral to subhedral and up to 1.5 mm in size. Epidote is euhedral against biotite but in contact with plagioclase and quartz shows highly embayed, wormy contacts that are almost myrmekitic (Fig. 3d). Some epidote crystals exhibit anhedral cores of allanite. The boundary between allanite and epidote is sharp. Epidotes are not deformed, occasionally are twinned on (100) and often contain inclusions of apatite or titanite. Blebs of hornblende completely included within the epidote are also present, a texture thought to be the result of hornblende resorbtion into the melt from which epidote then crystallized (Zen & Hammarstrom 1984).

Magmatic epidote has been described by many authors (e.g. Cornelius 1915; Zen & Hammarstrom 1984, 1986; Moench 1986; Tulloch 1986; Zen 1988; Dawes & Evans, 1991; Vihnal et al. 1991; Keane & Morrison 1997). The textural features here described for this mineral fit most of the criteria presented by these authors and are evidence of a magmatic origin for the epidote in the mafic–intermediate rocks of the Guitiriz massif.

The Puebla de Parga facies host is a medium- to coarse-grained granite with K-feldspar phenocrysts in a matrix of quartz, plagioclase, microcline, biotite and muscovite. Accessory minerals

are zircon, apatite and rutile. Secondary minerals are muscovite, chlorite, sericite, opaque and potassic feldspar. Myrmekitic textures are abundant.

Elongated microcline occurs as large phenocrysts up to 2 cm in length defining the magmatic foliation. They contain abundant inclusions of partially reabsorbed plagioclase and biotite. Plagioclase occurs as optically zoned and strongly twinned euhedral to subhedral crystals of variable size, with compositions ranging from An_2 to An_{22}. They are locally altered to sericite or muscovite. Biotite occurs as subhedral crystals containing abundant inclusions, mainly of zircon. Subhedral muscovite occurs usually associated with biotite.

The hybrid rocks, with fine to coarse grain size, consist of plagioclase, biotite, quartz, microcline $\pm$ muscovite $\pm$ epidote $\pm$ titanite $\pm$ hornblende $\pm$ allanite, with highly variable modal amounts. Accessory minerals always present in hybrid rocks are apatite and zircon.

Various disequilibrium textures constitute frequently cited evidence for magmatic hybridization. Hibbard (1991) proposed the concept of 'textural assemblage' whereby no single texture could be used unequivocally, but a combination of textures taken together may constitute good evidence for mixing. The criteria of particular interest to this purpose in the hybrid rocks of the Guitiriz hybridization zone (and in minor amount in the more mafic rocks) include: (1) the boxy cellular plagioclase crystals (Fig. 3e), likely resulting from a relative high growth rate and a low nucleation rate; the undercooled environment provided by the transfer of heat from the hot to the cool magma may be ideal for the boxlike cellular growth of plagioclase deriving from the more mafic system (Hibbard 1995); (2) the quartz xenocrysts with ferromagnesian rich outer zones (Fig. 3f); the latter would have formed through partial resorbtion of early formed quartz phenocrysts, which then underwent syntaxial growth enclosing biotite and hornblende; and (3) the acicular apatite as inclusion in all other major minerals, which suggests rapid growth from undercooled magma in a system chemically capable of crystallizing this mineral.

Mineral chemistry

All mineral analyses were performed with a Cameca SX100 microprobe (Clermont-Ferrand, France). Beam conditions were 20 kV and 40 nA for allanite and epidote and 15 kV, 15 nA for the rest of the silicates.

Biotite and plagioclase from all rock types were analysed. Hornblende, epidote and allanite were present only in the more mafic and some hybrid

Table 1. *Representative analyses of biotite*

	M-38 Mafic	M-38 Mafic	M-28 Hybrid	M-37 Hybrid	M-30-2 Hybrid	M-54 Granite	M-53 Granite	M-57 Granite
SiO_2	37.91	37.88	35.00	37.27	36.97	35.64	34.99	34.80
TiO_2	1.51	1.63	3.13	2.06	2.66	3.15	3.23	2.94
Al_2O_3	16.31	16.35	17.78	15.62	16.58	17.85	17.18	17.26
FeO_T	13.67	13.41	21.84	17.20	16.46	20.99	21.95	21.53
Cr_2O_3	0.00	0.10	0.00	0.00	0.00	0.03	0.03	0.01
MgO	15.55	15.06	8.04	12.50	11.65	7.38	7.61	7.35
MnO	0.34	0.28	0.30	0.30	0.23	0.25	0.29	0.21
CaO	0.00	0.00	0.06	0.00	2.19	0.05	0.01	0.01
Na_2O	0.08	0.04	0.07	0.06	0.06	0.01	0.04	0.05
K_2O	10.02	10.54	9.69	10.18	9.96	9.51	9.66	9.99
FeO	12.45	12.57	20.62	15.68	15.37	20.00	20.63	20.27
Fe_2O_3	1.34	0.92	1.34	1.67	1.20	1.09	1.46	1.39
Total	95.51	95.39	96.02	95.34	96.86	94.95	95.14	94.28

rocks, but they were analysed to ensure the magmatic origin of epidote and to constrain pressure–temperature (P–T) conditions during magmatic crystallization.

Biotite

Representative analyses of biotite from all rock types are listed in Table 1. Structural formulae have been calculated on the basis of 11 oxygens, Fe^{3+} was estimated following regression methods described by De Bruiyn *et al.* (1983). Biotite from the granite host, hybrid and more mafic rocks show minor variations in $X_{Fe} = Fe/(Fe + Mg)$: between 0.579 and 0.638 in the granite, between 0.344 and 0.318 in the more mafic rocks, and between 0.404 and 0.531 in the hybrid rocks. Biotite from all types of rocks show calc-alkaline affinity in terms of FeO_t–MgO–Al_2O_3 contents (Abdel-Rahman 1994). Only some rocks of the host granite show peraluminous affinity (Fig. 4). In the Al_t v. Mg diagram (Nachit *et al.* 1985), biotite analyses plot mostly in the calc-alkaline fieldbut close to the subalkaline–calc-alkaline boundary.

Amphibole

Representative analyses of hornblende are presented in Table 2. Structural formulae have been calculated on the basis of 23 oxygens and Σcats–Ca–Na–K = 13 following the method of Holland & Blundy (1994). Amphibole compositions display a range of Si per formula unit from 6.7 to 7.2 and are magnesian hornblende in the IMA classification (Leake 1978). The values of $Mg/(Mg+Fe^{2+})$ range between 0.7 and 0.8 suggesting crystallization conditions at high fO_2 (Pe-Piper 1988).

Epidote

Representative analyses of epidote and allanite are listed in Table 3. Normalization is based on eight cations and H_2O was calculated by stoichiometry. All Fe assumed to be ferric except for allanite where the following formulae were used: $Fe^{2+} = Zr + Ti + 2Th + \Sigma REEs + Y - Mg - Mn$ and $\Sigma Fe - Fe^{2+} = Fe^{3+}$. All Mn assumed to be divalent for allanite and trivalent for epidote. A site includes REEs, Ca, Sr, Ba and Th. Traverses were made along some epidote crystals showing no evidence for any zonation and no systematic variations in pistacite content. In addition, analyses were performed both on allanite core and epidote rim in order to determine whether this epidote was enriched in REE relative to epidote lacking an allanite core, finding that both are indistinguishable. The average pistacite component ($Ps = Fe^{3+}/(Fe^{3+} + Al)$) in the

Table 2. *Representative analyses of amphibole*

	M-38 Mafic	M-38 Mafic	M-38 Mafic	M-38 Mafic
SiO_2	48.05	48.54	47.29	46.83
TiO_2	0.87	0.67	0.00	0.72
Al_2O_3	8.01	7.23	8.73	9.12
FeO_T	0.00	0.02	0.04	0.08
Cr_2O_3	12.29	12.38	13.30	12.80
MgO	13.87	14.57	13.45	13.48
MnO	0.40	0.46	0.45	0.45
CaO	12.12	12.31	12.28	12.35
Na_2O	0.93	0.89	1.06	0.97
K_2O	0.60	0.57	0.83	0.81
FeO	9.54	8.49	9.48	9.26
Fe_2O_3	3.06	4.33	4.24	3.94
Total	97.13	97.65	97.42	97.61

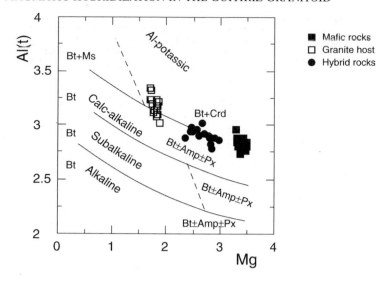

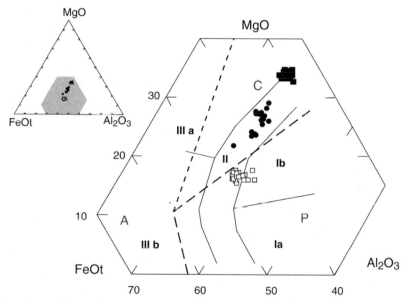

Fig. 4. Al_t v. Mg (Nachit *et al.* 1985) and FeO–MgO–Al_2O_3 (Abdel-Rahman 1994). Biotites from the three main rock types (mafic, felsic and hybrid rocks) are plotted.

mixing zone of Guitiriz is Ps_{26}, far from the values between Ps_0 to Ps_{24} reported for epidotes formed from alteration of plagioclase, and the values between Ps_{36} to Ps_{48} reported for those formed by alteration of biotite. But in the range between Ps_{25} to Ps_{29} reported for epidotes exhibiting magmatic textures (Tulloch 1979,

1986). In addition, Tulloch (1986) and Zen (1988) related the composition of magmatic epidote to epidote breakdown curves under different fO_2 conditions (Liou 1973); they indicate that, within probable fO_2 limits for magmas (i.e. between the HM and NNO buffers), epidote should range only from Ps_{25} to Ps_{33}.

Table 3. *Representative analyses of epidote and allanite*

	Epidote				Allanites	
	M-38 Mafic	M-37 Hybrid	M-30-2 Hybrid	M-39 Hybrid	M-37 Hybrid	M-37 Hybrid
SiO_2	37.74	37.64	37.87	37.181	33.33	32.99
TiO_2	0.13	0.09	0.15	0.16	0.89	0.88
Al_2O_3	24.10	23.74	23.53	22.98	17.19	17.02
Fe_2O_3	12.05	12.40	13.32	13.73	7.74	7.05
FeO	—	—	—	—	6.03	6.42
MgO	0.02	0.03	0.04	0.03	1.20	1.05
MnO	0.17	0.18	0.16	0.14	0.30	0.35
CaO	23.45	23.27	23.48	23.13	13.70	13.75
Na_2O	0.01	0.00	0.04	0.00	0.00	0.00
K_2O	0.00	0.00	0.08	0.01	0.00	0.00
ThO_2	—	—	—	—	0.01	0.42
Y_2O_3	—	—	—	—	0.10	0.00
Gd_2O_3	—	—	—	—	0.16	0.15
Sm_2O_3	—	—	—	—	0.31	0.26
Nd_2O_3	—	—	—	—	3.00	3.02
Pr_2O_3	—	—	—	—	0.96	0.97
Ce_2O_3	—	—	—	—	8.84	8.73
La_2O_3	—	—	—	—	4.17	4.36
H_2O	—	—	—	—	1.23	1.20
Total	97.67	97.35	98.66	97.36	99.15	98.63
Mol %	24.13	24.97	26.54	27.67	21.76	20.39

Whole-rock composition

Major and trace elements

For major and minor element whole rock analyses, measures were made with a ICP-AES ARL 3410 Minitorch® (Fisons) after decomposition of the sample with $Li_2B_4O_7$ flux in a Claisse Fluxy® fusion apparatus. Calibration lines were constructed using CRPG, CCRMP and GSJ standards and a pure $Li_2B_4O_7$ solution as the zero similar to the procedure outlined by Ortega & Menéndez (1998). The REE were measured by ICP-AES after ion-exchange separation following methods modified of Cantagrel & Pin (1994). Calibration lines were constructed using a mixed REE solution prepared from Aldrich 1000 μg ml^{-1} solutions as a high point and a pure 1 M HNO_3 as a low point. Precision and accuracy are equal or better than 10% as described in Cantagrel & Pin (1994) and reported in Ortega & Menéndez (1998).

Eleven samples representative of all compositional and textural rock types were analysed. Table 4 contains whole rock chemical analyses for studied mafic–intermediate, hybrid rocks and their host granite. Mafic–intermediate and hybrid rocks have SiO_2 contents between 61 and 68 wt% and low MgO concentrations (1.7–4 wt%). Samples studied are metalumious to weakly peraluminous (A/CNK ranges from 0.88 to 1.05) and show calc-alkaline affinity in terms of alkali/lime index and AFM relations

(Fig. 5a). In the $K_2O–SiO_2$ diagram, samples studied plot in the high-K calc-alkaline suites field (Fig. 5b).

The Harker diagrams show mostly linear trends with negative slope (TiO_2, Fe_2O_3, MgO, MnO, P_2O_5, Zr, Cr, REE) whereas some elements (Al_2O_3, Na_2O, K_2O) do not display significant variation with silica (Fig. 6). In terms of trace element composition these rocks are characterized by high contents in Sr (963–262), Zr (263–120) and high and variable contents in Ba (1200–350). Sample M-31-1 shows rather different chemical features, with low contents in Ca, Ba and Sr. Those differences could be explained by means of a minor modal amount in feldspars and a mayor modal amount in biotite. Sample M-54 has very high contents in Na. The presence of potassic feldspar phenocrysts replaced by albite could be related with later albitization processes. For this sample Ba and Sr are significantly depleted whereas other elements do not show large modification.

Rare earth element and Zr contents show a good correlation with silica. Chondrite-normalized rare earth element patterns (Fig. 7) are characterized by a significant LREE enrichment, and fractionated patterns that decrease with silica, whereas the Eu*/Eu anomaly increases. Sample M-38 of the more mafic rock

Table 4. *Whole-rock major and trace analyses for the studied rocks*

	M-38 Mafic	M-31-1 Hybrid	M-30-1 Hybrid	M-26-2 Hybrid	M-26-1 Hybrid	M-25 Hybrid	M-37 Hybrid	M-34 Hybrid	M-30-2 Hybrid	M-54 Granite	M-53 Granite
SiO_2	61.16	61.75	64.22	64.56	64.81	65.39	66.38	67.08	67.66	68.59	69.13
TiO_2	0.99	1.03	0.76	0.72	0.68	0.61	0.59	0.61	0.52	0.44	0.38
Al_2O_3	16.80	17.74	16.66	16.32	15.88	16.57	16.55	16.69	16.29	17.08	15.49
Fe_2O_3	5.39	6.69	4.65	4.62	3.87	3.55	3.09	3.31	3.09	3.64	2.28
MgO	3.93	3.31	2.78	3.27	2.43	2.10	1.94	2.33	1.70	1.40	0.86
MnO	0.09	0.06	0.06	0.07	0.05	0.04	0.04	0.05	0.04	0.04	0.03
CaO	6.22	3.20	3.88	4.69	4.79	3.87	3.84	4.12	3.64	2.72	1.75
Na_2O	3.00	3.47	3.33	3.14	3.58	3.25	3.47	3.19	3.07	5.53	3.51
K_2O	2.73	3.55	2.79	2.68	3.46	3.38	5.10	3.64	3.58	1.74	4.61
P_2O_5	0.63	0.45	0.40	0.49	0.36	0.43	0.36	0.33	0.35	0.29	0.25
Total	100.94	101.25	99.58	100.56	100.91	99.19	01.36	101.35	99.94	101.47	99.29
Ba	1216.00	368.00	379.00	477.00	796.00	614.00	1034.00	810.00	1009.00	153.00	570.00
Sr	963.00	440.00	620.00	807.00	819.00	608.00	794.00	687.00	670.00	224.00	263.00
Rb	163.00	279.00	197.00	187.00	169.00	195.00	192.00	178.00	179.00	209.00	241.00
Cr	48.80	144.30	57.20	34.90	43.10	27.50	33.20	60.90	32.00	21.60	12.80
Cu	24.50	7.80	15.20	—	14.70	—	14.00	16.10	8.00	4.20	—
Ni	49.50	26.10	9.90	29.50	13.80	22.50	18.30	4.40	7.20	12.80	—
V	306.00	29.80	41.80	71.30	12.30	56.60	52.60	106.60	91.80	36.40	20.60
Y	17.20	15.50	15.10	14.40	13.70	15.00	12.90	13.00	10.30	13.60	8.40
Zn	105.00	126.00	128.00	101.00	78.90	85.10	143.00	76.80	87.60	133.00	113.00
Zr	263.00	244.00	229.00	204.00	182.00	182.00	154.00	172.00	153.00	136.00	120.00
La	92.76	46.39	54.12	78.58	52.66	58.51	41.39	41.15	39.80	31.75	26.54
Ce	173.00	88.63	104.80	144.90	100.90	109.10	84.13	81.00	77.84	60.55	51.96
Nd	75.81	44.20	49.39	59.82	45.22	48.32	41.31	38.23	37.08	26.80	23.94
Sm	10.58	8.19	7.44	8.65	6.75	7.95	7.04	6.45	5.55	5.08	4.57
Eu	2.40	1.30	1.38	1.61	1.37	1.49	1.29	1.23	1.25	0.89	0.89
Gd	5.75	5.50	4.66	5.05	4.16	5.16	4.37	3.88	3.55	3.40	2.94
Er	1.77	1.43	1.56	1.41	1.32	1.36	1.34	1.39	1.09	1.31	0.89
Dy	3.26	3.47	3.34	2.73	2.73	3.16	3.05	2.70	2.26	2.81	1.91
Yb	1.38	1.25	1.32	1.28	1.20	1.19	1.20	1.13	0.98	1.13	0.70
Lu	0.21	0.19	0.20	0.19	0.17	0.18	0.17	0.17	0.16	0.16	0.11

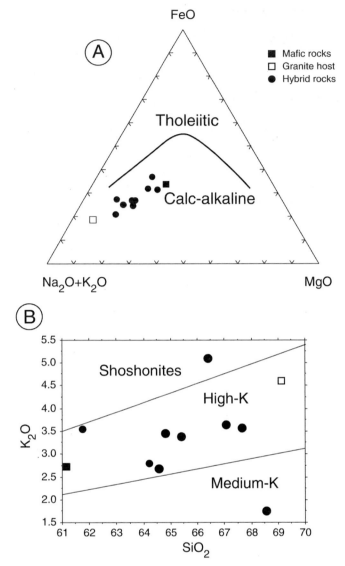

Fig. 5. Classification of granites and mafic–intermediate related rocks from Guitiriz: (**a**) AFM (A: $Na_2O + K_2O$, F: FeO_t, M: MgO), boundary between calc-alkaline and tholeitic fields after Irvine and Baragar (1971). (**b**) K_2O v. SiO_2 (subdivisions after Gill 1981).

analysed has the highest REE content ($\Sigma REE = 383$ ppm) and steepest normalized pattern (La/Lu = 45), with no significant Eu/Eu* anomaly (Eu/Eu* = 0.94). The granitic rocks have low REE contents ($\Sigma REE = 121$ ppm) and moderately fractionated patterns (La/Lu = 27). The hybrid rocks show intermediate characteristics in terms of REE contents ($\Sigma REE = 318–140$), fractionation patterns (La/Lu = 43–20), and Eu anomaly (Eu/Eu* = 0.86–0.75).

Sr isotopes

Sample dissolution (in PFA Teflon beakers) and chemical separation for whole rock Sr isotope analyses was performed following modified procedures after Pin & Bassin (1992). Sr was loaded onto a double Re–Ta assembly of outgassed filaments and measured on a Finnigan MAT 262 thermal ionization mass spectrometer at the University of País Vasco (Spain). Replicate analysis of the NBS 987 Sr standard during runs gave $^{87}Sr/^{86}Sr$ 0.710241 ± 12. X-ray fluorescence

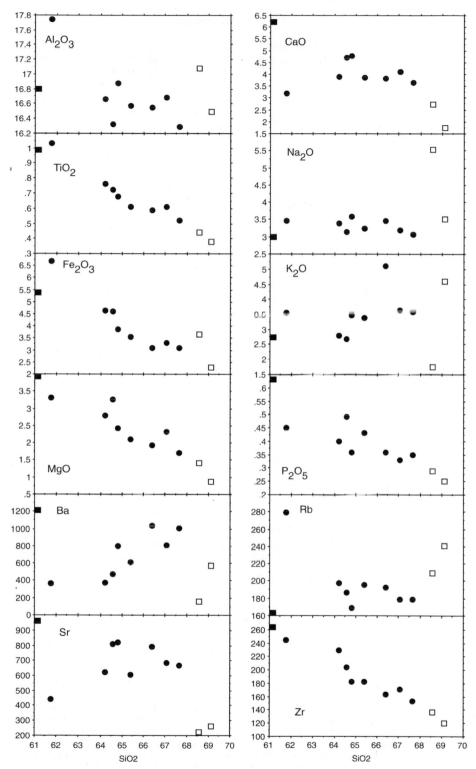

Fig. 6. Harker diagrams for major and some representative trace elements.

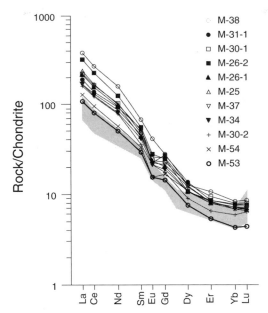

Fig. 7. REE chondrite-normalized diagram (Evensen *et al.* 1978) for the rock types of the Guitiriz hybridization zone.

(XRF) data on whole-rock powders for Rb and Sr were obtained at the University of Oviedo (Spain) using a Philips PW1480 spectrometer fitted with a Sc/Mo X-ray tube; estimated errors are <1% for these elements.
Sr isotope data are given in Table 5. The initial isotopic composition has been calculated assuming an age of 320 Ma (Bellido *et al.* 1990). Sr initial isotope ratios range from 0.7042 for sample M-38 of the more mafic rock analysed to 0.7104 for sample M-53 of host granite. All the data points fit a straight line when plotting $^{87}Sr/^{86}Sr$ versus the reciprocal of Sr concentration (Fig. 8, or a hyperbolic trend if plotted v. Sr).

Discussion

Geochemical interpretation

While major elements do not display a good correlation, the field relationships and the petrological and mineralogical features point to magma mixing as the more obvious interpretation to account for chemical variation of samples studied. Perfect magma mixing between varying proportions of felsic and mafic magmas would result in linear trends on binary geochemical plots, the statistical correlation being function of: (i) the degree to which perfect (homogeneous, closed system, ...) mixing occurred, (ii) the homogeneity of each end member (mafic and felsic), (iii) the degree to which subsequent processes (fractionation, ...) have occurred, and (iv) the accuracy in analytical data (Neves & Bouchez 1995; D'Lemos 1996).

Attempting to model magma mixing between the more mafic sample analysed (M-38) and granite host (M-53) was done following the methods proposed by Langmuir *et al.* (1978) and Fourcade & Allègre (1981). The results of mixing tests to produce hybrid rock compositions indicate that participation of felsic component ranges from 36 to 82%, with correlation coefficients (R) > 0.925 (Table 3). Taking the participation in felsic component as an index, the correlation coefficients are better for most trace elements, particularly rare earths and high field strength elements. The lack of correlation for some major elements (Al_2O_3, Na_2O and K_2O) with silica or the felsic component, may be explained taking into account the absence of chemical variation between the end terms of the mixing, and/or non-ideal crystal exchange, and/or the high mobility of Na_2O and K_2O as demonstrated experimentally by Watson & Jurewicz (1984). Notwithstanding, the mixing model appears reasonable since the degree of chemical homogenisation is controlled by the slowest

Table 5. *Whole-rock Sr isotope analyses for the studied rocks*

Sample	Rb	Sr	$^{87}Rb/^{86}Sr$	$^{87}Sr/^{86}Sr \pm 2\sigma$	$(^{87}Sr/^{86}Sr)_i$
M-38	163	963	0.4893	0.706465 (08)	0.7042
M-31-1	279	440	1.8345	0.715252 (08)	0.7069
M-30-1	197	620	0.9193	0.709337 (16)	0.7052
M-26-2	187	807	0.6716	0.707890 (07)	0.7048
M-26-1	169	819	0.5975	0.707412 (07)	0.7047
M-25	195	608	0.9291	0.710135 (09)	0.7059
M-37	192	794	0.7015	0.708022 (12)	0.7048
M-34	178	687	0.7514	0.708402 (12)	0.7050
M-30-2	179	670	0.7727	0.708852 (09)	0.7053
M-54	209	224	2.7106	0.722776 (12)	0.7104
M-53	241	263	2.6569	0.721633 (14)	0.7095

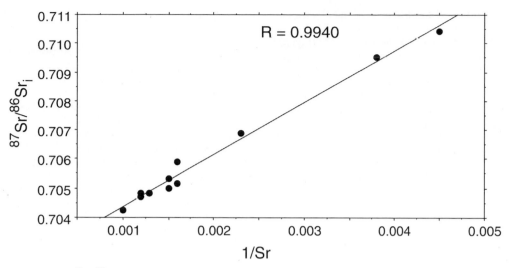

Fig. 8. Plot of $^{87}Sr/^{86}Sr$ versus the reciprocal of Sr concentration.

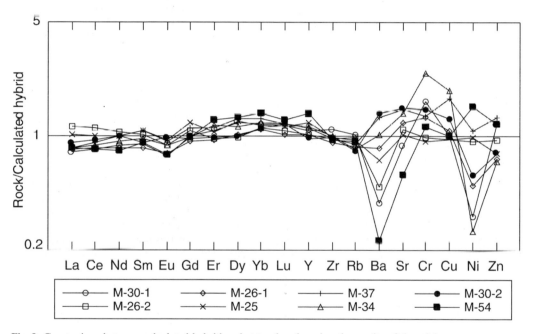

Fig. 9. Comparison between calculated hybrid and natural rocks using the results of the mixing test.

moving magmatic components, i.e. Si and non-alkali elements (Baker 1991; Chekhmir & Epel'baum 1991).

Using the proportions deduced from mixing tests we have calculated the trace element contents for theoretical hybrid rock compositions (Fig. 9). The agreement between theoretical and measured contents is remarkable for most elements, notably for most immobile elements (REE, Y, Zr). The greater discrepancies are for Ba and transition elements (Ni, Cr, Cu, Zn) that may be interpreted by concurrent factors as analytical errors, loss or gain of mobile elements or inhomogeneity of parent materials. In this sense, some trace elements (Cu, Ni) occur in such low concentration that large analytical uncertainties are most likely. On the other hand, variable contents for some elements (Rb, Sr, Ba) in the granite rocks (Bellido *et al.* 1990) would imply that the mixing process involved a heterogeneous parent, hence the mixed-magma series would not result necessarily in linear chemical trends (cf. D'Lemos 1996). Late stage mobilization of litophile elements might have as well contributed to some variation. In fact, the less mobile elements (REE, Y, Zr) show the best adjustment to the mixing model.

The above hypothesis appears to be supported by the Sr isotope data for the samples analysed. In effect, the goodness of fit to a straight line (Fig. 8) attests for the validity of a two-component mixing model and of the assumption that neither the strontium concentrations nor the $^{87}Sr/^{86}Sr$ ratios were modified after mixing had occurred (Faure 1986; Albarède 1996).

To validate this hypothesis, a mixing test have been made also for Sr isotope data considering the same end members and the percent of felsic component previously determined. All calculated values are very close to the theoretical values ($R^2 = 0.924$), except for sample M-54. The incipient albitisation processes in this sample could explain its low Sr contents and high Sr isotope ratio.

However, it is difficult to estimate the primary isotopic characteristics of the end-members involved. Sample M-38 of the more mafic rock analysed shows an initial Sr isotopic ratio 0.7042, providing evidence for the contribution of mantle-derived gabbroic magma. Yet, taking into account textural features, such as the incorporation of potassic feldspar crystals, and the major element composition, one must conclude that sample M-38 is already an evolved rock, with low Mg number (Mg No. = 58), and very low contents in transition elements (Cr = 49 ppm, Ni = 50 ppm). It is likely, therefore, that a mantle reservoir with an initial $^{87}Sr/^{86}Sr$ isotopic ratio <0.7042 was involved in the production of the primary mafic magma. The crustal end-member should have an initial Sr isotopic ratio >0.7104 (Bellido *et al.* 1990, reported initial Sr isotopic ratios between 0.7137 and 0.716 for the whole granite sample set which includes more evolved granite types of the Friol facies). Both magmas should have intruded at the same time as field and petrographical data account. The system must have cooled rapidly enough so that isotopic homogenisation was not attained. This suggests that the mafic–intermediate rocks did not reside in their final granitic melt for long before cooling of the whole system.

Emplacement conditions and epidote significance

In order to estimate the intrusion conditions, the geobarometers of Hammarstrom & Zen (1986), Hollister *et al.* (1987), Johnson & Rutherford (1989), Thomas & Ernst (1990), all based on the linear variation of the Al content of hornblende with pressure of crystallisation, have been applied. The average values obtained are 3.3, 3.3, 2.6 and 1.4 kbar, respectively. The equilibration temperature obtained for amphibole and plagioclase following the method of Holland & Blundy (1994) is *c.* 670 °C, as an average. Using this value for temperature, an average emplacement pressure of 2.5 kbar has been estimated by using the geobarometer of Anderson & Smith (1995). Finally, the method proposed by Ague (1997) based on the equilibrium $2Ab + 2An + Phl + Tr = 2Parg + 6Qz + Kfs$, with temperatures obtained as described previously, yielded estimated crystallisation conditions for the Guitiriz hybridization zone of *c.* 700 °C, 2.5 kbar. As it may be observed, all the values are below 5 kbar, the epidote low pressure stability limit at water saturated conditions and fO_2 buffered by NNO for a magma of tonalitic composition (Schmidt & Thompson, 1996).

Pressure and temperature from the Al in hornblende barometer and amphibole-plagioclase thermometer for the Guitiriz massif can compared to those of regional metamorphism in the study area. Arenas (1991) estimated peak metamorphic conditions near Guitiriz during D2-deformation phase at *c.* 600 °C and 5 kbar. Since the Guitiriz granitoid is synkinematic with respect to D3 (Aranguren *et al.* 1996), which is a retrometamorphic phase with a progressive decrease in P and T (Arenas 1991), the conditions of intrusion for the Guitiriz massif are clearly below the estimated low pressure stability limit of magmatic epidote.

The petrogenetical meaning of magmatic epidote has been widely discussed. For many authors the presence of magmatic epidote excludes a shallow intrusion level for a tonalitic or granodioritic magma. Crawford & Hollister (1982) estimated a minimum pressure of 6 kbar for the crystallization of magmatic epidote on the basis of experimental results by Liou (1973). Zen & Hammarstrom (1984) estimated as well, a minimum pressure between 6 and 8 kbar for magmatic epidote on the basis of experimental results by Naney (1983). More recently, Schmidt & Thompson (1996) set the limit of epidote + biotite stability field in the pressure interval 5 to 9 kbar for tonalite + H_2O and $fO_2 = NNO$.

However, epidotes like those here studied, exhibiting magmatic textures have been described in localities where the Al-hornblende geobarometer indicates pressures below or overlapping the epidote low pressure stability limit (Moench 1986). Tulloch (1986) also questioned the idea that magmatic epidote stability in calc-alkaline plutons required high pressures. Yet, Brandon *et al.* (1996) have proposed that the presence of magmatic epidote is not, by itself, sufficient to establish a high-pressure solidification of granite or to invoke lower-pressure stability limits for magmatic epidote beyond those observed in experiments. Using experimental measurements of epidote dissolution, Brandon *et al.* (1996) concluded that the presence of this mineral implies a rapid mechanism of magma transport. The magma transport rates deduced are consistent with calculated rates for fracture propagation in dykes, excluding the mechanism of diapiric magma ascent for these systems (Clemens & Mawer 1992; Koyaguchi 1985).

In the present case, the hypothesis of a rapid magma transport by dikes from the deep crust, is in agreement with the emplacement mechanism proposed by Aranguren (1994) and Aranguren *et al.* (1996). Moreover, the mafic–intermediate rocks studied are geographically related to a strong positive gravity anomaly which was interpreted by Aranguren (1994) and Aranguren *et al.* (1996) as one of the feeding zones of the Guitiriz massif reaching a deep >4 km. However, the accuracy of this estimation depends on the representability in depth of the samples taken at the surface. Since the mafic rocks studied here were not used by Aranguren (1994) and Aranguren *et al.* (1996) for the gravity calculations due to the lack of outcrops, the depths calculated may have been overestimated.

In any case, the hypothesis of a feeding zone in this area is also supported by the existence of lineations with higher plunge angles than those in the rest of the massif, as well as strong linear anisotropy (Aranguren 1994). In consequence, the proposed feeding zone might be interpreted as to correspond to a fracture developed in an extensional regime. The rapid ascent of the basic magma through this fracture should have favoured the preservation of magmatic epidote. The perfect crystal faces of magmatic epidote enclosed in biotite, but reacted where the same crystal is in contact with quartz or plagioclase, could be the result of epidote-melt reaction during magma ascent and low pressure crystallization of the magma (cf. Brandon *et al.* 1996).

Petrogenetic model

The *c.* 100 m wide contact between mafic–intermediate and granitic rocks results from interaction between contemporaneous magmas. Physical mingling (and mixing) is clearly recognised in this outcrop as evidenced by the presence of complete gradational contacts or large K-feldspar megacryst in the more mafic rocks. Incorporation of xenocrysts or infiltration of extraneous melts produced mineral scale disequilibria and growth textures such as boxy cellular plagioclase and ocellar quartz in mafic–intermediate and hybrid rocks (Fig. 3).

To be effective, the hybridization process requires that liquidus temperatures, volatile contents and vicosities of end member magmas will be similar (Sparks & Marshall 1986; Frost & Mahood 1987). In the Guitiriz hybridization zone the more mafic magma probably was already relative evolved and included a high water content. This would have resulted in low viscosity and temperature contrast with the granitic magma, leading to extensive physical mixing of the two magmas. Mixing between both magmas should have occurred near solidus temperature, leading to a restricted mixing process in space and time that precluded attainment of isotopic equilibrium for Sr.

The similarity of magmatic fabrics in both granite host and hybrid and mafic–intermediate rocks, and the geographical relationship between the mafic–intermediate bodies and the feeding zone of the host granite, suggest that this zone may also have acted as a conduit for the ascent of the mafic magma. The presence of epidote, exhibiting textural and compositional features indicative of a magmatic origin in the more mafic and some hybrid rocks, also supports the hypothesis of a conduit implicated in the hybridization process. Taking into account that the pressure conditions calculated for the emplacement are below the epidote stability limit, preservation of magmatic epidote in the Guitiriz massif required a rapid transport mechanism from deeper zones. Hence, a feeding zone appears to be the appropriate scenario to account for all the features described in the Guitiriz hybridization zone. The data obtained for the pressure of emplacement also indicate that mixing between both magmas could not have occurred at depth within the feeding zone.

Conclusions

The field relationships, and the petrographic and mineralogical evidence, substantiated by major and rare earth element geochemistry, suggest a two-component mixing model to account for the origin of hybrid rocks studied from the Guitiriz granite massif. Trace element behaviour points, however, towards a more complex process. Yet, the lack of linear trends for some trace elements could be explained by subsequent diffusion of mobile elements or non-ideal crystal exchange.

The Sr isotope data defining a hyperbolic trend support this model and, furthermore, attest a simple binary mixing between a mantle-derived magma and a crustal derived one. A mantle reservoir with an initial $^{87}Sr/^{86}Sr$ isotopic ratio <0.7042 was probably involved in the production of the primary gabbroic magma, whereas the crustal end-member must have an initial Sr isotopic ratio <0.7104. It is suggested that both magmas rose through a common feeding zone developed in an extensional regime in which favourable conditions for magma mixing occurred.

We are grateful to R. Alonso and P. Bezares (Department of Analytical Chemistry, UPV) for ICP-AES analytical facilities. To E. Ariño (University of Oviedo, Spain) who performed Rb–Sr XRF analyses. Discussions and critical reading by J. I. Gil Ibarguchi greatly contributed to an improvement of the previous manuscript. We are grateful to L. G. Corretgé, J. de la Rosa and Λ. Castro for helpful suggestions to improve the manuscript. Financial support was given by Project UPV-130.310-EB 034/96.

References

ABDEL-RAHMAN, A. M. 1994. Nature of biotites from alkaline, calc-alkaline and peraluminous magmas. *Journal of Petrology*, **35**, 524–541.

AGUE, J. J. 1997. Thermodynamic calculation of emplacement pressures for batholithic rocks, California: Implication for the aluminum-in-hornblende barometer. *Geology*, **25**, 563–566.

ALBARÈDE, F. 1996. *Introduction to geochemical modeling*. Cambridge University Press, Cambridge.

ANDERSON, J. L. & SMITH, D. R. 1995. The effects of temperature and fO_2 on the Al–in hornblende barometer. *American Mineralogist*, **80**, 549–559.

ARANGUREN, A. 1994. *Estructura y cinemática del emplazamiento de los granitoides del dominio del domo de Lugo y del Antiforme del Ollo de Sapo*. Laboratorio Xeológico de Laxe, Serie Nova Terra, **10**.

——, TUBIA, J. M., BOUCHEZ, J. L. & VIGNERESSE, J. L. 1996. The Guitiriz granite, Variscan belt of northern Spain: extension-controlled emplacement of magma during tectonic escape. *Earth Planetary Science Letters*, **139**, 165–176.

ARENAS, R. 1991. Opposite *P, T, t* paths of the Hercynian metamorphism between the upper units of the Cabo Ortegal Complex and their substratum northwest of the Iberian Massif. *Tectonophysics*, **191**, 347–364.

BAKER, D. 1991. Interdiffusion of hydrous dacitic and rhyolitic melts and the efficacy of rhyolite contamination of dacitic enclaves. *Contributions to Mineralogy and Petrology*, **106**, 462–473.

BELLIDO, F., GARCÍA GARZÓN, J. & REYES, J. 1990. Estudio petrológico y geocronológico Rb–Sr de los granitoides de Friol y Puebla de Parga NO de Lugo. *Boletín Geológico Minero de España*, **101**, 621–631.

BRANDON, A. D., CREASER, R. A. & CHACKO, T. 1996. Constrains on rates of Granitic Magma Transport from Epidote Dissolution Kinetics. *Science*, **271**, 1845–1848.

CANTAGREL, F. & PIN, C. 1994. Major, minor and rare-earth element determinations in 25 rock standards by ICP-Atomic Emission Spectrometry. *Geostandards Newsletter*, **18**, 123–138.

CAPDEVILA, R., CORRETGÉ, G. & FLOOR, P. 1973. Les granitoïdes varisques de la Meseta iberique. *Bulletin Société Géologie de France*, **15**, 209–228.

CASTRO, A., DE LA ROSA, J. D., FERNÁNDEZ, C. & MORENO-VENTAS, I. 1995. Unstable flow, magma mixing and magma-rock deformation in a deep-seated conduit: the Gil-Márquez Complex, south-west Spain. *Geologische Rundschau*, **84**, 359–374.

CHEKHMIR, A. S. & EPEL'BAUM, M. B. 1991. Diffusion in magmatic melts: new study. *In*: PERCHURK L. L. & KUSHIRO, I. (eds) *Physical chemistry of magmas*. Springer-Verlag, New York, 99–119.

CLEMENS, J. D. & MAWER, C. K. 1992. Granitic magma transport fracture propagation. *Tectonophysics*, **204**, 339–360.

CORNELIUS, H. P. 1915. Geologische Beobachtungen im Gebiet des Forno-Gletschers Engadin. *Centralblatt für Mineralogie Geologie und Paläontologie*, **8**, 246–252.

CORRETGÉ, L. G. 1983. Las rocas graníticas y granitoides del macizo ibérico. *In*: *Libro Juvilar, J. M. Ríos, Geología de España*, Instituto Geológico y Minero de España, 569–592.

CRAWFORD, M. L. & HOLLISTER, L. S. 1982. Contrast of metamorphic and structural histories across the Work Channel lineament, Coast Plutonic Complex, British Columbia. *Journal of Geophysical Research*, **87**, 3849–3860.

DAWES, R. L. & EVANS, B. W. 1991. Mineralogy and geotermobarometry of magnetic epidote-bearing dikes, Front Range, Colorado. *Geological Society of America Bulletin*, **103**, 1027–1031.

DE BRUIYN, H., VAN DER WESTHUIZEN, W. A. & SCHOCH, A. E. 1983. The estimation of FeO, F and H_2O + by regression in microprobe analyses of natural biotite. *Journal of Trace Microprobe Technologies*, **1**, 399–413.

D'LEMOS, R. S. 1996. Mixing between granitic and dioritic crystal mushes, Guernsey, Channel Island, UK. *Lithos*, **38**, 233–257.

DIDIER, J. & BARBARIN, B. 1991. *Enclaves and granite petrology*. Developments in Petrology, **13**. Elsevier, Amsterdam.

DíEZ BALDA, M. A., VEGAS, R. & GONZÁLEZ LODEIRO, F. 1990. Central-Iberian Zone. Authochthonous Sequences. Structure. *In*: DALLMEYER, R. D. &

MARTÍNEZ GARCÍA, E. (eds) *Pre-mesozoic geology of Iberia*. Springer-Verlag, Berlin Heidelberg.

EVENSEN, M. M., HAMILTON, P. J. & O'NIONS, R. K. 1978. Rare earth abundances in chondritic meteorites. *Geochimica et Cosmochimica Acta*, **42**, 1199–1212.

FAURE, G. 1986. *Principes of isotope geology*, 2nd edition. John Wiley & Sons, New York.

FOURCADE, S. & ALLÈGRE, J. C. 1981. Trace elements behavior in granite genesis: a case studied the calc-alkaline pluton assotiation from Querigut Complex (Pyrénées, France). *Contributions to Mineralogy and Petrology*, **76**, 177–195.

FROST, T. P. & MAHOOD, G. A. 1987. Field, chemical and physical constraints on mafic–felsic magma interaction in the Lamarck Granodiorite, Sierra Nevada, California. *Bulletin of Geological Society of America*, **99**, 272–291.

GALÁN, G., PIN, C. & DUTHOU, J. L. 1996. Sr–Nd isotopic record of multi-stage interactions between mantle-derived magmas and crustal components in a collision context-The ultramafic–granitoid association from Vivero (Hercynian belt, NW Spain). *Chemical Geology*, **131**, 67–91.

GILL, B. J. 1981. *Orogenic andesites and plate tectonic*. Springer-Verlag.

HAMMARSTROM, J. M. & ZEN, E. 1986. Aluminium in hornblende: An empirical igneous geobarometer. *American Mineralogist*, **71**, 1297–1313.

HIBBARD, M. J. 1991. Textural anatomy of twelve magma-mixed granitoid systems. *In:* DIDIER, J. & BARBARIN, B. (eds) *Enclaves and granite petrology*. Developments in Petrology, **13**. Elsevier, Amsterdam.

—— 1995. *Petrography to petrogenesis*. Prentice-Hall, Englewood Cliffs, New Jersey.

HOLLAND, T. & BLUNDY, J. 1994. Non-ideal interactions in calcic amphiboles and their bearing on amphibole–plagioclase thermometry. *Contributions to Mineralogy and Petrology*, **116**, 433–447.

HOLLISTER, L. S., GRISSOM, G. C., PETERS, E. K., STOWELL, H. H. & SISSON, V. B. 1987. Confirmation of the empirical correlation of the Al in hornblende with pressure of solidification in calc-alkaline plutons. *American Mineralogist*, **72**, 231 239.

IGLESIAS, M. & CHOUKROUNE, P. 1980. Shear zones in the Iberian arc. *Journal of Structural Geology*, **2**, 63–68.

IRVINE, T. N. & BARAGAR, W. R. A. 1971. A guide to the chemical classification of common volcanic rocks. *Canadian Journal Earth Sciences*, **8**, 523–548.

JANOUSEK, V., ROGERS, G. & BOWES, D. R. 1995. Sr–Nd isotopic constrains on the petrogenesis of the Central Bohemian Pluton, Czech Republic. *Geologische Rundschau*, **84**, 520–534.

JOHNSON, M. C. & RUTHERFORD, M. J. 1989. Experimental calibration of the aluminium-in-hornblende geobarometer with aplication to Long Valley caldera (California) volcanic rocks. *Geology*, **17**, 837–841.

JULIVERT, M., FONTBOTÉ, J. M., RIBEIRO, A. & CONDE, L. 1972. *Mapa tectónico de la Península Ibérica y Baleares*. Instituto Geológico y Minero de España, Madrid.

KEANE, S. D. & MORRISON, J. 1997. Distinguishing magmatic from subsolidus epidote: laser probe oxygen isotope compositions. *Contributions to Mineralogy and Petrology*, **126**, 265–274.

KOYAGUCHI, T. 1985. Magma mixing in a conduit. *Journal of Geological and Geothermal Research*, **25**, 365–369.

LANGMUIR, C. H., VOCKE, R. D., HANSON, G. N. & HART, S. R. 1978. A general mixing equation with applications to Icelandic basalts. *Earth and Planetary Science Letters*, **37**, 380–392.

LEAKE, B. E. 1978. Nomenclature of amphibole. *Canadian Mineralogist*, **16**, 501–520.

LIOU, J.G. 1973. Synthesis and stability relations of epidote, $Ca_2Al_2FeSi_3O_{12}(OH)$. *Journal of Petrology*, **14**, 381–413.

MARTÍNEZ, F. J., CARRERAS, J., ARBOLEYA, M. L. & DIETSCH, C. 1996. Structural and metamorphic evidence of local extension along the Vivero fault coeval with bulk crustal shortening in the Variscan chain (NW Spain). *Journal of Structural Geology*, **18**, 61–73.

MOENCH, R. H. 1986. Comment on 'Implications of magmatic epidote-bearing plutons on crustal evolution in the acreted terranes of northwestern North America' and 'Magmatic epidote and its petrologic significance'. *Geology*, **14**, 187–188.

MORENO-VENTAS, I., ROGERS, G. & CASTRO, A. 1995. The role of hybridization in the genesis of Hercynian granitoids in the Gredos Massif, Spain: inferences from Sr–Nd isotopes. *Contributions to Mineralogy and Petrology*, **120**, 137–149.

NACHIT, H., RAZAFIMAHEFA, N., STUSSI, J. M. & CARRON, J. P. 1985. Composition chimique des biotites et typologie magmatique des granitoïdes. *Comptes Rendus de l'Académie des Sciences de Paris*, **301**, 813–818.

NANEY, M. T. 1983. Phase equilibria of rock-forming ferromagnesian silicates in granitic systems. *American Journal of Science*, **283**, 993–1033.

NEVES, S. P. & BOUCHEZ, A. 1995. Successive mixing and mingling of magmas in a plutonic complex of Northeast Brazil. *Lithos*, **34**, 275–299.

ORTEGA, L. A. & MENÉNDEZ, M. 1998. Desarrollo de un método para la determinación de elementos mayoritarios, traza y tierras raras mediante ICP de emisión óptica. *Boletín de la Sociedad Española de Mineralogía*, **21**, 168–169.

PE-PIPER, G. 1988. Calcic-amphiboles of amfic rocks of the Jeffers Brooks pluonic complex, Nova Scotia, Canada. *American Mineralogist*, **76**, 993–1006.

PIN, C. & BASSIN, C. 1992. Evaluation of a strontium-specific extraction chromatographic method for isotopic analysis in geological materials. *Analytica Chimica Acta*, **269**, 249–255.

SCHMIDT, M. W. & THOMPSON, A. B. 1996. Epidote in calc-alkaline magmas: An experimental study of stability, phase relationships, and the role of epidote in magmatic evolution. *American Mineralogist*, **81**, 462–474.

SPARKS, R. S. J. & MARSHALL, L. A. 1986. Thermal and mecanical contraints on mixing between mafic and silicic magmas. *Journal of Volcanology and Geothermal Research*, **29**, 99–124.

THOMAS, W. N. & ERNST, W. G. 1990. The aluminium content of hornblende in calc-alkaline granitic rocks: A mineralogic barometer calibrated experimentally to 12 kbars. *In*: SPENCER, R. J. & I.-MING CHOU (eds) *Fluid mineral interactions: a tribute to H. P. Eugster*. Geochemical Society Special Publications, **2**, 59–63.

TULLOCH, A. J. 1979. Secondary Ca–Al silicates as low-grade alteration products of granitoid biotite. *Contributions to Mineralogy and Petrology*, **69**, 105–117.

—— 1986. Comment on 'Implications of magmatic epidote-bearing plutons on crustal evolution in the acreted terranes of northwestern North America' and 'Magmatic epidote and its petrologic significance'. *Geology*, **14**, 187–188.

VERNON, R. H., ETHERIDGE, M. A. & WALL, V. J. 1988. Shape and microstructure of granitoid enclaves: indicators of magma mingling and flow. *Lithos*, **22**, 1–11.

VIHNAL, C. R., McSWEEN, H. Y. & SPEER, J. A. 1991. Hornblende chemistry in southern Appalachian granitoids: Implications for aluminum hornblende thermobarometry and magmatic epidote stability. *American Mineralogist*, **76**, 176–188.

WATSON, E. B. & JUREWICZ, S. R. 1984. Behavior of alcalies during diffusive interaction of granitic xenoliths with basaltic magma. *Journal of Geology*, **92**, 121–131.

ZEN, E-AN 1988. Tectonic significance of high-pressure plutonic rocks in the western cordillera of North America. *In*: ERNST, W. G. (ed.) *Metamorphism and crustal evolution of the Western United States*. Rubey volume VII. Prentice-Hall, Englewood Cliffs, New Jersey, 41–67.

—— & HAMMARSTROM, J. M. 1984. Magmatic epidote and its petrologic significance. *Geology*, **12**, 515–518.

—— & —— 1986. Reply on the comments on 'Implications of magmatic epidote-bearing plutons on crustal evolution in the acreted terranes of northwestern North America' and 'Magmatic epidote and its petrologic significance' by A. J. Tulloch & by R. H. Moench. *Geology*, **12**, 515–518.

ZORPI, M. J., COULON, C., ORSINI, J. B. & COCIRTA, C. 1989. Magma mingling, zoning and emplacement in calc-alkaline granitoid plutons. *Tectonophisics*, **157**, 315–329.

Index

Page numbers in *italics* refer to Figures and page numbers in **bold** refer to Tables